Mass Facility

Post-Translational Modifications of Proteins

"Only connect....."

(E. M. Forster. Frontispiece to *Howard's End*)

Post-Translational Modifications of Proteins

Editors

John J. Harding, Ph.D.
Senior Research Scientist
Nuffield Laboratory of Ophthalmology
Oxford University
Oxford, England

M. James C. Crabbe, M.A., B.Sc., M.Sc., Ph.D., D.Sc., C.Chem., F.R.S.C.
Reader in Microbiology, University of Reading, and
Supernumerary Fellow, Wolfson College
Oxford, England

CRC Press
Boca Raton Ann Arbor London

Library of Congress Cataloging-in-Publication Data

Post-translational modifications of proteins / editors, John J.
  Harding, M. James C. Crabbe.
    p.   cm.
  Includes bibliographical references and index.
  ISBN 0-8493-4171-X
  1. Post-translational modification.   I. Harding, John J., Ph. D.
  II. Crabbe, M. James C.
  QH450.6.P68   1991
  574.19′245--dc20
                                                                    91-28200
                                                                      CIP

Direct all inquiries to CRC Press, Inc., 2000 Corporate Blvd., N.W., Boca Raton, Florida 33431.

© 1992 by CRC Press, Inc.

International Standard Book Number 0-8493-4171-X

Library of Congress Card Number 91-28200

Printed in the United States of America   1 2 3 4 5 6 7 8 9 0

PREFACE

Many polypeptides are chemically modified *in vivo* before they become fully functional proteins. Other proteins have their functions impaired by later modifications. Both types of post-translational modification are considered in this book.

Proline and lysine are hydroxylated in collagen and a few other proteins. Kivirikko and colleagues describe the occurrence and function of the hydroxylated residues, and the three enzymes required for their formation, including recent advances in our knowledge of lysyl hydroxylase.

Reactions between sugars and proteins may lead to diabetic complications and age-related changes in proteins. These reactions and their damaging effects are discussed by Bucala and colleages with emphasis on the later changes that yield brown and fluorescent products.

Van Heyningen's chapter reviews ADP-ribosylation of proteins outside the nucleus, a process which may be important for normal function but is mostly known as an activity of bacterial toxins.

Monod, Wyman, and Changeux's concerted model for allosteric control of enzyme activity has been well exemplified by studies on glycogen phosphorylase, as described in the chapter by Oikonomakos et al. The authors detail the extensive contributions of X-ray crystallography to our knowledge of this enzyme where phosphorylation is the key post-translational event in control of catalysis. The crystal structures of both R and T states are pictured, together with studies on binding to the allosteric site, the glycogen storage site, the inhibitor site, and the catalytic site.

Fast Atom Bombardment Mass Spectrometry (FAB-MS) is a very powerful technique that has been used to uncover the glycan structures in glycoproteins. The chapter by Dell et al. details many of the important areas where FAB-MS has been applied to the structure and biosynthesis of various oligosaccharide moieties in proteins.

The current understanding of the structural diversity and function of glycoprotein oligo-saccharides and of the mechanism and control of their assembly is reviewed in the chapter by Hemming.

THE EDITORS

John J. Harding B.A., Ph.D. is Senior Research Biochemist at the Nuffield Laboratory of Ophthalmology, University of Oxford. He graduated in Natural Sciences from the University of Cambridge in 1961, and obtained a Ph.D. from London University for studies of the chemistry of collagen. He moved to Oxford in 1967 to work with the leading ocular biochemist, Antoinette Pirie, and has continued to study lens proteins and how they change in cataract, prevention of cataract, and the structure of collagen. He is the author of over a hundred scientific publications, including a book on cataract, and two chapters in "Advances in Protein Chemistry" — the most recent on non-enzymic post-translational modification of proteins.

He is a member of the Biochemical Society, International Society for Eye Research, British Diabetic Association, Association for Eye Research, and a European Community concerted action on ageing (EURAGE). He is an editor of Experimental Eye Research.

Dr. Harding has presented lectures at many international meetings on four continents. He was an invited speaker at the International Congress of Biochemistry in Prague.

M. James C. Crabbe M.A., B.Sc., M.Sc., Ph.D., D.Sc., C. Biol., M.I.Biol., C.Chem., F.R.S.C., F.R.S.A., is Reader in Microbiology in the Department of Microbiology at the University of Reading, and Supernumerary Fellow at Wolfson College, Oxford.

Dr. Crabbe graduated in Biochemistry from the University of Hull in 1971, and obtained his M.Sc. and Ph.D. degrees from the University of Manchester. On moving to Oxford to work with John Harding on a Wellcome Trust Fellowship, he was elected Guy Newton Stipendiary Junior Research Fellow at Wolfson College, and later Research and Governing Body Fellow at the College. He was a Senior Fellow in the Nuffield Laboratory of Ophthalmology at the University of Oxford, and lecturer in Biochemistry at New and St. Peter's Colleges, Oxford. After positions in the Dyson Perrins Laboratory and the Department of Biochemistry at the University of Oxford, he moved in 1988 to an Endowed Readership at the University of Reading. He was awarded the degree of D.Sc. from the University of Manchester in 1990.

Dr. Crabbe is a member of the Royal Society of Chemistry, the Royal Society for the Encouragement of Arts, Manufactures and Commerce, the Institute of Biology, the Biochemical Society, the Blair Bell Research Society, the International Society for Clinical Enzymology, the Connective Tissue Society, the International Society for Eye Research, the Histamine Society, the Association for Eye Research, the Association for Research in Vision and Ophthalmology, the Molecular Graphics Society, and the Society of General Microbiology. He is an editor of *Computers and Chemistry*.

Dr. Crabbe has received the Association for Eye Research International Award, and a Science and Technology Media Fellowship from the British Association for the Advancement of Science. In 1991 he was invited to be a member of the Academic Audit Unit for British Universities.

Dr. Crabbe has been the recipient of research grants from the National Institutes of Health, the Research Councils, Charitable Organisations, and Industry. He has published over 110 scientific articles and books, and been an invited speaker at many National and International meetings. He is perhaps best known for his molecular modelling software package Desk Top Molecular Modeller (DTMM — Oxford University Press), which has been highly acclaimed in the International Press. His research interests include molecular modelling applications, enzymology, enzyme kinetics, mechanisms of cataract formation and prevention, ageing and glycosylation, and diabetes and its complications.

CONTRIBUTORS

K. Ravi Acharya
Biochemistry Department
University of Bath
Bath, England

Richard Bucala
Laboratory of Medical Biochemistry
Rockefeller University
New York, New York

Anthony Cerami
Laboratory of Medical Biochemistry
Rockefeller University
New York, New York

Anne Dell
Department of Biochemistry
Imperial College
London, England

Frank W. Hemming
Department of Biochemistry
University of Nottingham Medical School
Queen's Medical Centre
Nottingham, England

Simon van Heyningen
Department of Biochemistry
University of Edinburgh
Edinburgh, Scotland

Louise N. Johnson
Laboratory of Molecular Biophysics
University of Oxford
Oxford, England

Kari I. Kivirikko
Department of Medical Biochemistry
University of Oulu
Oulu, Finland

Roy A. McDowell
Department of Biochemistry
Imperial College
London, England

Raili Myllylä
Department of Medical Biochemistry
University of Oulu
Oulu, Finland

Nikos G. Oikonomakos
Institute of Biological Research
National Hellenic Research Foundation
Athens, Greece

Taina Pihlajaniemi
Department of Medical Biochemistry
University of Oulu
Oulu, Finland

Mark E. Rogers
M-Scan Inc.
West Chester, Pennsylvania

Helen Vlassara
Laboratory of Medical Biochemistry
Rockefeller University
New York, New York

TABLE OF CONTENTS

Chapter 1

HYDROXYLATION OF PROLINE AND LYSINE RESIDUES IN COLLAGENS AND OTHER ANIMAL AND PLANT PROTEINS

Kari I. Kivirikko, Raili Myllylä, and Taina Pihlajaniemi

TABLE OF CONTENTS

I. INTRODUCTION

4-Hydroxyproline, 3-hydroxyproline and hydroxylysine (Figure 1) are found in animals almost exclusively in collagens and a few other extracellular proteins with collagen-like amino acid sequences. 4-Hydroxyproline is also found in plants, primarily in a family of cell-wall glycoproteins, the best studied among which are the extensins. All three amino acids are formed as cotranslational and posttranslational modifications by the hydroxylation of peptide-bound proline or lysine residues by three separate enzymes, prolyl 4-hydroxylase (EC 1.14.11.2, procollagen-proline, 2-oxoglutarate 4-dioxygenase), prolyl 3-hydroxylase (EC 1.14.11.7, procollagen-proline, 2-oxoglutarate 3-dioxygenase), and lysyl hydroxylase (EC 1.14.11.4, procollagen-lysine, 2-oxoglutarate 5-dioxygenase). Prolyl 4-hydroxylase is widely distributed in nature, as it has now been identified in a number of vertebrate, invertebrate, and plant sources. The two other enzymes are also found in a variety of animal sources, but they appear to be absent from plants.

Prolyl 4-hydroxylase plays a central role in collagen synthesis, as the 4-hydroxyproline residues formed in the reaction are an absolute requirement for the folding of the newly synthesized procollagen polypeptide chains into triple-helical molecules. This crucial role of 4-hydroxyproline in collagens makes prolyl 4-hydroxylase a potential target for pharmacological modulation of the excessive collagen formation found in patients with various fibrotic disorders and has prompted a vast amount of research into the enzyme. In plant proteins the 4-hydroxyproline residues are frequently *O*-glycosylated and may form intermolecular cross-links which contribute to cell-wall assembly and stability. The hydroxylysine residues of animal proteins have two important functions: their hydroxyl groups serve as sites of attachment for carbohydrate units, and they are essential for the stability of certain collagen cross-links. The role of the 3-hydroxyproline residues is unknown at present.

Prolyl 4-hydroxylase, prolyl 3-hydroxylase, and lysyl hydroxylase belong to the group of 2-oxoglutarate dioxygenases, the other members of which include ε-*N*-trimethyl-L-lysine hydroxylase and γ-butyrobetaine hydroxylase, two enzymes involved in the biosynthesis of L-carnitine, together with thymine 7-hydroxylase, pyrimidine deoxyribonucleoside 2'-hydroxylase and deoxyuridine 1'-hydroxylase. Prolyl 4-hydroxylase is one of the earliest discovered and best investigated of these enzymes, and has often been used as a model for studying the other 2-oxoglutarate dioxygenases.

The aim of this chapter is to review current information on the occurrence, functions, and formation of 4-hydroxyproline, 3-hydroxyproline, and hydroxylysine in animal and plant proteins, with special emphasis on the corresponding enzymes. Since the investigations of Stetten and Shoenheimer[1] and Stetten,[2] it has been known that free 4-hydroxyproline is not incorporated into collagen but that the hydroxyproline of collagen is derived from proline which is hydroxylated in a peptide-bound form. The literature on the topics covered in this chapter is now voluminous, and this review will therefore concentrate mainly on more recent discoveries. For more detailed accounts and more complete references, especially on the early work, the reader is referred to several previous reviews on the three hydroxy amino acids, their formation, and the corresponding enzymes in animals[3-10] and plants.[3,7,11-13] Reviews are also available on methodological aspects,[14,15] which will not be discussed in detail here.

II. OCCURRENCE OF 4-HYDROXYPROLINE, 3-HYDROXYPROLINE, AND HYDROXYLYSINE IN ANIMAL AND PLANT PROTEINS

A. ANIMAL PROTEINS

The vast majority of 4-hydroxyproline, 3-hydroxyproline, and hydroxylysine in vertebrate proteins is found in various collagens. 4-Hydroxyproline and in most but not all cases,

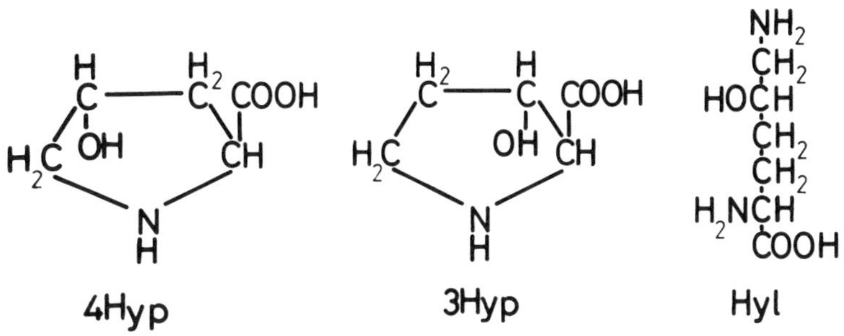

FIGURE 1. The structures of 4-*trans*-hydroxy-L-proline, 3-*trans*-hydroxy-L-proline, and 5-hydroxy-L-lysine.

TABLE 1
Vertebrate Proteins and Peptides Reported to Contain 4-Hydroxyproline (4-Hyp), 3-Hydroxyproline (3-Hyp), or Hydroxylysine (Hyl)[a]

Protein or peptide	4-Hyp	3-Hyp	Hyl
Collagens	+	+	+
C1q of complement	+	−	+
Acetylcholinesterase	+	−	+
Pulmonary surfactant apoproteins	+	−	−
Mannose-binding proteins	+	−	+
Macrophage scavenger receptors	+[b]	−	+[b]
Conglutinin (bovine)	+	−	+
Elastin	+	−	−
[Hydroxyproline]-lysyl-bradykinin	+	−	−
[Hydroxyproline[9]] luteinizing hormone-releasing hormone			
Somatostatin-28 (anglerfish)	−	−	+

[a]   For references, see appropriate paragraphs in the text.
[b]   The presence of these residues is likely on the basis of the cDNA-derived amino acid sequences.

hydroxylysine, are also found in the collagen-line domains of a few other proteins: The C1q subcomponent of complement, acetylcholinesterase, pulmonary surfactant apoproteins, mannose-binding proteins, types I and II macrophage scavenger receptors, and bovine conglutinin (Table 1). 4-Hydroxyproline but not hydroxylysine, is likewise present in elastin, whereas 3-hydroxyproline has not been identified in any other protein except collagens (Table 1). A single 4-hydroxyproline residue is also found in hydroxyproline-lysyl-bradykinin, and hydroxyproline luteinizing hormone-releasing hormone, while a single hydroxylysine residue is found in anglerfish somatostatin-28 (Table 1).

Collagens are a family of closely related, although chemically distinct, extracellular matrix macromolecules. All collagen molecules consist of three polypeptide chains, called α-chains. Each of these is coiled into a left-handed helix, and the three helical chains are twisted around each other into a right-handed triple helix. In some collagen types all three α-chains of the molecule are identical, while in other types the molecule contains two, and in some cases even three different α-chains. In all collagen types the polypeptide chains also have noncollagenous domains of varying sizes. The triple-helical regions of the molecule characteristically have a repeating triplet amino acid sequence, -Gly-X-Y-, where X and Y denote amino acids other than glycine. For detailed information on collagen structure and types, the reader is referred to a recent book[16] and a number of reviews.[17-21]

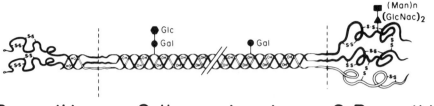

N-Propeptides   Collagen domain   C-Propeptides

FIGURE 2.   Structure of a type I procollagen molecule. Glc denotes glucose; Gal, galactose; Man, mannose; and GlcNac, *N*-acetylglucosamine. (From Prockop, D. J. and Kivirikko, K. I., *N. Engl. J. Med.*, 311, 376, 1984. With permission.)

Fourteen distinct collagen types, containing altogether 28 different α-chains, have now been characterized, and preliminary data are available on the existence of additional types.[19-23] Some of these collagens have a distinct distribution, e.g., types II, IX, X, and XI are found primarily or exclusively in cartilage and type IV collagen is found only in basement membranes. On the other hand, some are present in most extracellular matrices. The molecules of most collagen types are synthesized in precursor forms, procollagen molecules (Figure 2), which differ from the final collagen molecules in that they have additional propeptide extensions at both their amino- and carboxy-terminal ends. The polypeptide chains of these precursor molecules are called pro-α-chains (for reviews on collagen synthesis, see References 24 to 27).

4-Hydroxyproline, 3-hydroxyproline, and hydroxylysine content values have now been determined for the triple-helical (i.e., collagenous) domains of most collagen polypeptide chains (Table 2).[16,19,28-33] The 4-hydroxyproline content shows only relatively small, though distinct, differences between collagens in most cases, whereas 3-hydroxyproline and hydroxylysine vary markedly. In the case of 3-hydroxyproline, the values range from 0 to over 10 resides per 1000 amino acids, and in the case of hydroxylysine from about 5 to 70. Additional variations are found in the content of the three amino acids within the same collagen type in different tissues and even in the same tissue in many physiological and pathological states (for References, see 8). The 4-hydroxyproline content of a given collagen type again varies only within narrow limits, apparently because of the function of 4-hydroxyproline (Section III), whereas large variations are found in 3-hydroxyproline and hydroxylysine. For example, the hydroxylysine content of type I collagen varies from about 6 to 17 residues per 1000, and the 3-hydroxyproline content of type IV collagen from about 1 to 20.[8]

The subcomponent C1q of complement is composed of 18 polypeptide chains, six A, six B, and six C chains. Each of the three types of polypeptide chains is approximately 225 amino acid residues long and contains a region of 78 to 81 amino acid residues of a collagen-like -Gly-X-Y- repeating sequence located close to the amino-terminus (Figure 3B). The carboxy-terminal portion of each chain has approximately 136 amino acid residues of a noncollagen-like amino acid sequence, which forms a globular "head" group (for reviews, see References 34, 35). The collagen-like sequences have high contents of 4-hydroxyproline and hydroxylysine.[34-36]

Acetylcholinesterase consists of 12 catalytic subunits arranged in three tetrameric groups (Figure 3A), and a triple-helical collagen-like tail structure that is thought to anchor the enzyme to the basement membrane.[37] This collagen-like domain again has high 4-hydroxyproline and hydroxylysine.[37-38] Additional vertebrate proteins so far reported to contain similar collagenous domains are pulmonary surfactant apoproteins,[39-42] mannose-binding proteins,[43-45] and bovine conglutinin.[49] The presence of 4-hydroxyproline and hydroxylysine in the macrophage scavenger receptors is likely on the basis of their cDNA-derived amino

TABLE 2

4-Hydroxyproline (4-Hyp), 3-Hydroxyproline (3-Hyp), and Hydroxylysine (Hyl) in the Triple-Helical Domains of Various Collagen Chains

Collagen type and chain		Residues/1000 amino acid residues[a]					
		4-Hyp	3-Hyp	Hyp + Pro	Hyl	Hyl + Lys	Ref.
I:	α1	114	1	233	10	37	19
	α2	105	1	220	12	32	19
II:	α1	96	1	203	18	38	19
III:	α1	126	0	233	6	35	19
IV:[b]	α1	118	1[c]	207	49	56	19
	α2	107	1[c]	181	39	44	19
	α3	n.r.	n.r.	n.r.	n.r.	n.r.	
	α4	n.r.	n.r.	n.r.	n.r.	n.r.	
	α5	n.r.	n.r.	n.r.	n.r.	n.r.	
V:	α1	109	3	230	35	55	19
	α2	109	3	209	24	42	19
	α3	92	2	193	43	58	19
VI:	α1	75	n.r.	167	48	59	30
	α2	62	n.r.	166	67	83	30
	α3	54	n.r.	138	48	69	30
VII:	α1	84	0	165	41	59	19
VIII:	α1 }[d] α2 }[d]	126	9	235	22	45	32
IX:	α1[e]	112	n.r.	230	33	53	29
	α2[e]	109	n.r.	212	46	58	29
	α3	n.r.	n.r.	n.r.	n.r.	n.r.	
X:	α1	112	n.r.	248	35	55	31
XI:	α1	100	2	223	37	56	33
	α2	96	2	210	40	57	33
	α3	104	2	224	21	36	33
XII:	α1	n.r.	n.r.	n.r.	n.r.	n.r.	
XIII:	α1	n.r.	n.r.	n.r.	n.r.	n.r.	
XIV:	α1	n.r	n.r.	n.r.	n.r.	n.r.	

[a] n.r. = not reported.

[b] Type IV collagen forms at least two types of molecule that contain altogether four different α-chains.[20,22,53]

[c] Type IV collagen from placental basement membrane; in most studies on other basement membranes the values range from about 5 to 10 residues/1000.

[d] The value shown is the mean for two 50,000 molecular weight fragments of the triple-helical protein. The 3-hydroxyproline content is based on the finding that 6.6% of the hydroxyproline is present as the 3-isomer.[28]

[e] Calculated for two of the three collagenous domains of these chains.

acid sequences, but has not been established directly.[46-48] All the types I and II macrophage scavenger receptor[46-48] proteins contain 4-hydroxyproline, but at least some of the pulmonary surfactant apoproteins contain no hydroxylysine.[33-40]

Elastin is the main protein component of the elastic fibers, and differs from the proteins discussed so far in that it has no triple-helical collagen-like domain. Nevertheless, the elastin polypeptide chain has repeating -Gly-X-Y- sequences, which contain 4-hydroxyproline but no hydroxylysine.[50,51] The 4-hydroxyproline content of elastin may vary greatly, usually being about 10 to 15 residues per 1000 amino acids,[51] but ranging up to about 50 residues per 1000 in special situations.[52]

4-Hydroxyproline has been identified in vertebrate collagens and other proteins containing this amino acid essentially exclusively in the Y positions of the repeating -X-Y-Gly-

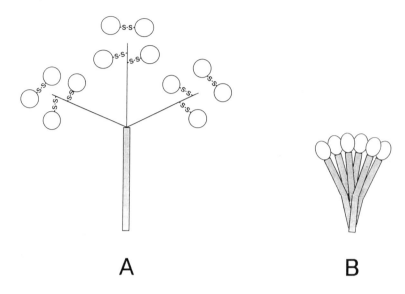

A   **B**

FIGURE 3.   Structures of acetylcholinesterase (A) and the subcomponent C1q of complement (B). ▨ Denotes triple-helical collagen-like structures. (Part A modified from Rosenberry, T. L., Barnett, P., and Mays, C., *Meth. Enzymol.*, 82, 325, 1982. With permission. Part B from Reid, K. B. M., *Biochem. Soc. Trans.*, 11, 1, 1983. With permission.)

sequences[3-10,16-21] (the repeating -Gly-X-Y- sequences are written here as -X-Y-Gly- due to the properties of the three hydroxylases, see Section V). The only reported exceptions are position 8 in the B chain of the human C1q subcomponent of complement and position 35 in the C chain,[35,36] and the extreme carboxy-terminal triplet of the collagen domain in the α3 chain of type IV collagen,[53] in which 4-hydroxyproline has been identified in a sequence -X-4Hyp-Ala-. Hydroxylysine is likewise found almost exclusively in the Y positions of the repeating -X-Y-Gly- sequences, but in some collagens the short nontriple-helical sequences at the ends of the α-chain contain one sequence of -X-Hyl-Ser- or -X-Hyl-Ala-.[8,16] Although 4-hydroxyproline and hydroxylysine are usually found only in proteins containing repeating -X-Y-Gly- sequences, the human kinin mixture in plasma, urine, and ascitic fluid contains both lysyl bradykinin and small amounts of hydroxyproline-lysyl-bradykinin with a single 4-hydroxyproline residue in the sequence -Pro-4Hyp-Gly-.[54,55] The hydroxyproline luteinizing hormone-releasing hormone also contains a single 4-hydroxyproline residue in the sequence -Arg-4Hyp-Gly-.[56] Correspondingly, anglerfish somatostatin-28 contains a single hydroxylysine residue in the sequence -Trp-Hyl-Gly-.[57,58] 3-Hydroxyproline has been identified in collagens only in the sequence -Gly-3Hyp-4Hyp-Gly-.

The 4-hydroxylation of proline residues in the Y positions is not always complete, and thus the collagen in a single tissue may contain molecules in which a given Y position is occupied by a 4-hydroxyproline residue and others in which the same position is occupied by proline.[8,16-20] The degree of hydroxylation of many lysine residues in the Y positions is likewise incomplete, and a number of lysine residues in the Y positions of some collagens are not hydroxylated at all.[8,16-20] Such variation in the extent of hydroxylation of the proline and lysine residues explains the differences that are found in 4-hydroxyproline and hydroxylysine content within the same collagen type from various sources. This variation also partly explains the differences found in the amounts of these two amino acids between various collagen types, but additional factors involved are differences in the total amounts of proline and lysine residues incorporated per 1000 amino acids (sum of 4-hydroxyproline and proline or hydroxylysine and lysine in Table 2), and differences in the distribution of the incorporated proline and lysine residues between the X and Y positions. The variability in 3-hydroxyproline content within a given collagen type and between various collagen types is similarly due to differences in the extent of 3-hydroxylation of proline residues in the X positions of the -Pro-4Hyp-Gly- sequences.

Most invertebrate collagens resemble the vertebrate collagens and have a roughly similar 4-hydroxyproline content in their triple-helical domains.[59,60] Two exceptions are earthworm cuticle collagen and *Ascaris* cuticle collagen which provide striking examples of variation in 4-hydroxyproline content, about 95% of the incorporated proline residues being found as 4-hydroxyproline in the former, while only about 5% of the proline residues are hydroxylated in the latter.[59,60] Earthworm cuticle collagen also differs from the others in that most, if not all, of the 4-hydroxyproline residues are confined to the X positions of the repeating -X-Y-Gly- sequences.[59,60]

B. PLANT PROTEINS

In contrast to the detailed information on many of the 4-hydroxyproline-containing proteins in animals, our knowledge about the corresponding proteins in plants is limited.[13] Since the pioneering work of Lamport, it has been known that 4-hydroxyproline is a major amino acid in plant cell wall hydrolysates.[61] Most of the hydroxyproline-rich plant proteins characterized so far fall into three groups: extensins, 4-hydroxyproline-rich lectins, and arabinogalactan proteins.[11-13]

Extensins are a family of insoluble plant cell wall glycoproteins that contain characteristic -Ser-4Hyp-4Hyp-4Hyp-4Hyp- sequences, with most of the 4-hydroxyproline residues *O*-glycosylated by oligoarabinosides and many of the serine residues galactosylated.[11,13] The insoluble extensins are synthesized in the form of soluble monomers that become insoluble in the cell wall, possibly by the formation of isodityrosine and other cross-links.[13] Circular dichroism studies (see Reference 13) suggest that all of the amino acids are in the poly(L-proline) II helix conformation (Section III).

A number of extensin genes have recently been cloned and sequenced.[62-64] A carrot extensin gene contains three repetitive sequences (with slight alterations) that account for 257 amino acids out of the 306 amino acid polypeptide: (1) -Ser-Pro-Pro-Pro-Pro-Lys-His-; (2) -Ser-Pro-Pro-Pro-Pro-Thr-Pro-Val-Tyr-Lys-Tyr-Lys-; and (3) -Ser-Pro-Ala-Pro-Val-His-His-Tyr-Lys-Tyr-Lys-.[13,62] The second of these sequences has previously been reported in proteolytic fragments of a tomato extensin in the form of -Ser-4Hyp-4Hyp-4Hyp-4Hyp-Thr-4Hyp-Val-Tyr-Lys- with the 4-hydroxyproline residues arising from the hydroxylation of the corresponding proline residues.[11] The amino acid sequences in which 4-hydroxyproline is found in plant extensins are thus distinctly different from the repeating -X-4Hyp-Gly-triplets in animal proteins.

4-Hydroxyproline-rich lectins have so far been isolated only from a single plant family, the Solanaceae.[13] The best studied of these lectins is potato lectin, a glycoprotein with a molecular weight of about 50,000.[13,65] This consists of at least two dissimilar domains, a binding-site domain which is rich in cysteine and glycine and depends on disulfide bridges for its stability, and a 4-hydroxyproline-rich domain (see Reference 65). The latter is highly glycosylated with oligoarabinosides linked to 4-hydroxyproline and single galactose residues linked to serine.[65,66] This 4-hydroxyproline-rich domain is extensin-like and is in the poly(L-proline) II conformation.[65]

Arabinogalactan proteins are found in most higher plants and in many of their secretions.[12,13] They are a group of macromolecules containing large arabinogalactan side chains *O*-glycosidically linked to polypeptides rich in 4-hydroxyproline, serine and alanine. 4-Hydroxyproline residues contain linkages to both the arabinose and the galactose of the carbohydrate side chains.[12] Serine and threonine residues contain additional carbohydrate units. Little detailed information is available on the structure of the protein core.

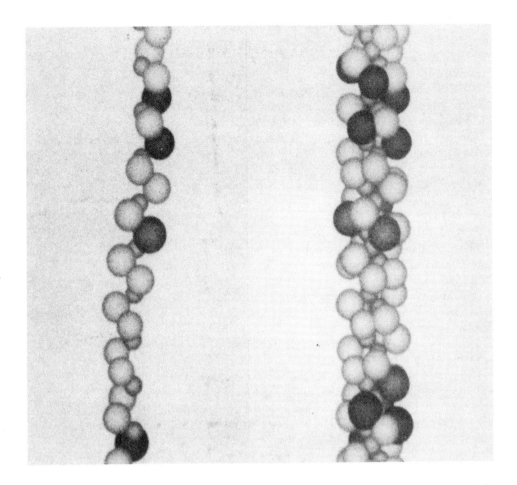

FIGURE 4.   A space-filling model of the collagen triple helix showing one chain (left) and three chains (right). Each sphere denotes an amino acid residue. The small spheres arranged up the center are glycine residues, while the large spheres on the outside can be any amino acid. (Reprinted by permission of the publisher from Piez, K. A., in *Extracellular Matrix Biochemistry,* Piez, K. A. and Reddi, A. H., Eds., Elsevier, New York, 1984, 1. Copyright 1984 by Elsevier Science Publishing Co., Inc.)

4-Hydroxyproline is also found in the cell wall proteins of the green algae. *Chlamydomonas reinhardii*, for example, is composed of concentric layers of 4-hydroxyproline-rich glycoproteins.[67] Algal cell wall glycoproteins containing as much as 62% 4-hydroxyproline have been reported.[68]

III. FUNCTIONS OF 4-HYDROXYPROLINE, 3-HYDROXYPROLINE, AND HYDROXYLYSINE

The hydroxyl groups of the 4-hydroxyproline residues in animal proteins are not substituted in any way, and for a long time it was not understood why collagens contain so much 4-hydroxyproline. The triple-helical conformation of the collagens is dependent on a number of requirements.[17,18,69] The presence of glycine, the smallest amino acid, in every third position is an absolute requirement, as the amino acid in this position occupies the restricted space in which the three helical chains come together in the center of the triple helix (Figure 4). The triple-helical conformation also depends on the presence of proline. Its ring structure prevents rotation about the $N-C_\alpha$ bond in a polypeptide chain, the rotation about the $C_\alpha-C=O$ bond also being restricted. In fact, poly(L-proline) forms a highly stable left-handed single-chain poly(L-proline) II helix with three residues per turn that is very similar to the helix of a collagen α-chain.[17,18,69] Gustavson suggested in the mid-1950s that 4-hydroxyproline residues may have an additional stabilizing function, perhaps because their hydroxyl groups allow the formation of additional hydrogen bonds among adjacent molecules in the fibers.[70] This suggestion was nevertheless discounted because the thermal stabilities of various collagens were found to be more closely proportional to their total imino acid content (proline plus 4-hydroxyproline) than to their 4-hydroxyproline content (for References, see 6).

It was only in 1973 that the hydroxyl group of 4-hydroxyproline residues was shown to have a crucial role in stabilizing the collagen triple helix under physiological conditions.[71-75] Initial evidence for this role of 4-hydroxyproline was obtained in experiments with synthetic peptides having structures of (Pro-Pro-Gly)$_{10}$ and (Pro-4Hyp-Gly)$_{10}$.[74] The presence of 4-hydroxyproline was found to increase the thermal stability of their helices, the midpoint of thermal transition from helix to coil (T_m) being about 35°C higher for the latter peptide than for the former.[74] Similar conclusions were derived from experiments on the thermal denaturation of molecules consisting of three nonhydroxylated type I procollagen pro-α-chains. The T_m of such molecules was found to be only 24°C, a value about 15°C lower than that for molecules consisting of hydroxylated pro-α-chains.[71-73] Investigations into the influence of proline hydroxylation on the thermal stability of a variably 4-hydroxylated collagen fragment also supported this role for 4-hydroxyproline.[75] These findings have subsequently been confirmed and extended (for References, see 6, 8).

It is clear from the above that nonhydroxylated procollagen polypeptide chains cannot form triple-helical molecules at 37°C, and almost complete 4-hydroxylation of proline residues in the Y positions of -X-Y-Gly- triplets is required for the formation of a molecule that is stable at that temperature. It is probable that 4-hydroxyproline has a similar function in the other proteins with a collagen-like triple helix, whereas its role in elastin is unknown.

The means by which 4-hydroxyproline stabilizes the collagen triple helix is not known. The most likely possibility is that the hydroxyl group forms an additional bond, via a water molecule, either between two polypeptide chains or within the same chain in the collagen triple helix.[17,69,76-78] Four such postulated mechanisms are shown in Figure 5. Two of these schemes (b and c) are not possible when the X position is occupied by a proline residue and are therefore not plausible as explanations for the total body of experimental data.[17,77,78]

The 4-hydroxyproline residues in plant proteins are frequently O-glycosylated and thus have an important function as sites of attachment for carbohydrate units (Section II.B). In the case of extensins and 4-hydroxyproline-rich lectins, the oligosaccharide side chains stabilize the conformation of the polypeptide chains, as removal of the oligosaccharides decreases both the amount of poly(L-proline) II helix and the thermal stability of this conformation.[13,65] Several other functions have also been suggested for the 4-hydroxyproline-linked carbohydrate side chains in the various plant proteins, but they must currently be regarded as speculative.[12,13]

The hydroxylysine residues of collagen have two important functions: their hydroxyl groups serve as sites of attachment for carbohydrate units, and they are essential for the stability of the intermolecular collagen cross-links. The carbohydrate is present partly in the form of the monosaccharide galactose and partly the disaccharide glucosylgalactose, the structure of the disaccharide with its peptide attachment being 2-O-α-D-glucopyranosyl-O-β-D-galactopyranosylhydroxylysine, thus involving an unusual α1→2-O-glycosidic bond between glucose and galactose.[79,80] The extent of glycosylation of the hydroxylysine residues and the ratio of galactosylhydroxylysine to glucosylgalactosylhydroxylysine vary markedly between the genetically distinct collagen types and within the same collagen type from various sources.[16,80]

The intermolecular collagen cross-links are formed by condensation between an aldehyde derived from a lysine or hydroxylysine residue and the ε-amino group of a second lysine,

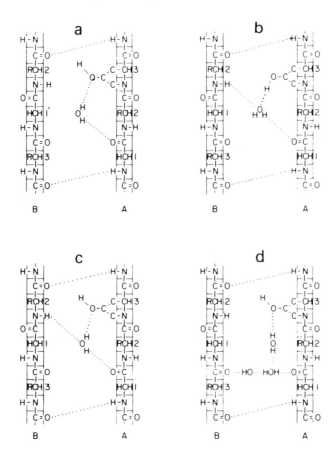

FIGURE 5.   Schemes suggested for the mechanism of the stabilization
of the collagen molecule by 4-hydroxyproline residues. (From Suzuki, F.,
Fraser, R. D. B., and MacRae, T. P., *Int. J. Biol. Macromol.*, 2, 54,
1980. With permission.)

hydroxylysine or glycosylated hydroxylysine residue. The cross-links formed from a hydroxylysine-derived aldehyde are more stable than those formed from a lysine-derived aldehyde, possibly owing to migration of the double bond to a stable oxo form.[17,27,81-83] Hydroxylysine-derived cross-links are converted with time to trivalent 3-hydroxypyridinium residues consisting of either three residues of hydroxylysine or two residues of hydroxylysine and one of lysine.[81-84] The significance of lysine hydroxylation for the stabilization of collagen cross-links is clearly demonstrated by the profound changes in the mechanical properties of certain tissues that are seen in patients with a genetic deficiency in lysyl hydroxylase activity (Section X).

The hydroxylysine residues in the subcomponent Clq of complement[34-36] and mannose-binding proteins[44] are also glycosylated, mostly by the disaccharide glucosylgalactose, whereas the presence of hydroxylysine-linked carbohydrate units has not been reported for acetylcholinesterase. The cross-links derived from hydroxylysine appear to be unique to collagens.

The role of the 3-hydroxyproline residues is unknown at present.

IV. MOLECULAR PROPERTIES OF THE HYDROXYLASES

A. ANIMAL PROLYL 4-HYDROXYLASES

Prolyl 4-hydroxylase was first purified to near homogeneity from chick embryos[85,86] and the skin of newborn rats[87] by conventional procedures. Two affinity column procedures were subsequently developed that allowed the production of large quantities of pure enzyme.[88,89] The first is based on the affinity of the enzyme for its polypeptide substrate,[88] and the second for its polypeptide inhibitor.[89] Prolyl 4-hydroxylase has now been purified to homogeneity by these procedures or their modifications[14,15] from several vertebrate sources,[88-93] including human tissues.[94,95]

The active prolyl 4-hydroxylase in vertebrates is a tetramer ($\alpha_2\beta_2$) with a molecular weight of about 240,000[86,88,89] and consisting of two types of monomer with molecular weights of about 64,000 (α-subunit) and 60,000 (β-subunit).[87-90,94] Electron microscopy shows the monomers to be rod-shaped, and the tetramer to consist probably of two interlocked V-shaped dimers.[96] Intrachain disulfide bonds seem to be essential for the monomers to maintain the native structure necessary for their association, whereas the existence of interchain disulfide bonds does not seem likely.[6,97,98] The monomers produced by dissociation of the enzyme tetramer do not possess any hydroxylase activity.[88,98,99] All attempts at reconstituting an active enzyme tetramer from the dissociated monomers *in vitro* have proved unsuccessful.[88,98,99]

Amino acid analyses of pure prolyl 4-hydroxylase from chick embryos,[88,89] rat skin,[91] and human tissues[94] indicate that there is a great similarity between the enzyme proteins from all these sources. Distinct differences are found between the α-subunit and β-subunit in their amino acid composition and the peptide fragments produced by a number of proteinases.[91,97] Recent molecular cloning and nucleotide sequencing of the two types of subunit have in fact indicated that these subunits have entirely different amino acid sequences (Section VIII). Prolyl 4-hydroxylase, like the other two hydroxylases of collagen synthesis, is a glycoprotein, most or all of the carbohydrate being located on the α-subunit.[91,97,100,101] The role of this carbohydrate is unknown and its removal does not influence the enzyme activity.[101]

Two forms of the α-subunit have been found in prolyl 4-hydroxylase from various sources such as chick embryo fibroblasts, L929 cells, newborn mice and human liver.[101,103] In chick embryo fibroblasts the larger α-subunit, termed α', contains two asparagine-linked high mannose oligosaccharides, each with eight mannose units, whereas the smaller α-subunit, termed α, contains a single asparagine-linked oligosaccharide with seven mannose units.[101] Both oligosaccharides can be cleaved with endo-β-N-acetylglucosaminidase H and completely digested with α-mannosidase to yield mannosyl-N-acetylglucosamine.[101] Biosynthetic findings and other data suggest that the two forms of the α-subunit represent two highly homologous gene products in which slight differences in amino acid sequence may result in the use of two glycosylation sites on the α'-subunit but only one on α.[101,102] Molecular cloning demonstrates the presence of only one gene for the α-subunit in the human genome and indicates that two slightly different forms of the α-subunit arise by alternative splicing of the products of one gene (Section VIII.A).

Nonvertebrate prolyl 4-hydroxylases have been partially characterized at a low level of purification from the muscle layers of *Ascaris lumbricoides*[104] and the subcuticular epithelium of earthworms,[105] and one such enzyme has been purified more extensively from another nematode *Panagrellus silusiae*.[106] A high degree of purification has been reported for the enzyme from the foot of the marine mussel *Mytilus edulis*.[107] The *Panagrellus* and *Mytilus* enzymes appear to be similar to the vertebrate prolyl 4-hydroxylase in their molecular properties, although slight differences are found in the molecular weights of the subunits.[106,107]

B. PLANT PROLYL 4-HYDROXYLASES

Prolyl 4-hydroxylases have been partially characterized from many plant species.[108-116] A high extent of purification has been obtained for the enzyme from suspension-cultured cells of the French bean, *Phaseolus vulgaris*.[113] The molecular weight of this enzyme by gel filtration is similar to that of the vertebrate prolyl 4-hydroxylase tetramer, and the French bean enzyme may consist of two types of subunit with molecular weights of about 65,000.[113]

A similar molecular weight for the native enzyme is also found in the case of two vascular plants,[116] suggesting that higher-plant prolyl 4-hydroxylases may resemble the vertebrate enzyme in their basic structure.

The prolyl 4-hydroxylase found in multicellular and unicellular green algae such as *Chlamydomonas, Enteromorpha* and *Volvox* differs distinctly from the vertebrate and higher-plant prolyl 4-hydroxylases in that it appears to be a monomer.[115,116] It has a molecular weight of about 65,000 and is antigenically related to the α-subunit of the vertebrate enzyme.[116] Nevertheless, the algal enzyme activity is partially inhibited by a monoclonal antibody to the β-subunit of human prolyl 4-hydroxylase, suggesting that the algal enzyme monomer may also contain sequences that correspond to those in the β-subunit of the vertebrate enzyme.[115] One report claims that the algal enzyme is a tetramer consisting of subunits with molecular weights ranging from about 65,000 to 60,000.[114] Unfortunately the molecular weight was not determined for the native enzyme, and therefore it seems possible that some of the polypeptides present in the dissociated preparation may in fact have been derived from a contaminating protein rather than the presumed enzyme tetramer.

C. LYSYL HYDROXYLASE

Lysyl hydroxylase is very similar to prolyl 4-hydroxylase in its catalytic properties (Sections V and VI), but it was demonstrated at an early stage that the two enzyme activities are derived from separate proteins.[85,117,118] Lysyl hydroxylase was first purified several hundred- or thousand-fold from chick embryo extract by conventional procedures.[118,119] Subsequenty two affinity column procedures were developed,[120,121] and the enzyme was purified to homogeneity from chick embryos[121] and human placental tissues.[122]

The active lysyl hydroxylase is probably a dimer (α_2) with a molecular weight of about 190,000 in gel filtration and consisting of only one type of monomer with a molecular weight of about 80,000 to 85,000.[121-123] An additional enzyme form that has an apparent molecular weight of about 550,000 in gel filtration is found under some conditions.[118,119] This is likely to be a polymeric form of the enzyme dimer. Lysyl hydroxylase is a glycoprotein[119,120,123] containing asparagine-linked carbohydrate units, which, unlike the case of prolyl 4-hydroxylase, are essential for maximal enzyme activity.[123]

Some data have been interpreted as suggesting that lysyl hydroxylase may have collagen type-specific or tissue-specific isoenzymes. Large differences are found in the extent of lysine hydroxylation between genetically distinct collagen types and within the same collagen type from different tissues (Section II.A). The deficiency in lysyl hydroxylase activity present in the type VI variant of the Ehlers-Danlos syndrome leads to hydroxylysine deficiency, which also shows a wide variation in degree from tissue to tissue (Section X). Nevertheless, the molecular and catalytic properties of lysyl hydroxylase purified from a number of sources[121-123] and the catalytic properties determined in crude extracts from cells synthesizing various genetically distinct collagen types[124] are essentially identical. Other data indicate that differences in the extent of lysine hydroxylation between collagen types can be explained by regulation of a single lysyl hydroxylase and thus do not require the presence of isoenzymes.[125,126] Immunological characterization of chick embryo lysyl hydroxylase similarly does not support the suggestion that various tissues have collagen type-specific or tissue-specific forms.[127]

Circumstantial evidence has been presented for the existence of a separate lysyl hydroxylase specific to the lysines present in the sequences -X-Lys-Ser- or -X-Lys-Ala- in the short noncollagenous domains at the ends of the α-chains in some collagens (Section II.A). This evidence is based on the finding that highly purified lysyl hydroxylase acting on lysine residues in the collagenous -X-Lys-Gly- triplets did not hydroxylate these lysines.[128] No direct data are available on the presence or properties of this presumed second lysyl hydroxylase, however. It may be noted that the only lysyl hydroxylase characterized so far has been found to hydroxylate lysine residues in arginine-rich histone, which does not contain any -X-Lys-Gly- sequences but does contain the sequences discussed above (Section V.C).

D. PROLYL 3-HYDROXYLASE

The formation of 3-hydroxyproline and 4-hydroxyproline residues has been shown to be catalyzed by separate enzymes.[129,130] Prolyl 3-hydroxylase has been purified to about 5000-fold from an ammonium sulfate fraction of chick embryo extract by a procedure involving one affinity chromatography,[131] but it has not been isolated as a homogeneous protein.

The molecular weight of prolyl 3-hydroxylase from chick embryos[131] and rat kidney cortex[130] is about 160,000 in gel filtration, but the subunit structure is currently unknown. Prolyl 3-hydroxylase is probably a glycoprotein, as its activity is inhibited by concanavalin A, and as the enzyme binds to columns of this lectin and can be eluted with a buffer containing α-D-mannoside.[131]

V. PEPTIDE SUBSTRATES

The proline or lysine residue to be hydroxylated must be in the peptide linkage, since none of the enzymes acts on the corresponding free amino acid.[108,129,132,133] As discussed below, the minimum sequence requirement for hydroxylation by a vertebrate prolyl 4-hydroxylase is fulfilled by an -X-Pro-Gly- triplet and that for hydroxylation by lysyl hydroxylase by an -X-Lys-Gly- triplet. In some cases, however, both enzymes also hydroxylate certain other sequences. Prolyl 3-hydroxylase appears to require a -Pro-4Hyp-Gly- triplet. The interaction with a vertebrate prolyl 4-hydroxylase and with lysyl hydroxylase is further influenced by the amino acid in the X position of the triplet that is hydroxylated, and the interaction with all these animal enzymes is affected by other nearby amino acids, the peptide chain length, and the peptide conformation. Plant prolyl 4-hydroxylases appear to require primarily that the proline residue to be hydroxylated should be in a peptide with a poly(L-proline) II helix conformation, although other factors may also influence the reaction.

A. ANIMAL PROLYL 4-HYDROXYLASES

Early studies demonstrated that the tripeptides Gly-Pro-Pro, Gly-Pro-Ala and Pro-Gly-Pro do not act as substrates for a vertebrate prolyl 4-hydroxylase, whereas the tripeptides Pro-Pro-Gly and Ala-Pro-Gly do.[133-138] The vasoactive peptide bradykinin, with the structure H-Arg-Pro-Pro-Gly-Phe-Ser-Pro-Phe-Arg-OH, also serves as a substrate,[139] whereas this property is lost if the glycine is replaced by alanine, sarcosine, or 5-amino valine.[140] Poly-tripeptides with the structure (Pro-Pro-Gly)ₙ are good substrates, with only the proline residues preceding glycine being hydroxylated.[133,141] (Polytripeptides synthesized as [Gly-X-Y]ₙ or [Y-Gly-X]ₙ are referred to here as [X-Y-Gly]ₙ to simplify the comparisons.) Similarly, polytripeptides with the structure (Ala-Pro-Gly)ₙ are good substrates, whereas those with the structure (Pro-Ala-Gly)ₙ are either very poor substrates or completely inactive.[136,142] Experiments on the hydroxylation of a biologically prepared 4-hydroxyproline-deficient protein substrate also demonstrated that only proline residues preceding glycine are hydroxylated.[143] All these findings agree with the presence of 4-hydroxyproline in collagens and other proteins containing this amino acid in the Y position of -X-Y-Gly- triplets (Section II.A), and indicate that the minimum sequence requirement for a vertebrate prolyl 4-hydroxylase is fulfilled by an -X-Pro-Gly- triplet. Nevertheless, glycine can well be replaced by β-alanine, an amino acid which is larger than glycine but does not contain a side chain.[144] The few cases of -X-4Hyp-Ala- sequences in the subcomponent C1q of complement and type IV collagen (Section II.A) also indicate that the requirement for an -X-Pro-Gly- sequence is not absolute, and that the glycine can in some instances be replaced by alanine. In agreement with these data, a low but significant rate of hydroxylation of the tetrapeptides -Pro-Pro-Ala-Pro- and -Pro-Pro-Gln-Pro was reported recently.[145]

Prolyl 4-hydroxylase from the subcuticular epithelium of earthworms differs distinctly in its specificity from all other animal prolyl 4-hydroxylases studied so far in that (Pro-Ala-Gly)$_n$ is a much better substrate than (Ala-Pro-Gly)$_n$.[105] This property agrees with the unusual occurrence of 4-hydroxyproline in the X positions of the -X-Y-Gly- sequences in earthworm cuticle collagen.

The rate of hydroxylation of a proline residue by a vertebrate prolyl 4-hydroxylase is clearly dependent on the amino acid in the X position of the -X-Pro-Gly- triplet to be hydroxylated. Exact comparison of the values reported in different papers is not possible, however, as other properties of the peptides and the reaction conditions used also vary. Data obtained in experiments with polytripeptides of the structure (X-Pro-Gly)$_n$ suggest that proline in the X position may give a particularly high maximal reaction velocity *(V)*, while alanine, leucine, arginine, valine and glutamate give a lower *V* in this order.[6,8,136,142,146-148] The reaction velocities found with (Arg-Pro-Gly)$_5$ and (Glu-Pro-Gly)$_5$, for example, are about one third and one twentieth, respectively, of the velocity observed with (Pro-Pro-Gly)$_5$,[6] and a similar difference has been found in the comparison between (Glu-Pro-Gly)$_3$ and (Pro-Pro-Gly)$_3$.[147] The presence of glycine or sarcosine in the X position of the (X-Pro-Gly)$_n$ polytripeptides completely prevents hydroxylation.[136,142,148] The data obtained with (X-Pro-Gly)$_n$ polytripeptides nevertheless do not necessarily hold for a single -Gly-Pro-Gly- sequence present among other types of -X-Pro-Gly- triplet, as various collagens do contain a few -Gly-4Hyp-Gly- sequences.[16]

Amino acids in other parts of the peptide also influence the hydroxylation of an -X-Pro-Gly- triplet. The K_m for Arg-Gly-(Pro-Pro-Gly)$_5$, for example, is only about one-half of that for (Pro-Pro-Gly)$_5$.[146] Similarly, insertion of an additional arginine residue into the amino-terminal end of bradykinin (see above) reduces the K_m to less than half compared with the insertion of glycine.[140] The insertion of glutamic acid into polytripeptides appears to have no effect on binding to the enzyme, as the K_m values for (Pro-Pro-Gly)$_5$, Glu-Gly-(Pro-Pro-Gly)$_5$ and (Glu-Pro-Gly)$_5$ are all similar,[6,16] although the *V* for (Glu-Pro-Gly)$_5$ is very low (see above). Nevertheless, the addition of a glutamic acid residue to the amino-terminal end of bradykinin decreases both the K_m and *V* values.[140] Several other alterations in the amino acid sequence of bradykinin also influence the interaction with prolyl 4-hydroxylase, thus demonstrating an effect of amino acids outside the triplet to be hydroxylated.[140] It is not known, however, to what extent the effects found with this short peptide containing only a single -X-Pro-Gly- triplet may be related to those occurring with long peptides having glycine in every third position.

The chain length of the peptide substrate has a marked effect on the K_m, whether expressed as the molar concentration of the X-Pro-Gly- triplets or of the peptide, in that this value decreases with increasing chain length.[135,141,146] Experiments performed with peptides of the structure (Pro-Pro-Gly)$_n$ (Table 3) and (Ala-Pro-Gly)$_n$[136] demonstrate a similar effect. Interestingly, the chain length of the peptide does not seem to influence the *V* of the reaction (Table 3). Type I protocollagen, a biologically prepared protein consisting of non-hydroxylated pro-α-chains with molecular weights of about 150,000,[24-26] has a particularly low K_m (Table 3), about 0.4% of that for (Pro-Pro-Gly)$_{20}$, but the *V* for this protein is not significantly different from that for Pro-Pro-Gly-NHCH$_3$ (Table 3).

The conformation of the peptide substrate has a crucial effect, in that the triple-helical conformation of the collagenous peptides completely prevents hydroxylation.[150-152] In collagens and related peptides a minimal structural requirement for proline 4-hydroxylation appears to be a poly(L-proline) II helix (Section III) followed by a β-turn.[145,153,154] The poly(L-proline) II structure may aid effective interaction at the substrate binding subsites, while the β-turn may be essential at the catalytic site of the enzyme.[145] In elastin and related

TABLE 3
Effect of Peptide Chain Length on the Michaelis Constant (K$_m$) and Maximal Velocity (V) of the Hydroxylation of Random-Coil Forms of (Pro-Pro-Gly)$_n$ by Prolyl 4-Hydroxylase

Peptide substrate	Relative V (%)	K$_m$ of -X-Pro-Gly- (μM)	Ref.
Pro-Pro-Gly	No data	20,000	6
Acetyl-Pro-Pro-Gly	No data	23,000	6
Pro-Pro-Gly-NHCH$_3$	100	44,000	6
(Pro-Pro-Gly)$_5$	100	1,750	146
(Pro-Pro-Gly)$_{10}$	100	280	146
(Pro-Pro-Gly)$_{20}$	105	50	146
Type I protocollagen	100	0.2	149

From Kivirikko, K. I. and Myllylä, R., *The Enzymology of Post-Translational Modification of Proteins*, Freedman, R. B. and Hawkins, H. C., Eds., Academic Press, London, 1980, 53. With permission.

peptides, a β-strand structure may be interchangeable with the poly(L-proline) II conformation.[145]

A further aspect of the peptide substrate is that prolyl 4-hydroxylase hydroxylates (Pro-Pro-Gly)$_5$ and (Pro-Pro-Gly)$_{10}$ asymmetrically, in that the fourth[156] or ninth[149] triplet from the amino-terminal end, respectively, is hydroxylated more readily than any other. These findings suggest that the enzyme has an asymmetrical active site at which binding subsites are not located symmetrically with the catalytic subsite.[149] Thus the hydroxylation of a given -X-Pro-Gly- triplet may be influenced by amino acids far outside the triplet.

B. PLANT PROLYL 4-HYDROXYLASES

Plant prolyl 4-hydroxylases do not act on Pro$_2$, Pro$_3$ or Pro$_4$, whereas Pro$_6$ and Pro$_n$ (n>6) are readily hydroxylated.[109,110] Pro$_5$ is hydroxylated at 15°C, but this property is lost at 30°C.[110] Poly(L-proline) forms a unique type II conformation in aqueous solutions (Section III). This conformation is found in Pro$_n$ (n≥6) at 30°C and Pro$_5$ at 15°C but not 30°C.[110] These data thus indicate that plant prolyl 4-hydroxylases recognize primarily the poly(L-proline) II helix.[110] Although reactions with model peptides of the structure Pro$_n$ require the presence of at least five proline residues, even a single proline residue appears to be hydroxylated provided that it is present in a peptide with poly(L-proline) II conformation. This is demonstrated by the presence of 4-hydroxyproline in extensins in a sequence -Ser-4Hyp-4Hyp-4Hyp-4Hyp-Thr-4Hyp-Val-Tyr-Lys- (Section II.B), in which a single proline residue in the sequence -Thr-Pro-Val- has been hydroxylated. Unlike the case of the vertebrate prolyl 4-hydroxylase, the peptide chain length does not seem to influence the reaction, provided that the peptide is long enough to form the poly(L-proline) II helix.[109,110,115] No data are yet available on the influence of other nearby amino acids on the rate of hydroxylation of a proline residue in the appropriate conformation.

The specificity of plant prolyl 4-hydroxylases for poly(L-proline) and their natural substrates is not absolute, however, as a low hydroxylation rate has been found with some other proline-containing peptides. An early work suggested that prolyl 4-hydroxylase from carrot disks can hydroxylate nonhydroxylated type I procollagen pro-α-chains,[108] although it was unknown how the rate of this reaction was related to that obtained with the natural plant substrate. Subsequent papers have reported low rates of hydroxylation of (Pro-Pro-Gly)$_5$ and (Pro-Pro-Gly)$_{10}$ under conditions in which these peptides exist in random-coil forms,[109,112,114,115] whereas virtually no hydroxylation has been found with triple-helical (Pro-Pro-Gly)$_{10}$.[115] A low hydroxylation rate has also been reported for bradykinin with the structure H-Arg-Pro-Pro-Gly-Phe-Ser-Pro-Phe-Arg-OH.[112,114]

C. LYSYL HYDROXYLASE

Lysyl hydroxylase does not hydroxylate the tripeptide Lys-Gly-Pro, whereas the tripeptide Ile-Lys-Gly serves as a substrate.[157] Lysine vasopressin, with the structure Cys-Tyr-Phe-Gln-Asn-Cys-Pro-Lys-Gly-NH$_2$,[157] lysine-rich histone,[158] which contains -X-Lys-Gly- sequences, and polypeptides with the structure (X-Lys-Gly)$_n$ also act as substrates.[157,159] These data indicate that the minimum sequence requirement for lysyl hydroxylase is fulfilled by an -X-Lys-Gly- triplet. Nevertheless, lysyl hydroxylase also acts on arginine-rich histone,[158] which does not contain any -X-Lys-Gly- sequence, but does contain sequences such as -X-Lys-Ser-, -X-Lys-Ala-, and -X-Lys-Thr-. This agrees with the presence of hydroxylysine in the sequences -X-Hyl-Ser- and -X-Hyl-Ala- in the short noncollagenous domains at the ends of the α-chains in some collagens (Section II.A). It is not known at present whether these collagen sequences are hydroxylated *in vivo* by a separate lysyl hydroxylase specific to them (Section IV.C).

As in the case of the vertebrate prolyl 4-hydroxylase, the reaction with lysyl hydroxylase appears to be affected by the amino acid sequence around the lysine residue, the peptide chain length, and the peptide conformation. Comparisons of Ile-Lys-Gly, (Ile-Lys-Gly)$_2$, (Ile-Lys-Gly)$_3$-Phe, (Ile-Lys-Gly)$_5$-Phe, (Ala-Lys-Gly)$_{10}$, lysine vasopressin, and four synthetic peptides with more complex amino acid sequences as substrates demonstrate the influence of both the amino acid sequence and the chain length.[157,159,160] The K$_m$ values for (Ile-Lys-Gly)$_2$, (Ile-Lys-Gly)$_3$-Phe, and (Ile-Lys-Gly)$_5$-Phe, for example, are 4 mM, 1 mM, and 0.14 mM, respectively, all these values being expressed as mM concentrations of -X-Lys-Gly- units.[157,159] As in the case of the vertebrate prolyl 4-hydroxylase, the chain length of the peptide appears to influence only the K$_m$, whereas the V of the reaction seems to be unaffected.[157,159] The K$_m$ for type I protocollagen, a protein consisting of nonhydroxylated pro-α-chains, is particularly low, about 10^{-8} M, when expressed in terms of the molar concentration of peptide chains.[161]

The conformation of the peptide substrate has a marked effect in that triple-helical conformation of the peptide prevents hydroxylation.[161,162] By analogy with vertebrate prolyl 4-hydroxylase, it has been suggested that one conformational requirement for lysine hydroxylation may be a β-bend involving the lysine residue.[163]

A surprising aspect of the peptide substrate is that denatured 4-hydroxyproline and hydroxylysine-deficient collagen from the cuticle of *Ascaris* worms does not serve either as a substrate or an inhibitor, even though it contains about 40 lysine residues per 1000 amino acids and is a good substrate for vertebrate prolyl 4-hydroxylase.[157] Peptides prepared from this collagen by cyanogen bromide cleavage likewise fail to act as either substrates or inhibitors of lysyl hydroxylase.[157] It has also been found that a 99 amino acid peptide fragment from the α1-chain of rat skin collagen is a particularly poor substrate, even though it contains two -Ala-Lys-Gly- sequences that are virtually nonhydroxylated *in vivo*.[162] This finding cannot be explained by an inability of the -Ala-Lys-Gly- sequence to serve as a substrate, since other collagens contain many hydroxylated -Ala-Lys-Gly- sequences,[16] and the polytripeptide (Ala-Lys-Gly)$_{10}$ acts as a substrate for the enzyme.[159] The data on *Ascaris* cuticle collagen and the α1-chain peptide fragment thus suggest that some amino acid sequences around the Y position lysine residues may effectively prevent interaction with lysyl hydroxylase.

D. PROLYL 3-HYDROXYLASE

Type I protocollagen, a protein containing no 3-hydroxyproline or 4-hydroxyproline (see above), does not serve as a substrate for prolyl 3-hydroxylase.[130] Nevertheless, when this same protein is first fully hydroxylated with prolyl 4-hydroxylase to convert all the proline residues in the Y positions of the -X-Y-Gly- triplets to 4-hydroxyproline, the protein will then act as a good substrate for prolyl 3-hydroxylase.[130] 3-Hydroxyproline has been identified in collagens only in the sequence -Gly-3Hyp-4Hyp-Gly- (Section II.A), and thus the main or only substrate sequence for prolyl 3-hydroxylase appears to be a -Pro-4Hyp-Gly- triplet.

As in the cases of vertebrate prolyl 4-hydroxylase and lysyl hydroxylase, the 3-hydroxylation of various -Pro-4Hyp-Gly- triplets is likely to be influenced by other nearby amino acids.[129] Longer peptides are similarly better substrates for this enzyme than shorter ones, and a triple-helical conformation in the peptide completely prevents hydroxylation.[129] The K_m for fully 4-hydroxylated protocollagen as a substrate is very low, about $10^{-8} M$.[164]

VI. COSUBSTRATES, ACTIVATORS, AND REACTION MECHANISMS

The prolyl 4-hydroxylases, lysyl hydroxylase and prolyl 3-hydroxylase all require Fe^{2+}, 2-oxoglutarate, O_2 and ascorbate. The 2-oxoglutarate is stoichiometrically decarboxylated during hydroxylation, with one atom of the O_2 molecule being incorporated into the succinate while the other is incorporated into the hydroxyl group formed on the proline or lysine residue. Ascorbate is a highly specific requirement, but it is not consumed stoichiometrically, and the enzymes can catalyze their reactions for a number of catalytic cycles in the absence of this vitamin. As described below, the enzymes also catalyze an uncoupled decarboxylation of 2-oxoglutarate without subsequent hydroxylation of the peptide substrate, and ascorbate serves as an alternative oxygen acceptor in the uncoupled decarboxylation cycles. The reactions catalyzed by these enzymes can thus be summarized as shown for prolyl 4-hydroxylase in Figure 6.

A. COSUBSTRATES AND ACTIVATORS

The kinetic constants of the enzymes for their cosubstrates are very similar (Table 4).[115,131,134,165,166] The values shown with synthetic peptide substrates have been determined from initial velocity data obtained by varying two of the reaction components, and may hence be more correct than those determined with biologically prepared polypeptide substrates. As indicated for vertebrate prolyl 4-hydroxylase, the synthetic substrate values are accordingly slightly higher (Table 4).

The nonheme Fe^{2+} of the enzymes[3-8,167] is loosely bound, and the activity of most but not all[168] enzyme preparations is completely dependent on added Fe^{2+}. Kinetic studies indicate a binding of Fe^{2+} at thermodynamic equilibrium before the binding of 2-oxoglutarate (Section VII),[166,169] and the iron is not released from the enzymes between the catalytic cycles under turnover conditions.[166,169-171] In the case of vertebrate prolyl 4-hydroxylase it has been demonstrated that the enzyme binds two Fe^{2+} atoms per tetramer, i.e., one Fe^{2+} atom per pair of dissimilar subunits.[172] The Fe^{2+} is probably coordinated with the enzyme by three side chains,[173] some of which are likely to be cysteine residues.[8,173]

2-Oxoglutarate is an absolute and highly specific requirement.[174,175] Nevertheless, it can be replaced by 2-oxoadipinate in the reactions catalyzed by the vertebrate prolyl 4-hydroxylase[176] and lysyl hydroxylase,[177] although the K_m of 2-oxoadipinate is markedly higher. An early study suggested that carrot disk prolyl 4-hydroxylase can also utilize pyruvate and oxaloacetate,[108] but a strict specificity for 2-oxoglutarate has been found in subsequent experiments with plant prolyl 4-hydroxylases from *Vinca rosea*[109] and *Chlamydomonas reinhardii*.[115] Although the K_m values of the various hydroxylases for their cosubstrates are very similar (Table 4), the K_m for 2-oxoglutarate may be significantly higher in the case of lysyl hydroxylase than in that of the other enzymes.

The 2-oxoglutarate is stoichiometrically decarboxylated during hydroxylation.[109,139,157,175] In the absence of the peptide substrate, the enzymes catalyze an uncoupled

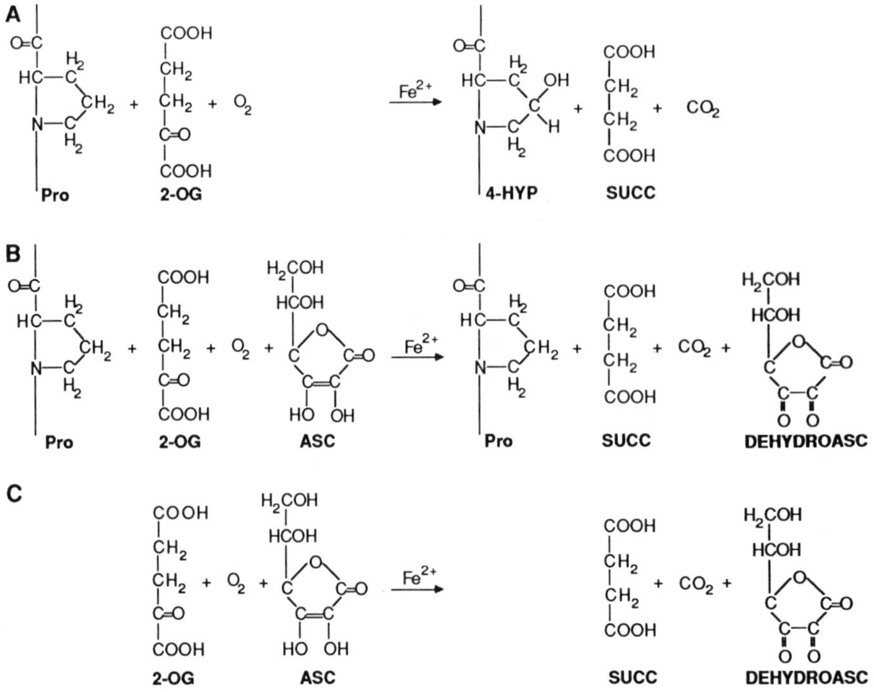

FIGURE 6.   Schematic representation of the reaction catalyzed by prolyl 4-hydroxylase. The 2-oxo-glutarate is stoichiometrically decarboxylated during the hydroxylation reaction, which does not need ascorbate (A). The enzymes also catalyze an uncoupled decarboxylation of 2-oxoglutarate without subsequent hydroxylation of the peptide substrate. Ascorbate serves as a stoichiometrically consumed alternative oxygen acceptor in the uncoupled reaction cycles, which may take place either in the presence (B) or absence (C) of the peptide substrate. Pro, proline; 4-Hyp, 4-hydroxyproline; 2-Og, 2-oxoglutarate; Succ, succinate; Asc, ascorbate; Dehydroasc, dehydroascorbate. (From Kivirikko, K. I., Myllylä, R., and Pihlajaniemi, T., *FASEB J.*, 3, 1609, 1989. With permission.)

TABLE 4
Apparent K_m Values for the Cosubstrates of the Hydroxylases

Enzyme and substrate	K_m for cosubstrate (μM)				
	Fe^{2+}	2-Oxoglutarate	O_2	Ascorbate	Ref.
Vertebrate prolyl 4-hydroxylase[a]					
Biological substrate	2	5	n.d.[b]	100	134
Synthetic substrate	4	22	43	300	165
Plant prolyl 4-hydroxylases[c]					
Synthetic substrate	28	30	n.d.[b]	200	115
Lysyl hydroxylase[a]					
Synthetic substrate	3	100	45	240	166
Prolyl 3-hydroxylase[a]					
Biological substrate	2	3	30	120	131

[a]   From chick embryos.
[b]   n.d. = not determined.
[c]   From *Chlamydomonas reinhardii*.

decarboxylation of 2-oxoglutarate (Figure 6),[115,166,169,178] the rate of which is about 1 to 10% of that found with a saturating concentration of the peptide substrate in the cases of the various hydroxylases.[115,166,169,179] Uncoupled decarboxylation requires the same cosubstrates as the complete reaction,[169,178] and its rate is distinctly increased by the presence of competitive peptide inhibitors.[142,178,179] The K_m values for Fe^{2+} and 2-oxoglutarate are similar in the complete and uncoupled reactions, whereas the K_m for O_2 in the uncoupled reaction is one order of magnitude higher.[180]

The oxygen of the hydroxyl group is derived from molecular oxygen,[181,182] the other atom of the O_2 molecule being incorporated into the succinate.[183] In the lysyl hydroxylase reaction the oxygen atoms present in an enzyme bound intermediate appear to be exchangeable with water, whereas such an exchange takes place only to a minor extent in the prolyl 4-hydroxylase reaction.[184] The first activated intermediate is probably a dioxygen unit bound to the Fe^{2+} of the enzyme,[173] while the final active intermediate performing the hydroxylation reaction is likely to be ferryl ion (Section VI.B).[173,180,185]

Ascorbate is an absolute requirement.[133,174,186,187] Early work suggested that it can be replaced to a significant extent by a number of other reductants,[87,174] but more recent studies indicate a high degree of specificity, only cysteine and dithiothreitol being able to replace it to a minor extent.[166,188] Nevertheless, ascorbate can be completely replaced by derivatives differing only in their side chain, such as D-isoascorbate and 5,6-O-isopropylidene L-ascorbate,[141,174,189] whereas modifications of the ring atoms that abolish the capacity to bind iron render the molecule inactive.[189] A low hydroxylation rate has also been found in certain cultured cells in the complete absence of ascorbate,[8,187] this being probably due to a microsomal reducing protein that contains cysteinyl-cysteine.[190,191] Ascorbate appears to react directly with the enzyme bound iron and thus to reduce it through an "inner sphere" mechanism.[188,189]

Ascorbate is not consumed stoichiometrically during hydroxylation (Figure 6), and the enzymes can complete many catalytic cycles at essentially a maximal rate in the absence of this vitamin (Figure 7).[166,188,192] Nevertheless, the hydroxylation then ceases rapidly, and ascorbate is needed to reactivate the system.[188,193] It has recently been demonstrated that the reaction requiring ascorbate is the uncoupled decarboxylation of 2-oxoglutarate, in which ascorbate is consumed stoichiometrically with the decarboxylation of 2-oxoglutarate (Figure 6 B,C).[179,180] The hydroxylases catalyze such uncoupled decarboxylation cycles even in the presence of a saturating concentration of their peptide substrates (Figure 6 B). Furthermore, certain peptides that do not become hydroxylated enhance the rate of the uncoupled reaction (see above). The biological peptide substrates of the prolyl hydroxylases and lysyl hydroxylase also contain many other sequences in addition to the substrate sequences. An interaction between some of the nonhydroxylatable sequences and the active sites of the enzymes is likely to lead to an uncoupled reaction cycle. It thus seems probable that the main biological function of ascorbate in these enzyme reactions, both *in vitro* and *in vivo*, is to serve as an alternative oxygen acceptor in the uncoupled reaction cycles.[179] The mechanisms of this function are discussed further in Section VI.B.

Even in the presence of optimal concentrations of all the cosubstrates, maximal activity is not obtained with the purified enzymes unless dithiothreitol, bovine serum albumin and catalase are added to the incubation solutions.[6,8,14] The stimulation provided by dithiothreitol agrees with the suggestion that the catalytic sites of the enzymes contain free thiol groups that are essential for the activity.[4,6,8] The action of bovine serum albumin is in part explained by a "protein effect" that can also be obtained with other proteins, but is in part more specific, probably due to the presence of many free thiol groups on this protein.[4,87] Catalase is likely to act partly by destroying peroxide, which is generated nonenzymatically by solutions of Fe^{2+}, O_2 and ascorbate,[141] and in part by a nonspecific protein effect.[87]

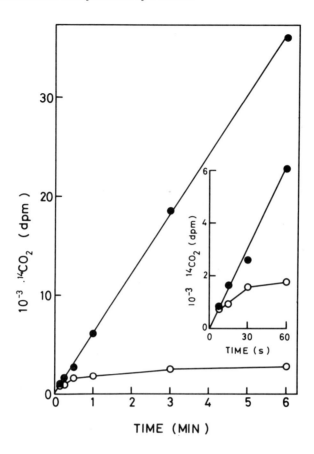

FIGURE 7. Hydroxylation-coupled decarboxylation of 2-oxo[^{14}C]glutarate by pure vertebrate prolyl 4-hydroxylase in the presence (●) or absence (○) of ascorbate. (From Myllylä, R., Kuutti-Savolainen, E.-R., and Kivirikko, K. I., *Biochem. Biophys. Res. Commun.*, 83, 441, 1978. With permission.)

B. REACTION MECHANISMS

Extensive kinetic studies have been performed to elucidate the mechanisms of the vertebrate prolyl 4-hydroxylase[165,169,188] and lysyl hydroxylase[166,194] reactions. These two reactions appear to have identical mechanisms. The kinetic and other[8,179] data are consistent with a reaction scheme (Figure 8 A) involving an ordered binding of Fe^{2+}, 2-oxoglutarate, O_2 and the peptide substrate to the enzyme in this order, and an ordered release of the hydroxylated peptide, CO_2, succinate, and Fe^{2+}, in which Fe^{2+} is not released between the catalytic cycles (see also Section VI.A) and in which the order of release of the hydroxylated peptide and CO_2 is uncertain. The more recent finding that the K_m for O_2 is lower in the complete reaction than in the uncoupled one,[180] would nevertheless be in better agreement with a scheme involving binding of O_2 before binding of the peptide substrate, and hence the order of binding of these two reactants must also be regarded as uncertain. Figures 8 B and C show the reaction schemes for the uncoupled reactions based on the kinetic[165,166,169,188,194] and other[179,195] data.

Three early suggestions for the mechanisms of these reactions are in disagreement with either the kinetic data or other facts subsequently learned concerning these enzymes. These mechanisms have been discussed in detail in previous reviews,[6,8,173] and will therefore not be described here.

A detailed stereochemical mechanism suggested for prolyl 4-hydroxylase[173] is in complete agreement with the above kinetic data. This mechanism assumes that the Fe^{2+} of a catalytic site is located in a pocket coordinated with the enzyme by three side chains (L_1-L_3

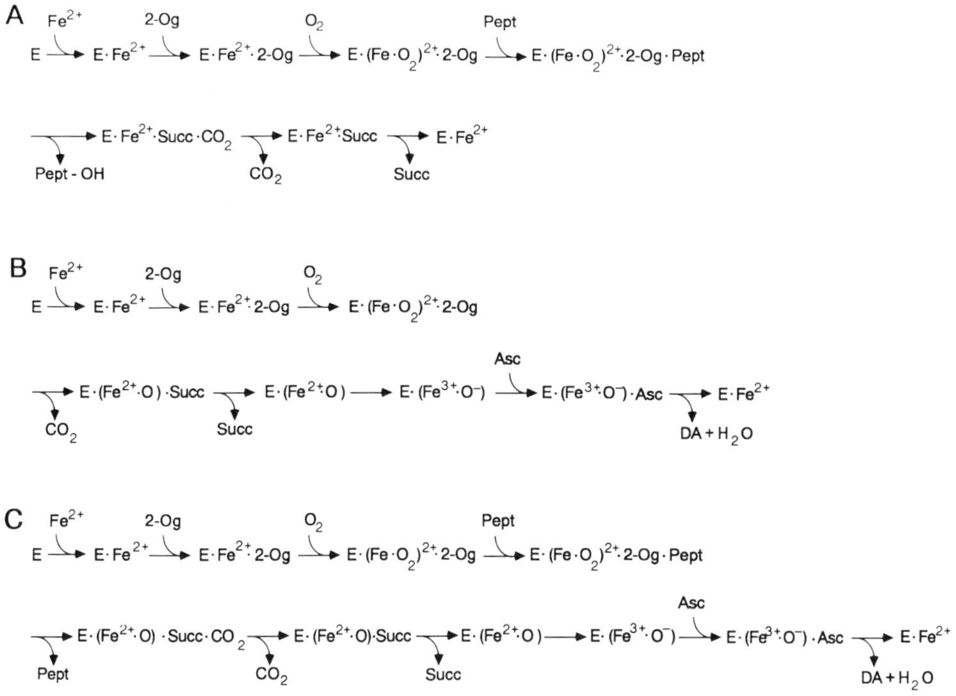

A

Fe^{2+} 2-Og O_2 Pept

$E \longrightarrow E \cdot Fe^{2+} \longrightarrow E \cdot Fe^{2+} \cdot 2\text{-Og} \longrightarrow E \cdot (Fe \cdot O_2)^{2+} \cdot 2\text{-Og} \longrightarrow E \cdot (Fe \cdot O_2)^{2+} \cdot 2\text{-Og} \cdot Pept$

$\longrightarrow E \cdot Fe^{2+} \cdot Succ \cdot CO_2 \longrightarrow E \cdot Fe^{2+} Succ \longrightarrow E \cdot Fe^{2+}$

Pept - OH CO_2 Succ

B

Fe^{2+} 2-Og O_2

$E \longrightarrow E \cdot Fe^{2+} \longrightarrow E \cdot Fe^{2+} \cdot 2\text{-Og} \longrightarrow E \cdot (Fe \cdot O_2)^{2+} \cdot 2\text{-Og}$

Asc

$\longrightarrow E \cdot (Fe^{2+} \cdot O) \cdot Succ \longrightarrow E \cdot (Fe^{2+}O) \longrightarrow E \cdot (Fe^{3+} \cdot O^-) \longrightarrow E \cdot (Fe^{3+}O^-) \cdot Asc \longrightarrow E \cdot Fe^{2+}$

CO_2 Succ $DA + H_2O$

C

Fe^{2+} 2-Og O_2 Pept

$E \longrightarrow E \cdot Fe^{2+} \longrightarrow E \cdot Fe^{2+} \cdot 2\text{-Og} \longrightarrow E \cdot (Fe \cdot O_2)^{2+} \cdot 2\text{-Og} \longrightarrow E \cdot (Fe \cdot O_2)^{2+} \cdot 2\text{-Og} \cdot Pept$

Asc

$\longrightarrow E \cdot (Fe^{2+} \cdot O) \cdot Succ \cdot CO_2 \longrightarrow E \cdot (Fe^{2+} \cdot O) \cdot Succ \longrightarrow E \cdot (Fe^{2+}O) \longrightarrow E \cdot (Fe^{3+} \cdot O^-) \longrightarrow E \cdot (Fe^{3+}O^-) \cdot Asc \longrightarrow E \cdot Fe^{2+}$

Pept CO_2 Succ $DA + H_2O$

FIGURE 8. Schematic representation of the mechanism for the vertebrate prolyl 4-hydroxylase reaction and lysyl hydroxylase reaction. The complete hydroxylation reaction is thought to proceed according to scheme (A), in which the order of binding of O_2 and the peptide substrate and the order of release of the hydroxylated peptide and CO_2 are uncertain. The uncoupled decarboxylation reaction is shown in schemes (B) and (C). Here the reactive iron-oxo complex is probably converted to $Fe^{3+} \cdot O^-$ (B,C). E denotes enzyme; 2-Og, 2-oxoglutarate; Pept, peptide substrate; Pept-OH, hydroxylated peptide substrate; Succ, succinate; Asc, ascorbate; DA, dehydroascorbate. Modified from schemes contained in References 15 and 218.

in Figure 9).[173] Binding of 2-oxoglutarate occurs via three distinct subsites. Subsite I is probably a positively charged side chain of the enzyme that ionically binds the C5 carboxyl group of the 2-oxoglutarate, subsite II consists of two *cis*-positioned equatorial coordination sites of the enzyme-bound Fe^{2+} and is chelated by the C1-C2 moiety, while subsite III involves a hydrophobic binding site in the C3-C4 region of the cosubstrate (Figure 8).[173,176] Molecular oxygen is thought to be bound end-on in an axial position, leading to a superoxide-like[196] structure with the noncoordinated oxygen atom sterically directed to facilitate a nucleophilic attack on the C2 of the 2-oxoglutarate. In order to allow for decarboxylation, the C2 of the 2-oxoglutarate undergoes rehybridization from its sp^2 hybridized planar oxo structure to an sp^3 hybridized tetrahedral transition state.[173] Decarboxylation will then occur simultaneously with cleavage of the O-O bond, and the original C2 of 2-oxoglutarate, which is now the C1 of succinate, will return to the sp^2 hybridization. At the same time, a highly reactive iron-oxo complex, ferryl ion, will be formed that acts as the active intermediate in oxygen transfer and subsequently hydroxylates the proline residue in the peptide substrate.[173] In the uncoupled reaction cycles, the reactive iron-oxo complex is probably converted to $Fe^{3+} \cdot O^-$ making the enzyme unavailable for new catalytic cycles until reduced by ascorbate.[179,180,195] Recent experimental data[176,177,189] are consistent with this hypothesis and suggest that the ascorbate binding site of prolyl 4-hydroxylase also contains the two *cis*-positioned equatorial coordination sites of the enzyme-bound iron, and is thus partially identical to the binding site of 2-oxoglutarate.[189]

A separate aspect of the reaction mechanism is the question of how the enzymes act on their biological substrates or any extended polypeptide with a number of residues that can

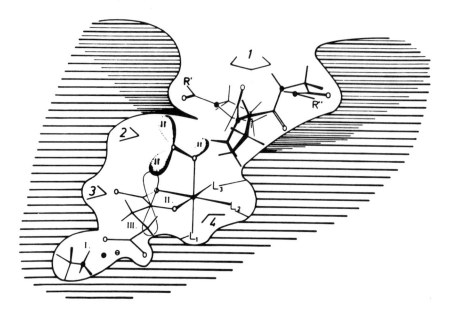

FIGURE 9.   Binding of the cosubstrates and peptide substrate at a catalytic site of vertebrate prolyl 4-hydroxylase. The iron (4) is thought to be located in a pocket coordinated with the enzyme by three side chains (L_1 to L_3). 2-Oxoglutarate (3) is bound to subsites I to III in its energetically most stable, staggered conformation. Molecular oxygen (2) is thought to be bound end-on in an axial position, leading to a superoxide-like structure. The peptide substrate (1) is sterically oriented to participate in the hydroxylation reaction stereospecifically. (From Majamaa, K., Hanauske-Abel, H. M., Günzler, V., and Kivirikko, K. I., *Eur. J. Biochem.*, 138, 239, 1984. With permission.)

be hydroxylated. Recent data suggest that the animal prolyl 4-hydroxylase $\alpha_2\beta_2$ tetramer has two active sites per pair of dissimilar subunits.[172,197-199] This suggestion is based on the findings that two Fe^{2+} atoms are bound per enzyme tetramer[172] and that each of the two α-subunits bears a peptide binding site[197] and a 2-oxoglutarate binding site.[198] The two large peptide binding sites probably act processively, which prevents dissociation of the enzyme-substrate complex between successive hydroxylations of a long peptide with multiple substrate sites (Figure 10).[199,201] Such a mechanism leads to a low K_m for a long peptide by overcoming the diffusional constraints on the association between the enzyme and the various substrate sites, and explains why long peptides are much better substrates than short peptides (Section V).[201] A similar explanation is likely to hold good for lysyl hydroxylase and prolyl 3-hydroxylase, which also show a preference for hydroxylation of long peptide substrates.

VII. INHIBITORS AND INACTIVATORS: IMPLICATIONS FOR THE DESIGN OF ANTIFIBROTIC DRUGS

The crucial role of collagen accumulation in fibrosis of the liver, lungs, and other organs has prompted attempts to develop drugs that inhibit its formation (for reviews, see References 202 and 203). The 4-hydroxylation of proline residues would seem an especially suitable target for chemical regulation, because inhibition of this reaction will prevent collagen triple helix formation and thus lead to a nonfunctional protein that is rapidly degraded. Many attempts have therefore been made to develop compounds that inhibit prolyl 4-hydroxylase, and in many cases also the two other collagen hydroxylases.

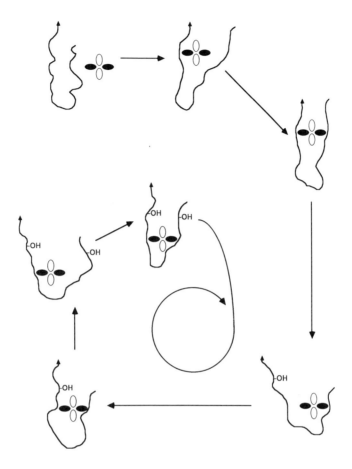

FIGURE 10. Processive action of the two peptide binding sites of the vertebrate prolyl 4-hydroxylase in the hydroxylation of a long polypeptide substrate. Only part of a long procollagen polypeptide chain is shown, with the arrowhead representing the rest of the molecule. The black ovals indicate the α subunits and the open ovals the β subunits of the prolyl 4-hydroxylase tetramer. (Modified from de Waal, A., thesis, University of Amsterdam, 1988).

A. INHIBITORS

A number of compounds inhibit the hydroxylases competitively with respect to the peptide substrate or some of the cosubstrates. Many peptides inhibit competitively with respect to the peptide substrate. In the case of vertebrate prolyl 4-hydroxylase these include polypeptides with the structure $(Pro-Ala-Gly)_n$,[135,136,142] and $(Gly-Pro-Gly)_n$,[136] poly(L-proline)[134,135,141] and several analogs of bradykinin.[204] The K_i of poly(L-proline), like the K_m of the peptide substrate, decreases with increasing chain length, poly(L-proline) with a molecular weight of 15,000 having a K_i of 0.02 μM.[205] The most potent inhibitor among the bradykinin analogs is glutamyl-3,4-dehydroprolylbradykinin, with a K_i of 100 μM.[204]

Many bivalent cations inhibit the enzymes competitively with respect to Fe^{2+}.[119] The most potent such inhibitor is Zn^{2+}, with a K_i of about 1 μM for vertebrate prolyl 4-hydroxylase[169,206] and lysyl hydroxylase,[194,206] and about 5 μM for plant prolyl 4-hydroxylase from *Chlamydomonas reinhardii*.[115] This cation can also be used for sensitive, selective inhibition of proline 4-hydroxylation, and to a lesser extent the other two collagen hydroxylations, in isolated cells and tissues,[207] and even in rapidly growing, young rats *in vivo*.[208] Furthermore, peroral zinc administration has been reported to have a direct, selective in-

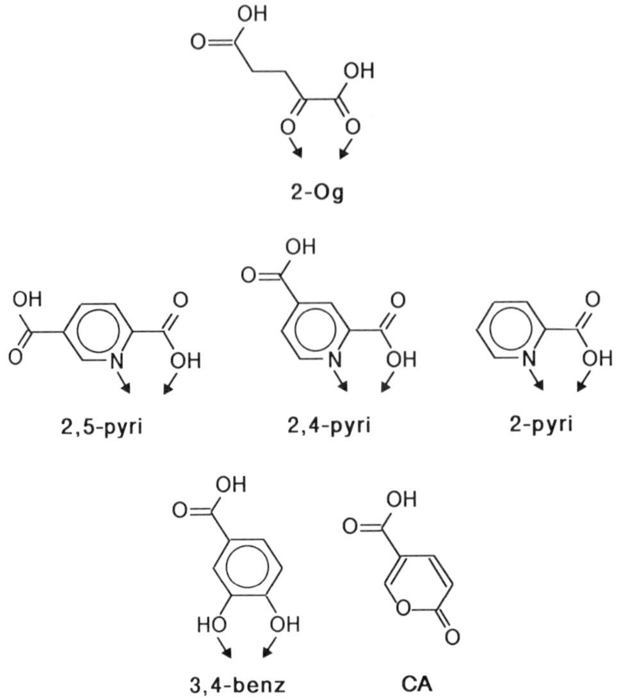

FIGURE 11.   Structures of certain compounds inhibiting or inactivating vertebrate prolyl 4-hydroxylase. Pyridine 2,5-dicarboxylate (2,5-pyri, K_i 0.8 μM), pyridine 2,4-dicarboxylate (2,4-pyri, K_i 2 μM), pyridine 2-carboxylate (2-pyri, K_i 25 μM) and 3,4-dihydroxybenzoate (3,4-benz, K_i 5 μM) inhibit it competitively with respect to 2-oxoglutarate (2-Og), 3,4-dihydroxybenzoate being also competitive inhibitor with respect to ascorbate. Coumalic acid (CA) inactivates the enzyme by a suicide mechanism, acting as a 2-oxoglutarate analog. The compounds chelate iron at the catalytic site of the enzyme in the manner indicated by the arrow-heads.

hibitory effect on hepatic collagen accumulation in rats with carbon tetrachloride-induced liver injury.[208]

Superoxide dismutase-active copper chelates inhibit the hydroxylases competitively with respect to oxygen, probably by dismutation of the activated form of oxygen at the catalytic site.[196] These compounds include Cu(acetylsalicylate)$_2$ and Cu(lysine)$_2$, both of which have a K_i of about 30 μM for vertebrate prolyl 4-hydroxylase and lysyl hydroxylase.[196] Superoxide dismutase does not inhibit vertebrate prolyl 4-hydroxylase, however,[4,169] probably because the activated form of the oxygen is sequestered in the active site pocket of the hydroxylase.

A number of aliphatic and aromatic compounds show competitive inhibition with respect to 2-oxoglutarate.[176,177,189,209] The most potent such inhibitors all have structural domains that can bind to the iron at the postulated subsite II of the 2-oxoglutarate site (Section VI.B.) and these compounds can also interact at subsites I and III. The two most effective such compounds found so far are pyridine 2,4-dicarboxylate and pyridine 2,5-dicarboxylate (Figure 11) with K_i values of 2 μM and 0.8 μM, respectively (Table 5).[176] Experiments with many related compounds have demonstrated that if the chelating moiety is destroyed by omission or shifting of the C2 carboxyl group or subsitution of the aromatic nitrogen, the inhibitory effect is markedly diminished.[176] Omission of the other carboxyl group thought to be bound at subsite I, or shifting of this group to a position other than 4 or 5, likewise greatly decreases the inhibitory potential.

The inhibition patterns found for the 2-oxoglutarate analogs with plant prolyl 4-hydroxylase and the two other collagen hydroxylases are basically similar to those found with

TABLE 5
K_i Values of Two Pyridine Dicarboxylates for the Hydroxylases and 2-Oxoglutarate Dehydrogenase

Enzyme	K_i (μM) for pyridine	
	2,4-dicarboxylate	2,5-dicarboxylate
Vertebrate prolyl 4-hydroxylase[a]	2	0.8
Plant prolyl 4-hydroxylase[b]	8	20
Lysyl hydroxylase[a]	50	150
Prolyl 3-hydroxylase[a]	3	15
2-Oxoglutarate dehydrogenase[c]	400	3000

[a] From chick embryos, References 176, 177.
[b] From *Chlamydomonas reinhardii*, Reference 115.
[c] From pig heart, Reference 177.

prolyl 4-hydroxylase, but several interesting differences in detail exist.[177] One systematic difference is that the K_i values of almost all the aliphatic and aromatic 2-oxoglutarate analogs are distinctly higher for lysyl hydroxylase than for the other three hydroxylases (Table 5).[115,177] This is probably related in part to the high K_m of 2-oxoglutarate in the case of lysyl hydroxylase (Table 4). Another definite difference is that the most potent inhibitor of vertebrate prolyl 4-hydroxylase is pyridine 2,5-dicarboxylate,[176] whereas for the three other hydroxylases it is pyridine 2,4-dicarboxylate.[115,177] There thus seem to be significant differences between the catalytic sites of the four enzymes. It is of particular importance that pyridine 2,4-dicarboxylate and pyridine 2,5-dicarboxylate are only very weak inhibitors of 2-oxoglutarate dehydrogenase (Table 5).[177] This enzyme differs from the four hydroxylases discussed here in that its reaction mechanism does not involve any metal ion. Therefore chelation of the enzyme-bound Fe^{2+} seems to be highly significant for the binding of various compounds at the 2-oxoglutarate sites of the four hydroxylases.[176,177] It would thus seem possible to develop potent prolyl 4-hydroxylase inhibitors that show a high degree of specificity.

3,4-Dihydroxybenzoate (Figure 11) also possesses the structural domains needed for binding at all three subsites of the 2-oxoglutarate binding site and is an effective inhibitor, with a K_i of 5 μM for vertebrate prolyl 4-hydroxylase[189] and 70 μM for plant prolyl 4-hydroxylase from *Chlamydomonas reinhardii*.[115] 3,4-Dihydroxybenzoate and related compounds differ from the pyridine derivatives, however, in their mode of inhibition with respect to ascorbate. The pyridine derivatives inhibit uncompetitively with respect to ascorbate, whereas 3,4-dihydroxybenzoate inhibits competitively with respect to this cosubstrate as well.[189] This agrees with the suggestion that an ascorbate binding site also contains the two *cis*-positioned equatorial coordination sites of the enzyme-bound iron (Section VI.B) and is therefore partially identical to the binding site for 2-oxoglutarate. The pyridine and dihydroxybenzoate inhibitors appear to react with different enzyme forms, as determined by the oxidation state of the iron atom at the catalytic site.

Most of the new prolyl and lysyl hydroxylase inhibitors display a low membrane permeability which makes them ineffective in cultured cells and *in vivo*. It has recently been demonstrated, however, that this problem can be solved by designing lipophilic proinhibitors.[210,211] These do not inhibit pure enzymes and are converted to the active inhibitors only inside the cell. One such proinhibitor is ethylpyridine-2,4-dicarboxylate,[210] which effectively inhibits 4-hydroxyproline formation, and thus collagen production, in chick embryo calvaria cultures. Inhibition by 50% is obtained at a 10 μM concentration,[210] whereas the actual inhibitor, pyridine 2,4-dicarboxylate, has poor cell membrane penetration and inhibits this system only in the mM range.[212] Similarly, ethyl 3,4-dihydrobenzoate at a 0.2 to 0.4 mM

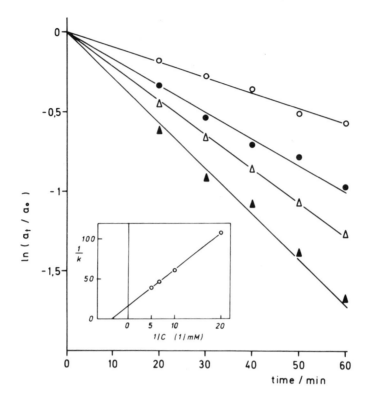

FIGURE 12.   Time-dependent inactivation of chick embryo prolyl 4-hydroxylase by an oxaproline-containing peptide with the structure acetyl-Pro-Oxaproline-Gly-benzylester. Concentrations of the inactivator were 50 (○), 100 (●), 150 (△), and 200 (▲) μM. The inset shows the secondary transform. a_o denotes initial enzyme activity; a_t, enzyme activity after a preincubation of t min. (From Günzler, V., Brocks, D., Henke, S., Myllylä, R., Geiger, R., and Kivirikko, K. I., *J. Biol. Chem.*, 263, 19498, 1988. With permission.)

concentration inhibits collagen production by 50% in cultures of normal human skin fibroblasts[213] and human keloid fibroblasts,[214] whereas 3,4-dihydroxybenzoate itself is a poor inhibitor. Furthermore, the *N*,*N*-ethylamide of pyridine 2,4-dicarboxylate inhibits hepatic collagen accumulation in rats with carbon tetrachloride-induced liver injury.[211] Further work will be needed, however, to demonstrate whether this approach can be used to develop effective drugs for the treatment of patients with fibrotic disorders.

Many other compounds, such as an anthraquinone glycoside P-1894B[215] and fibrostatin C,[216] also inhibit prolyl 4-hydroxylase, although their mode of action has not been elucidated in detail. P-1894B inhibits pure vertebrate prolyl 4-hydroxylase noncompetitively with respect to the peptide substrate and has a K_i of 1.8 μM.[215]

B. SUICIDE INACTIVATORS

A number of compounds have recently been identified that act as time-dependent (Figure 12) irreversible inactivators of vertebrate prolyl 4-hydroxylase and in most, but not all, cases presumably also the other hydroxylases discussed here.[195,217,218] These compounds can be divided into three groups depending on whether they react at the 2-oxoglutarate, ascorbate or peptide binding sites of the hydroxylases. Kinetic studies indicate that the compounds probably inactivate the enzymes by a suicide mechanism. All these compounds also become covalently bound at a catalytic site, as discussed in detail in Sections VIII.A. and B.

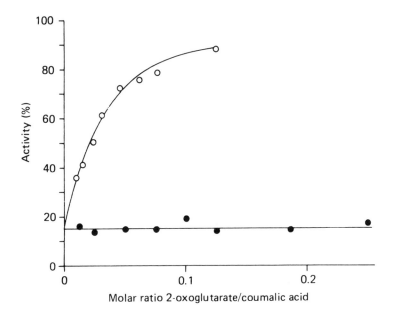

FIGURE 13. Prevention of coumalic acid inactivation of chick embryo prolyl 4-hydroxylase by 2-oxoglutarate. The cosubstrate was capable of preserving the enzyme activity only if added before (○) but not after (●) inactivation with coumalic acid. (From Günzler, V., Hanauske-Abel, H. M., Myllylä, R., Mohr, J., and Kivirikko, K. I., *Biochem. J.,* 242, 163, 1987. With permission.)

The first suicide inactivator to be discovered was coumalic acid (Figure 11), which appears to react in the manner of a 2-oxoglutarate analog.[217] It is not an effective inactivator, however, as quite a high concentration, about 2 mM, is required to inactivate chick embryo prolyl 4-hydroxylase by 50% in 1 h, and an even higher concentration is required for similar inactivation of chick-embryo lysyl hydroxylase.[217] This difference agrees with those found in the K_m values of these enzymes for 2-oxoglutarate and K_i values for 2-oxoglutarate analogs (Sections VI.A. and VII.A.). The high concentrations needed are probably explained by the lack of a functional group (Figure 11) able to bind to the Fe^{2+} at the catalytic site.

The suicide inactivation caused by coumalic acid can be prevented by the addition of 2-oxoglutarate before the inactivation period (Figure 13). Compounds that inhibit the enzymes competitively with respect to 2-oxoglutarate are even much more effective in rescuing the enzyme activity. A 5-μM concentration of pyridine 2,4-dicarboxylate, for example, gives 50% protection against the inactivation of chick embryo prolyl 4-hydroxylase by 8 mM coumalic acid.[217]

The second group of suicide inactivators consists of the anthracyclines doxorubicin and daunorubicin (Figure 14), which inactivate human and chick embryo prolyl 4-hydroxylase by 50% in 1 h at a 60-μM concentration.[218] This inactivation can be prevented by high concentrations of ascorbate or a low concentration of its competitive analogs, whereas 2-oxoglutarate and its competitive analogs offer no protection.[218] These and other data indicate that the anthracyclines probably react with the $Fe^{3+} \cdot O^-$ form of the enzyme in the uncoupled catalytic cycles and act as ascorbate analogs.[218] The inactivation of prolyl 4-hydroxylase by doxorubicin can readily be demonstrated in cultured human skin fibroblasts, but this compound also has other effects on collagen synthesis and proliferation of these cells.[219]

The third group of suicide inactivators consists of peptides in which the proline residue to be hydroxylated has been replaced by 5-oxaproline (Figure 15), a proline analog containing oxygen as part of its ring.[195] These inactivators appear to be specific to vertebrate prolyl

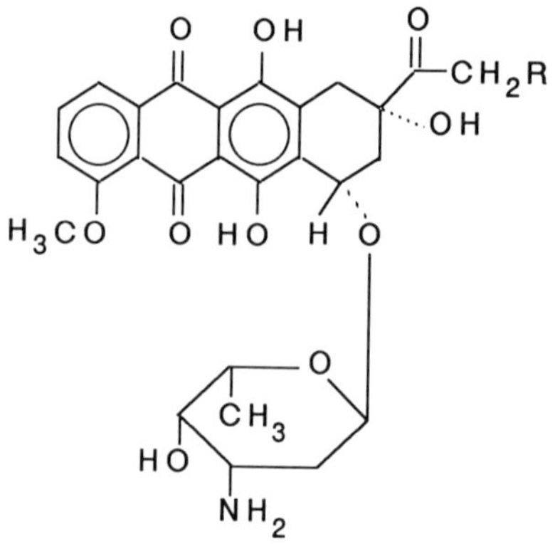

FIGURE 14.   Structures of doxorubicin (R = OH) and daunorubicin (R = H).

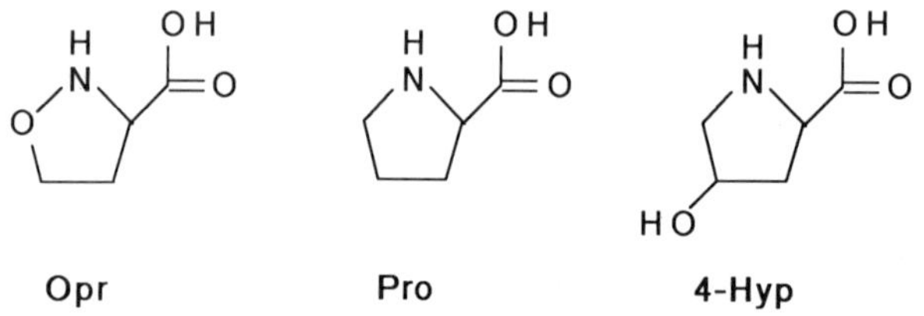

Opr   Pro   4-Hyp

FIGURE 15.   Structure of 5-oxaproline (Opr). Proline (Pro) and 4-hydroxyproline (4-Hyp) are also shown.

4-hydroxylase.[195] The most potent oxaproline-containing peptide tested so far has the structure benzyloxycarbonyl-Phe-Oxaproline-Gly-benzylester and inactivates human and chick embryo prolyl 4-hydroxylase by 50% in 1 h at a concentration of 0.8 μM.[195] Experiments with a number of related peptides have demonstrated that replacement of either of the two aromatic blocking groups by an aliphatic residue distinctly reduces the inactivating effect. Similarly, if the phenylalanine residue is replaced by alanine, valine, or glutamate, the potency is lost in this order, a glutamate-containing peptide giving no inactivation. Replacement of the glycine by other amino acids, including β-alanine, completely abolishes the inactivating effect.[195] The oxaproline peptides also act in cultured human skin fibroblasts, although the concentrations needed for a similar degree of prolyl 4-hydroxylase inactivation are about one order of magnitude higher than those required with the pure enzyme.[220]

VIII. DETAILED STRUCTURES OF THE SUBUNITS OF VERTEBRATE PROLYL 4-HYDROXYLASE AND LYSYL HYDROXYLASE

Recent molecular cloning of the subunits contained in the human and other vertebrate prolyl 4-hydroxylase $\alpha_2\beta_2$ tetramer and the lysyl hydroxylase α_2 dimer has made it possible to determine complete cDNA-derived amino acid sequences for these subunits. Data are currently also available on the structures of the respective genes and their location on human chromosomes. As discussed below, binding experiments with suicide inactivators and other compounds have indicated that the α-subunits probably contribute to most parts of the catalytic sites of the prolyl 4-hydroxylase tetramer, but some parts of the large catalytic sites may be cooperatively built up of both the α- and β-subunits. The β-subunit of prolyl 4-hydroxylase has been found to be a highly unusual multifunctional polypeptide.

TABLE 6
Search for Catalytic Sites of Vertebrate Prolyl 4-Hydroxylase by Binding Experiments

Compound	Reacts at	Binds to	Ref.
Coumalic acid	2-Oxoglutarate site	α-Subunit	217
[14]Doxorubicin	Ascorbate site	α- and β-Subunits	218
Z-Phe-Oxa-[14C]Gly-Oet[a]	Peptide site	α-Subunit	221
5-Azido-pyridine-2-carboxylic acid	2-Oxoglutarate site	α-Subunit	198, 222
N-(4-Azido-2-nitrophenyl)glycyl-(Pro-Pro-Gly)₅	Peptide site	α-Subunit	197
Inhibitory monoclonal antibodies	Catalytic site (?)	α- and β-Subunits	223, 224

[a] Benzyloxycarbonyl-Phe-Oxaproline-[14C]Gly-ethylester

A. THE CATALYTICALLY IMPORTANT α-SUBUNIT OF PROLYL 4-HYDROXYLASE

Various sets of binding results indicate that the α-subunits are likely to contribute most parts to the catalytic sites of the vertebrate prolyl 4-hydroxylase (Table 6). One set of data has been obtained in experiments with various suicide inactivators (Section VII.B.) that become covalently bound at the catalytic sites of the enzyme. When chick embryo prolyl 4-hydroxylase was incubated with coumalic acid and the result examined by SDS-poly-acrylamide gel electrophoresis, a very slight but distinct decrease was found in the mobility of the α-subunit, whereas the β-subunit was unaffected.[217] This suggests that coumalic acid became bound to the α-subunit, so that this subunit must contribute to the 2-oxoglutarate binding sites of the enzyme. When similar experiments were performed with a radioactive oxaproline-containing peptide, a strong radioactive labeling was found in the α-subunit, whereas no label could be detected in the β-subunit.[221] By contrast, experiments with [14-14C]doxorubicin, an analog of ascorbate, demonstrated radioactive labeling of both types of subunit, about 70% of the total radioactivity being found in the α-subunit.[218] Thus the β-subunits may also be involved in some parts of the ascorbate binding sites.[218]

Another approach, involving photoaffinity labeling with analogs of 2-oxoglutarate[198,222] and the peptide substrate,[197] has likewise demonstrated that the corresponding binding sites are both located on the α-subunits (Table 6). Nevertheless, as vertebrate prolyl 4-hydroxylase hydroxylates (Pro-Pro-Gly)₅ asymmetrically[156] and reacts preferentially with the fourth triplet from the amino-terminal end (Section V.A), the affinity label in N-(4-azido-2-nitro-phenyl)glycyl-(Pro-Pro-Gly)₅ is relatively far removed from the preferred hydroxylation site.[197]

An additional approach has involved examining the effect of monoclonal antibodies on the enzyme activity (Table 6).[223,224] Most of the inhibitory monoclonal antibodies react with the α-subunit of the enzyme, but some react with the β-subunit.[223,224] The latter finding agrees with the data obtained with [14C]doxorubicin and suggests that some parts of the β-subunits may also be involved in or located close to the catalytic sites.

Results regarding early evolutionary forms of prolyl 4-hydroxylase present in multicellular and unicellular green algae also support the suggestion that the α-subunits contribute to most parts of the catalytic sites of the vertebrate enzyme. As described in Section IV.B., the algal enzyme differs distinctly from the vertebrate and higher-plant prolyl 4-hydroxylases in being a monomer,[115,116] but its catalytic properties greatly resemble those of the tetrameric enzymes (Sections VI and VII). The algal enzyme monomer is antigenically related to the α-subunit of the vertebrate hydroxylase, but may also contain some sequences corresponding to those in the β-subunit (Section IV.B.).

Molecular cloning and nucleotide sequencing of the α-subunit of human prolyl 4-hy-droxylase has demonstrated that the final polypeptide is 517 amino acid residues in length

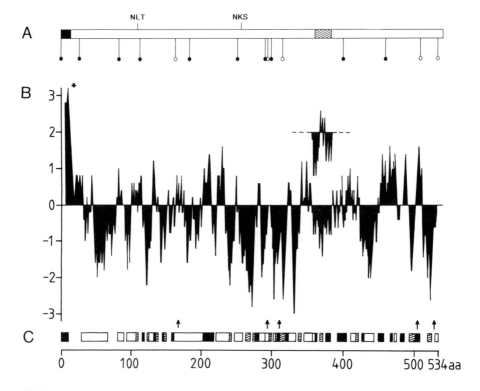

FIGURE 16.   Some characteristic features of the human prolyl 4-hydroxylase α-subunit polypeptide as deduced from the nucleotide sequences of the cDNAs. (A) Schematic representation of the polypeptide. The signal sequence is indicated by a black box and mature protein by an open box. The segment indicated by the shaded box is alternatively spliced in a mutually exclusive fashion. M indicates the first amino acid of the signal sequence, a methionine residue. NLT and NKS denote potential glyco-sylation sites with sequences -Asn-Leu-Thr- and -Asn-Lys-Ser-, respectively. Open circles indicate cysteine residues and filled circles methionines. (B) Hydropathy plot, calculated as the average hydro-pathy of seven consecutive amino acids. The hydropathy plot as predicted from the alternatively spliced region is shown for both sequence variants. (C) Secondary structure prediction: (□), α-helix; (■), β-pleated sheet; (▨), random coil configuration. (Based on data in References 225 and 227.)

and initially contains a signal peptide comprising an additional 17 amino acids.[225,226] This α-subunit sequence was found to be unique among the currently available data bank se-quences. Two types of cDNA clone were identified that differ over a stretch of 64 nucleotides (Figure 16) and were shown to be due to alternative splicing of the primary RNA transcripts of one gene.[225,227] Since no cDNA clone was found that either contained or lacked both 64 nucleotide stretches, this alternative splicing appears to take place in a mutually exclusive fashion. The two types of mRNA were present in equal or near equal amounts, which raises the possibility that the vertebrate prolyl 4-hydroxylase $\alpha_2\beta_2$ tetramer may contain one α-subunit of each type.

The human α-subunit contains two potential glycosylation sites for asparagine-linked oligosaccharides, an -Asn-Leu-Thr- sequence and an -Asn-Lys-Ser- sequence.[225] Neither of them is located in the differentially spliced segment of the human α-subunit (Figure 16), and thus the relation between the two differentially glycosylated forms of the chick α-subunit (Section IV.A.) and the alternative splicing of the RNA transcripts of the human α-subunit remains unknown.

The hydrophilicity/hydrophobicity plot indicates that the polypeptide is predominantly hydrophilic (Figure 16).[225,227] It contains only five cysteine residues (Figure 16), some of which are likely to be involved in the binding of Fe^{2+} at the catalytic site of the enzyme

(Section VI) and some in intrachain disulfide bonds (Section IV.A). Further research will nevertheless be needed to identify the cysteines required for these different functions.

The human gene coding for the α-subunit is located in the q21.3-23.1 region of chromosome 10, and is not physically linked to the gene for the β-subunit (Section VIII.B) or to any of the genes coding for the major collagen polypeptide chains.[228] The size of the transcribed part of this gene is over 70 kb, and characterization of its exon-intron organization is now in progress.[227]

B. THE MULTIFUNCTIONAL β-SUBUNIT OF PROLYL 4-HYDROXYLASE

Molecular cloning and nucleotide sequencing of the β-subunit of human prolyl 4-hydroxylase have indicated that the final polypeptide consists of 491 amino acids and is synthesized in a form containing a signal peptide of 17 additional amino acids.[229] Surprisingly, these human sequences were found to be highly homologous to those reported[230] for rat protein disulfide isomerase. Furthermore, Southern blot analyses of human genomic DNA with various cDNA probes for the human β-subunit demonstrated the presence of only one gene coding for these sequences.[229] The data thus indicated that the β-subunit of prolyl 4-hydroxylase and the enzyme protein disulfide isomerase are products of the same gene.[229]

Protein disulfide isomerase (EC 5.3.4.1) catalyzes the rearrangement of disulfide bonds in various proteins *in vitro* and is likely to be the *in vivo* catalyst of disulfide bond formation in the biosynthesis of various secretory and cell surface proteins,[231-233] including procollagens.[234] Comparison of the β-subunit isolated from the prolyl 4-hydroxylase tetramer with pure protein disulfide isomerase at the protein level also indicated that these two proteins are identical with respect to many criteria such as the molecular weight of the polypeptide chains, peptide maps and recognition by various polyclonal and monoclonal antibodies.[235] Moreover, the β-subunit isolated from the prolyl 4-hydroxylase tetramer was found to have the same amount of protein disulfide isomerase activity as protein disulfide isomerase itself, and even the β-subunit when present in the intact prolyl 4-hydroxylase tetramer had one half of this activity.[235]

Work on both the cDNA[229] and protein[235] levels thus demonstrate that a single polypeptide is involved in two entirely different enzymatic functions (Figure 17). The polypeptide acts as protein disulfide isomerase both when present as a monomer and when incorporated into the intact prolyl 4-hydroxylase tetramer. In the latter case it also catalyzes the 4-hydroxylation of proline residues, contributing together with the α-subunit to the structures of the catalytic sites of the prolyl 4-hydroxylase tetramer. It had already been demonstrated that a number of cell types contain, in addition to the prolyl 4-hydroxylase tetramer, large amounts of a protein that is identical to the β-subunit monomer on a number of criteria.[4,6,8,91,103] It is now evident that one function of this protein is to act as the enzyme protein disulfide isomerase.

The protein disulfide isomerase activity of the β-subunit is probably not involved in the catalytic mechanism of the prolyl 4-hydroxylase reaction. This suggestion is based on the findings that algal prolyl 4-hydroxylase, an enzyme monomer antigenically related to the α-subunit of the vertebrate enzyme (see above), and lysyl hydroxylase, an α_2 dimer, do not have any protein disulfide isomerase activity,[236] even though they catalyze a similar hydroxylation event.

Recent data indicate that the β-subunit/protein disulfide isomerase polypeptide may have at least three additional functions (Figure 17). One of these was found in molecular cloning and nucleotide sequencing of a major human[237] and bovine[238] cellular thyroid hormone binding protein present in the endoplasmic reticulum which becomes specifically labeled upon culturing of various cells with N-bromoacetyl-3,3′,5-[^{125}I]triiodo-L-thyronine.[239] Sequencing of cDNA clones for this protein demonstrated that it is identical to the β-subunit/protein disulfide isomerase polypeptide. The physiological significance of this apparently highly specific binding of thyroid hormone to the β-subunit is currently unknown.

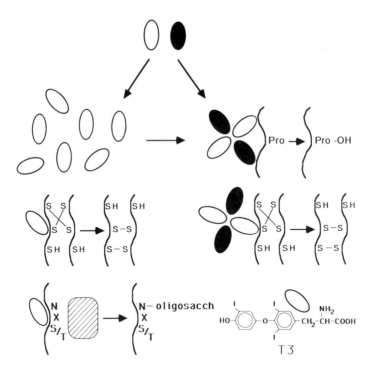

FIGURE 17.   Functions of the multifunctional β-subunit of the vertebrate prolyl 4-hydroxylase. The β-subunit (open oval) is synthesized in an excess over the α-subunit (black oval) and enters a β-subunit pool before being incorporated into the prolyl 4-hydroxylase tetramer. The α-subunit is incorporated into the tetramer immediately after synthesis. The β-subunit has protein disulfide isomerase activity both when present as the monomer and when present in the prolyl 4-hydroxylase tetramer. In the latter case it also catalyzes the 4-hydroxylation of proline residues (Pro→Pro-OH), contributing together with the α-subunit to the structure of the catalytic site of the prolyl 4-hydroxylase tetramer. The β-subunit also serves as a major cellular thyroid hormone (T$_3$) binding protein and as a glycosylation site (NXS/T) binding component of oligosaccharyl transferase *in vitro*. An additional function is to act as one of the two kinds of protein component of the microsomal triglyceride transfer protein complex, although this is not shown here. S-S indicates disulfide bonds and SH free thiol groups in cysteine residues. The wavy lines indicate polypeptides. (From Kivirikko, K. I., Helaakoski, T., Tasanen, K., Vuori, K., Myllylä, R., Parkkonen, T., and Pihlajaniemi, T., *Ann. N.Y. Acad. Sci.*, 580, 132, 1990. With permission.)

A further function proposed for the β-subunit polypeptide is to act *in vitro* as a glycosylation site binding protein component of oligosaccharyl transferase.[240] This enzyme catalyzes the transfer of the oligosaccharide unit from dolichylpyrophosphoryl oligosaccharide to the asparagine residue in the -Asn-X-Ser/Thr- sequences of the nascent polypeptide chains. The glycosylation site binding protein specifically recognizes the sequence -Asn-X-Ser/Thr-. The cDNA-derived amino acid sequence of this protein was found to be highly similar to that of the multifunctional β-subunit polypeptide.[240] Nevertheless, recent studies indicate that this polypeptide is not an essential component of the cotranslational N-glycosylation *in vivo*.[241,242]

The most recent function suggested for the cellular β-subunit polypeptide is to serve as one of the two kinds of protein component of the microsomal triglyceride transfer protein complex.[243,244] Many different data imply that the β-subunit is an important integral part of this protein complex, which appears to consist of one β-subunit and one 88,000 mol wt polypeptide.[244]

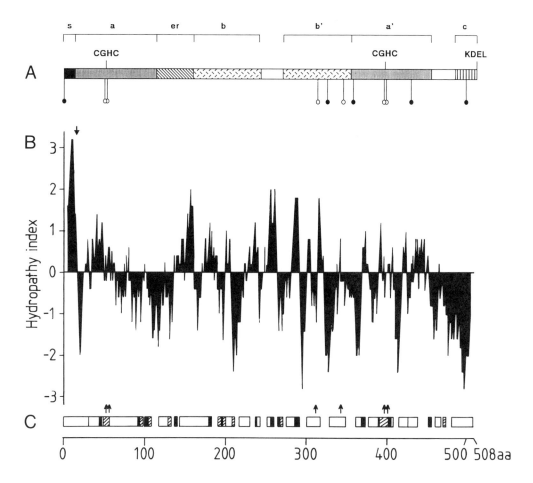

FIGURE 18. Schematic representation of the human β-subunit polypeptide. Symbols: (■), signal sequence (S); (▣), internally homologous regions a and a'; (⧖), internally homologous regions b and b'; (▨), a region homologous to the estrogen receptor (ER);[255] (▥), highly acidic C terminal end of the polypeptide (C). The two CGHC sequences indicate the catalytic sites of protein disulfide isomerase, -Cys-Gly-His-Cys- sequences.[232] KDEL denotes the carboxy-terminal sequence -Lys-Asp-Glu-Leu suggested as a requirement for the retention of a polypeptide within the lumen of the endoplasmic reticulum.[256,257] Open circles indicate cysteine residues and filled circles methionines. (B) Hydropathy plot, calculated as the average hydropathy of seven consecutive amino acids. (C) Secondary structure prediction: (□), α-helix; (■), β-pleated sheet; (▨), random coil configuration. Based on data in Reference 229.

The multifunctional β-subunit polypeptide has two pairs of internal repeats.[229,230,245,246] Region a is homologous to a', and region b to b' (Figure 18). The whole β-subunit polypeptide is homologous to a phosphoinositide-specific phospholipase C,[247] and regions a and a' of both proteins are homologous to the thioredoxins,[229,230,245,248-253] and to three thioredoxim-like repeats of a 72-kDa protein known as ERp 72.[254] A 43-amino acid segment of the β-subunit and the phospholipase, on the immediate carboxy-terminal side of region a, is further homologous to a segment of the estrogen-binding domain of the estrogen receptor (Figure 18) and to a lower degree to the corresponding segments of the progresterone and glucocorticoid receptors.[258] Regions a and a' both contain an active site for protein disulfide isomerase, a -Cys-Gly-His-Cys- sequence (Figures 18 and 19). The amino acid sequences around these two active sites in regions a and a' of the β-subunit polypeptide are identical, and they are highly homologous to the corresponding sequences in regions a and a' of the phosphoinositide-specific phospholipase C, as well as to the sequences around the active sites of thioredoxins from various sources (Figure 19).

```
β/PD1, region a , human    27–46    L L V E F Y A P W C G H C K A L A P E Y
                  chick    27–46    L L V E F Y A P W C G H C K A L A P E Y
          a', human   371–390    V F V E F Y A P W C G H C K Q L A P I W
              chick   371–390    V F V E F Y A P W C G H C K Q L A P I W
PI-PLC, region a ,   rat    24–43    M L V E F F A P W C G H C K R L A P E Y
          a',   rat   372–391    V L I E F Y A P W C G H C L N L E P K Y
Thioredoxin,       human    23–42    V V V D F S A T W C G P C K M I N P F F
                  chick    23–42    V V V D F S A T W C G P C K M I K P F F
              Spinach   28–47    S M V D F W A P W C G P C K L I A P V I
          C.anabaena   22–41    V L V D F W A P W C G P C R M V A P V V
              E.coli   23–42    I L V D F W A E W C G P C K M I A P I L
```

FIGURE 19. Comparison of amino acid sequences adjacent to the protein disulfide isomerase active sites in regions *a* and *a'* of the multifunctional human[229] and chick[245] β-subunit with corresponding sequences of a human phosphoinositide-specific phospholipase C[247] and the active sites of thioredoxins from spinach chloroplasts,[251] *Cyanobacterium anabaena*,[249] *Escherichia coli*,[250] chick[252] and human tissues.[253] The amino acids are given by single letter abbreviations, identical amino acid residues are boxed.

It has recently been proposed that proteins residing within the lumen of the endoplasmic reticulum have a carboxy-terminal sequence of -Lys-Asp-Glu-Leu, which is both necessary and sufficient for the retention of a protein within this cell organelle.[256,257] The multifunctional β-subunit polypeptide from all the species studied so far has this or a closely related carboxy-terminal sequence,[229,230,237,238,245,246,258-260] whereas the α-subunit does not.[225] One function of the β-subunit in the prolyl 4-hydroxylase tetramer would therefore seem to be to retain the enzyme within the lumen of the rough endoplasmic reticulum.

The human gene coding for the multifunctional β-subunit polypeptide is located in the long arm of human chromosome 17, at band 17q25.[261-263] The size of the transcribed part is about 16.5 kb, and it consists of 11 exons and 10 introns.[264] The chick gene for the β-subunit is similar in its exon-intron organization, but its introns are considerbly shorter, and hence the chick gene is only about 9 kb.[265] The two thioredoxin-like regions *a* and *a'* are coded by exons numbers 1, 2 and 8, 9, respectively, while regions *b* and *b'* are coded by exons 4 to 7, and the segment homologous to the estrogen receptor is coded by exon 3.[264] The codons for the two active sites of protein disulfide isomerase, each a -Cys-Gly-His-Cys-sequence, are located 12 bp from the beginning of exons 2 and 9. The last three amino acids coded by exons 1 and 9 at the respective exon-intron junctions, and the first nine coded by exons 2 and 9, including a broken codon for tyrosine, are identical. The data thus suggest that evolution of this gene has involved exon shuffling and duplication of a two-exon unit, in which the internal exon-intron junctions have remained entirely conserved.[264]

C. THE SUBUNIT OF LYSYL HYDROXYLASE

Recent molecular cloning and nucleotide sequencing of the chick lysyl hydroxylase subunit has demonstrated that the final polypeptide is 710 amino acid residues in length and initially contains a signal peptide of 20 additional amino acids.[266] The cDNA-derived amino acid sequence contains four sequences, -Asn-Ile-Ser-, -Asn-Lys-Ser-, Asn-Tyr-Thr-, and -Asn-Cys-Ser-, which may serve as attachment sites for aparagine-linked oligosaccharides. The hydrophilicity/hydrophobicity blot indicates that the polypeptide is mainly hydrophilic but also contains several hydrophobic regions. The carboxy-terminal sequence is -Ser-Phe-Ile-Asp-Pro and thus contains no retention signal typical of the luminal or transmembrane proteins of the endoplasmic reticulum. The polypeptide contains nine cysteine residues, at least one of which is likely to be involved in the binding of the iron atom at a catalytic site of the enzyme (Section VI).[266] The human lysyl hydroxylase subunit consists of 709 amino acid residues and a signal peptide of 18 additional amino acids.[267] The human coding sequences are 76% identical to the chick sequences at the amino acid level.[266,267] The carboxy-terminal region is especially well conserved, a 139 amino acid region, residues 570-709 (carboxy-terminus of the human polypeptide), being 94% identical between the two species and a 76 amino acid region, residues 621-697, 99% identical.[267] These comparisons suggest that lysyl hydroxylase may contain functionally significant sequences, especially in its carboxy-terminal region.

A surprising finding was that no significant homology was found between the primary structures of lysyl hydroxylase and the two types of subunit of prolyl 4-hydroxylase, in spite of the marked similarities between the kinetic properties of these two enzymes.[266,267] A computer-assisted comparison indicated only an 18% overall identity between chick lysyl hydroxylase and the α-subunit of chick prolyl 4-hydroxylase, and a 19% overall identity between chick lysyl hydroxylase and the β-subunit of chick prolyl 4-hydroxylase.[266] Closer inspection of the most homologous areas, however, revealed the presence of several regions of about 20 to 40 amino acids in which the identity between lysyl hydroxylase and one of the prolyl 4-hydroxylase subunits exceeds 30%, but no data are currently available to indicate whether any or all of these homologies represent mere chance rather than true evolutionary or functionally significant homologies.[266]

The human lysyl hydroxylase gene is located in the p36.2-36.3 region of chromosome 1[267] and is thus not physically linked to those for the α- and β-subunits of prolyl 4-hydroxylase (Section VIII.A,B). The size of the transcribed part of this gene is probably about 30 kb,[267] and characterization of its exon-intron organization is now in progress.

IX. INTRACELLULAR SITES OF HYDROXYLATION

Vertebrate prolyl 4-hydroxylase and lysyl hydroxylase and the multifunctional β-subunit of the vertebrate prolyl 4-hydroxylase have been identified to the cisternae of the rough endoplasmic reticulum by a variety of techniques.[102,119,239,240,268] Prolyl 4-hydroxylase and the multifunctional β-subunit reside within the lumen of the cisternae, whereas lysyl hydroxylase appears to be in part within the lumen or loosely bound to the inner membrane and in part tightly bound. The location of the vertebrate prolyl 4-hydroxylase tetramer and its β-subunit agrees with the presence of a -Lys-Asp-Glu-Leu sequence as the carboxy-terminus of the β-subunit polypeptide (Section VIII.B.). Prolyl 3-hydroxylase likewise appears to be located within the same cell compartment.[8,129] Evidence has recently been presented for the existence of active systems for the transport of 2-oxoglutarate[271] and ascorbate[272] into the cisternae of the endoplasmic reticulum through microsomal membranes.

The procollagen polypeptide chains are synthesized on membrane-bound ribosomes and pass through the membranes into the cisternae of the rough endoplasmic membrane while being synthesized (see References 6, 17, 24). Most of the 4-hydroxylation and 3-hydroxylation of proline residues and the hydroxylation of lysine residues occurs while the nascent polypeptide chains are growing on the ribosomes.[6,8,17,24,25] All these reactions are continued after the release of the complete polypeptide chains from the ribosomes, however,[6,8,17,24,125] until triple helix formation in the newly synthesized polypeptide chains prevents any further hydroxylation (Section V). As described in Section III, an almost complete 4-hydroxylation of proline residues in the Y positions of the -X-Y-Gly- sequences is required for the formation of a triple-helical molecule at body temperature. The triple-helical conformation is a requirement for a normal rate of procollagen secretion. If triple helix formation is prevented, the random coil polypeptide chains first accumulate within the cisternae of the rough endoplasmic reticulum and are then in part degraded and in part secreted at a delayed rate.[6,8,24,25]

Although most of the hydroxylation of proline and lysine residues usually occurs on nascent polypeptide chains, there are some conditions under which all the hydroxylation takes place only after the release of complete polypeptide chains into the cisternae of the rough endoplasmic reticulum. This has been demonstrated by incubating cells or tissues under conditions in which hydroxylation is inhibited and subsequently reversing the inhibition.[6,8] The cells synthesize nonhydroxylated polypeptide chains during the inhibition period, and these accumulate within the cisternae of the rough endoplasmic reticulum and are then hydroxylated during the reversal period.

One report suggests that hydroxylation of proline and lysine residues in the biosynthesis of a mannose-binding protein by rat hepatocytes may occur within the Golgi complex.[273] This suggestion is in distinct disagreement with all the data on cellular location of the corresponding enzymes and the presence of the -Lys-Asp-Glu-Leu sequence as the carboxyterminus of the β-subunit polypeptide (see above). Further work is obviously needed to resolve this controversy.

Plant prolyl 4-hydroxylases also appear to reside within the endoplasmic reticulum in free or loosely bound form, as a substantial part of the enzyme activity can be extracted after disruption of the membranes without any detergent.[108,109,111-114,274] Small amounts of enzyme activity are also found within the Golgi apparatus,[111-113] possibly owing to contamination of the Golgi fraction with endoplasmic reticulum during the isolation procedures.[112,113]

X. DEFICIENCY OF LYSYL HYDROXYLASE ACTIVITY IN THE TYPE VI VARIANT OF THE EHLERS-DANLOS SYNDROME

The Ehlers-Danlos syndrome is characterized by joint hypermobility, skin changes such as hyperextensibility, thinness and fragility, and other signs of connective tissue involvement. The disorder has been divided into ten subtypes on the basis of clinical, genetic, and other findings, but it is evident that even the subtypes are heterogeneous.[24,26,27,275,276]

Type VI of the Ehlers-Danlos syndrome includes all the general clinical characteristic of the syndrome and, as special features, kyphoscoliosis and ocular manifestations such as fragility of the ocular globe, spontaneous retinal detachment, microcornea, and slightly blue sclerae. The biochemical abnormality in most[277-281] but not all[281,282] patients is a deficiency in lysyl hydroxylase activity. The significance of lysine hydroxylation for the stability of collagen fibers is clearly demonstrated by the profound changes in the mechanical properties of certain tissues that are seen in these patients. Since hydroxylysine is required for the formation of the most stable intermolecular collagen crosslinks (Section III), the low hydroxylysine content probably explains the clinical manifestations in those patients with the type VI Ehlers-Danlos syndrome who have the corresponding enzyme deficiency. Nevertheless, as some patients with similar clinical findings do not have such a biochemical abnormality, the disease must be produced by a different molecular defect in these cases.

Detailed biochemical investigations have demonstrated that the lysyl hydroxylase deficiency is probably caused by different mutations of the enzyme protein in different families.[283,284] The mutant enzyme in one family has been reported to deviate from normal in properties such as an elevated apparent K_m for ascorbate,[283,284] but no abnormality in this K_m has been found in two other families.[122,285]

An interesting aspect of the type VI Ehlers-Danlos syndrome is that the deficiency in collagen hydroxylysine shows a wide variation in degree between tissues.[277] One explanation may be the existence of tissue-specific or collagen type-specific lysyl hydroxylase isoenzymes, but many observations argue against their existence (Section IV.C.). One report suggests that the residual activity of the mutant lysyl hydroxylase may be preferentially directed toward type IV collagen rather than type I,[286] but work on other patients does not support this suggestion.[122] Moreover, the polypeptide chains of a number collagen types synthesized by the affected fibroblasts have been shown to be deficient in hydroxylysine.[281,283,287,288] The most likely explanation for the variation in the degree of hydroxylysine deficiency is that cells in some tissues have an excess of lysyl hydroxylase activity and hence partial deficiency may be of no consequence.[126] Such an enzyme excess may also exist in many cultured cells, as the extent of hydroxylation of lysine residues in the collagens synthesized by cultured normal fibroblasts is higher than that found *in vivo*,[289] and as the deficiency in lysine hydroxylation in the collagens produced by the affected skin fibroblasts in culture is considerably milder than that present in the skin of the patients.[281,287]

XI. REGULATION

Vertebrate prolyl 4-hydroxylase activity has been found in both fibroblasts and a variety of nonfibroblastic cell types, including cells of nonmesenchymal origin.[4,8,290-292] The amounts of prolyl 4-hydroxylase activity varies markedly between cell types and even in the same cell type in different experimental situations. A good correlation usually exists between the amount of this enzyme activity and the rate of collagen synthesis in various cells and tissues in a large number of physiological and pathological states.[4,6,8] On this basis, assays of prolyl 4-hydroxylase activity have been used widely for estimating the rate of collagen synthesis in many experimental models and in patients with various diseases.[8]

There are some situations in which the amount of vertebrate prolyl 4-hydroxylase activity does not correlate with the rate of collagen synthesis, however. Some cells have prolyl 4-hydroxylase activity even though they do not produce any collagenous polypeptides, and some have relatively high enzyme activity even though only traces of collagenous polypeptides are synthesized.[8,291,292] In these cases prolyl 4-hydroxylase activity is clearly present in the cell in an excess. On the other hand, cultured fibroblasts have relatively low prolyl 4-hydroxylase activity during the logarithmic phase of growth, distinctly lower than during the stationary phase,[4,6,8] even though the rate of collagen synthesis is higher.[25] The amount of prolyl 4-hydroxylase activity is rate limiting for collagen production in such logarithmic phase cultures, the proline residues in the newly synthesized polypeptide chains being incompletely hydroxylated, so that many chains do not form triple-helical molecules and are subject to rapid intracellular degradation.[293,294]

The rates of synthesis of the two types of subunit of the vertebrate prolyl 4-hydroxylase are regulated differently. The α-subunit appears to be utilized to form the active tetrameric enzyme immediately after its synthesis, whereas the β-subunit is produced in an excess and enters a pool of the multifunctional β-subunit polypeptide (Figure 17) before being incorporated into the tetramer.[8,102,103,295,296] Regulation of the amounts of the active enzyme tetramer therefore appears to take place mainly through regulation of the synthesis of the α-subunit. The mechanisms by which the rate of synthesis of the α-subunit is regulated with alterations in the rate of collagen synthesis are unknown, however. The availability of cDNA[225] and genomic[227] clones for the α-subunit of human prolyl 4-hydroxylase will provide tools for research into such mechanisms.

A recent study indicates that differentiation of mouse F9 teratocarcinoma stem cells, which is known to be associated with a marked increase in the synthesis rates and mRNA concentrations of basement membrane proteins such as type IV collagen, is also associated with a marked increase of up to 50-fold in the concentrations of the mRNAs for the α- and β-subunits of prolyl 4-hydroxylase.[297] The levels of the mRNAs for the two types of prolyl 4-hydroxylase subunit have also been reported to vary during growth of chick embryo tendon fibroblasts, changes in the level of the α-subunit mRNA being similar to those in the mRNA concentrations for the $\alpha 1$ and $\alpha 2$ chains of type I collagen.[298] The addition of hydralazine to cultures of human skin fibroblasts rapidly increases concentration of the prolyl 4-hydroxylase α-subunit mRNA up to about 4-fold, and more slowly also the level of the β-subunit mRNA.[299] Further studies are needed to elucidate the mechanisms involved in these changes.

The human gene coding for the multifunctional β-subunit possesses several potential control elements.[264] The promoter region contains a TATA box, six CCAAT boxes, two highly GC-rich segments and several possible sites for the binding of the transcription factor

TABLE 7
Activity of Prolyl 4-Hydroxylase (4-PH), Prolyl 3-Hydroxylase (3-PH), and Lysyl Hydroxylase in Confluent Cultures of Human Cells Producing Various Collagens at Different Rates[a]

Cell type	Relative collagen synthesis (%)[b]	Main collagen type	Relative enzyme activity[c]		
			4-PH	3-PH	LH
Skin fibroblasts	6.3	I	100	100	100
Lung fibroblasts (WI-38)	5.2	I	92	143	116
SV-40 transformed WI-38[d]	1.0	I	34	123	62
Rhabdomyosarcoma cells (RD)	1.0	III	22	274	73
HT-1080 cells[e]	0.7[e]	IV	66	61	207
A-204 cell[f]	1.0	V	29	98	111

[a]   Based on References 125, 289, and 301.
[b]   Expressed either in terms of collagenase-sensitive [^{14}C]lysine-labeled protein as a percentage of total [^{14}C]lysine-labeled protein[289] or collagenase-sensitive [^{14}C]proline-labeled protein as a percentage of total [^{14}C]proline-labeled protein, the latter value corrected for the number of amino acids in collagen relative to those in noncollagen protein.[125]
[c]   The activities are expressed per unit soluble cell protein, that in human skin fibroblasts being taken as 100.
[d]   Simian virus 40-transformed WI-38 cells.
[e]   Human tumor cell line 1080. Incorporation of [^{14}C]proline per cell was 2.2 times that in skin fibroblasts, and hence the rate of collagen chain production per cell was about 30% of that in skin fibroblasts in spite of the much lower relative rate.[125]
[f]   Rhabdomyosarcoma cells.

Sp1. The first exon and the first intron contain further potential Sp1 binding sites, and the first intron also contains two additional long GC-rich regions and other possible control elements. The presence of these various elements may explain the efficient expression of the β-subunit gene in a variety of cell types,[264] but further work is needed to elucidate their functional significance.

Lysyl hydroxylase and prolyl 3-hydroxylase activities also vary markedly between cell types, and in the same cell type in different situations.[8,125,289,292,300,301] Even so, distinct differences are found in the proportions of the three collagen hydroxylase activities between confluent cultures of cells synthesizing either the same collagen type at varying rates or different collagen types, as shown in Table 7. The four sarcoma cell lines shown in Table 7 synthesize collagen at a much lower rate than the skin and lung fibroblasts, and also have a distinctly lower prolyl 4-hydroxylase activity. This agrees with the concept that prolyl 4-hydroxylase activity usually decreases and increases with changes in the rate of collagen synthesis (see above). The two other hydroxylases seem to be regulated less efficiently in relation to a decreased rate of collagen synthesis (Table 7), and therefore exist in a relative excess in all sarcoma cell lines in view of their low rate of collagen production. This is accompanied by an increase in the extent of hydroxylation of lysine residues and 3-hydroxylation of proline residues even in cases in which the respective enzyme activities are similar to those in normal human skin fibroblasts or slightly lower.[289,301]

Changes in concentrations of the cosubstrates needed by the three animal hydroxylases may also influence the rates of the reactions. Exclusion of any of the cosubstrates from isolated cells or tissues will inhibit the hydroxylations, for example.[6,8] Scurvy has been regarded as a classic example of the influence of ascorbate concentration, and it has been demonstrated on a number of occasions that hydroxylation of collagen produced by cultured cells and tissues is markedly impaired in the absence of ascorbate.[6,8] Recent work with scorbutic guinea pigs *in vivo* has demonstrated only relatively small decreases in collagen hydroxylation, however, suggesting that the impaired connective tissue formation seen in scurvy may also involve a humoral factor other than ascorbate.[302]

The extents of lysine hydroxylation and proline 3-hydroxylation, which are markedly incomplete in many collagen types (Section II.A.), are dependent not only on the amounts of active enzyme present, the rate of production of collagenous polypeptides, (see above) and the amounts of the cosubstrates needed by the enzymes, but also on the time available before the newly synthesized polypeptide chains fold into the triple-helical conformation.[6,8] This factor is critical, as none of the three collagen hydroxylases acts on triple-helical molecules (Section V). Large differences are found in the time needed for triple helix formation in the biosynthesis of various genetically distinct collagens,[17,24,25,80] and these contribute to differences in the extent of modification among the various collagen types. It has also been demonstrated that acceleration of triple helix formation by experimental manipulations of isolated chick embryo tendon[303] or cartilage[304] cells leads to a marked decrease in the extent of hydroxylation of lysine residues, whereas inhibition of triple helix formation leads to an increase in this[304,305] and in the 3-hydroxylation of proline residues.[306] Nevertheless, differences in the time needed for the triple helix formation are not the main reason for the high extent of modification of type II and IV collagens as compared to types I and III.[125,307] The main factors seem to be the amounts of the respective enzyme activities and, in the case of lysine hydroxylation, the availability of lysine residues in the Y positions of the -X-Y-Gly- triplets in the respective collagen polypeptide chains.[125] The latter factor is especially important in the case of type IV collagen, in which more than 85% of the lysine residues present in the triple-helical domains are hydroxylated.[125] This is possible only because most of the lysine residues initially incorporated into the polypeptide chains are located in Y positions.[16]

Only limited data are currently available on the regulation of plant prolyl 4-hydroxylases, but a rapid, marked induction of prolyl 4-hydroxylase has been found in the French bean in response to elicitors from a phytopathogenic fungus.[275,308] Prolyl 4-hydroxylase activity in the unicellular green alga *Chlamydomonas reinhardii* has been reported to vary with the life cycle, with a clear maximum at the time when the cells are dividing and depositing new walls around the daughter cells.[114] Further work is evidently needed to elucidate the mechanisms of such changes.

REFERENCES

1. **Stetten, M. R. and Schoenheimer, R.,** The metabolism of 1(-)proline studied with the aid of deuterium and isotopic nitrogen, *J. Biol. Chem.,* 153, 113, 1944.
2. **Stetten, M. R.,** Some aspects of the metabolism of hydroxyproline studied with the aid of isotopic nitrogen, *J. Biol. Chem.,* 181, 31, 1949.
3. **Kuttan, R. and Radhakrishnan, A. N.,** Biochemistry of the hydroxyprolines, *Adv. Enzymol.,* 37, 273, 1973.
4. **Cardinale, G. J. and Udenfriend, S.,** Prolyl hydroxylase, *Adv. Enzymol.,* 41, 245, 1974.
5. **Hayaishi, O., Nogaki, M., and Abbott, M. T.,** Oxygenases: Dioxygenases, in *The Enzymes,* Vol. 12, Boyer, P. D., Ed., Academic Press, New York, 1975, 119.
6. **Prockop, D. J., Berg, R. A., Kivirikko, K. I., and Uitto, J.,** Intracellular steps in the biosynthesis of collagen, in *Biochemistry of Collagen,* Ramachandran, G. N. and Reddi, A. H., Eds., Plenum Press, New York, 1976, 163.
7. **Adams, E. and Frank, L.,** Metabolism of proline and the hydroxyprolines, *Annu. Rev. Biochem.,* 49, 1005, 1980.
8. **Kivirikko, K. I. and Myllylä, R.,** Hydroxylation of prolyl and lysyl residues, in *The Enzymology of Post-translational Modification of Proteins,* Freedman, R. B. and Hawkins, H. C., Eds., Academic Press, London, 1980, 53.
9. **Kivirikko, K. I. and Myllylä, R.,** Post-translational processing of procollagens, *Ann. N.Y. Acad. Sci.,* 460, 187, 1985.

10. **Kivirikko, K. I., Myllylä, R., and Pihlajaniemi, T.**, Protein hydroxylation: prolyl 4-hydroxylase, an enzyme with four cosubstrates and a multifunctional subunit, *FASEB J.*, 3, 1609, 1989.
11. **Lamport, D. T. A.**, Structure, biosynthesis and significance of cell wall glycoproteins, *Recent Adv. Phytochem.*, 11, 79, 1977.
12. **Fincher, G. B. and Stone, B. A.**, Arabinogalactan-proteins; Structure, biosynthesis, and function, *Annu. Rev. Plant Physiol.*, 34, 47, 1983.
13. **Cooper, J. B., Chen, J. A., van Holst, G.-J., and Varner, J. E.**, Hydroxyproline-rich glycoproteins of plant cell walls, *Trends Biochem. Sci.*, 12, 24, 1987.
14. **Kivirikko, K. I. and Myllylä, R.**, Post-translational enzymes in the biosynthesis of collagen: Intracellular enzymes, *Methods Enzymol.*, 82, 245, 1982.
15. **Kivirikko, K. I. and Myllylä, R.**, Recent developments in post-translational modification: Intracellular processing, *Methods Enzymol.*, 144, 96, 1987.
16. **Mayne, R. and Burgeson, R. E., Eds.**, *Structure and Function of Collagen Types*, Academic Press, Orlando, 1987, 1.
17. **Bornstein, P. and Traub, W.**, The chemistry and biology of collagen in *The Proteins*, Vol. 4, Neurath, H., Hill, R. L., and Boeder, C. L., Eds., Academic Press, New York, 1979, 411.
18. **Piez, K. A.**, Molecular and aggregate structures of the collagens, in *Extracellular Matrix Biochemistry*, Piez, K. A. and Reddi, A. H., Eds., Elsevier, New York, 1984, 1.
19. **Burgeson, R. E. and Morris, N. P.**, The collagen family of proteins, in *Connective Tissue Disease, Molecular Pathology of the Extracellular Matrix*, Uitto, J. and Perejda, A. J., Eds., Marcel Dekker, New York, 1987, 3.
20. **Vuorio, E. and de Crombrugghe, B.**, The family of collagen genes, *Annu. Rev. Biochem.*, 59, 837, 1990.
21. **Shaw, L. M. and Olsen, B. R.**, FACIT collagens: diverse molecular bridges in extracellular matrices, *Trends Biochem. Sci.*, 16, 191, 1991.
22. **Pihlajaniemi, T., Pohjolainen, E.-R., and Myers, J. C.**, Complete primary structure of the triple-helical region and the carboxy-terminal domain of a new type IV collagen chain, $\alpha5(IV)$, *J. Biol. Chem.*, 265, 13758, 1990.
23. **Pihlajaniemi, T. and Tamminen, M.**, The $\alpha1$ chain of type XIII collagen consists of three collagenous and four non-collagenous domains and its primary transcript undergoes complex alternative splicing, *J. Biol. Chem.*, 265, 16922, 1990.
24. **Prockop, D. J., Kivirikko, K. I., Tuderman, L., and Guzman, N. A.**, The biosynthesis of collagen and its disorders, *N. Engl. J. Med.*, 301, 13 & 77, 1979.
25. **Kivirikko, K. I. and Myllylä, R.**, Biosynthesis of collagens, in *Extracellular Matrix Biochemistry*, Piez, K. A. and Reddi, A. H., Eds., Elsevier, New York, 1984, 88.
26. **Prockop, D. J. and Kivirikko, K. I.**, Heritable diseases of collagen, *N. Engl. J. Med.*, 311, 376, 1984.
27. **Kivirikko, K. I. and Kuivaniemi, H.**, Post-translational modifications of collagen and their alterations in heritable diseases, in *Diseases of Connective Tissue: The Molecular Pathology of the Extracellular Matrix*, Uitto, J. and Perejda, A. J., Eds., Marcel Dekker, New York, 1987, 263.
28. **Sage, H., Pritzl, P., and Bornstein, P.**, A unique, pepsin-sensitive collagen synthesized by aortic endothelial cells in culture, *Biochemistry*, 19, 5747, 1980.
29. **Reese, C. A., Wiedemann, H., Kühn, K., and Mayne, R.**, Characterization of highly soluble collagenous molecule isolated from chicken hyaline cartilage, *Biochemistry*, 21, 826, 1982.
30. **Jander, R., Rauterberg, J., and Glanville, R. W.**, Further characterization of the three polypeptide chains of bovine and human short-chain collagen (Intima collagen), *Eur. J. Biochem.*, 133, 39, 1983.
31. **Schmid, T. M. and Linsenmayer, T. F.**, A short chain (pro)collagen from aged endochondral chondrocytes, Biochemical characterization, *J. Biol. Chem.*, 258, 9504, 1983.
32. **Kapoor, R., Bornstein, P., and Sage, H.**, Type VIII collagen from bovine Descemet's membrane: structural characterization of a triple-helical domain, *Biochemistry*, 25, 3930, 1986.
33. **Eyre, D. and Wu, J.-J.**, Type XI or $1\alpha2\alpha3\alpha$ collagen, in *Structure and Function of Collagen Types*, Mayne, R. and Burgeson, R. E., Eds., Academic Press, Orlando, 1987, 261.
34. **Reid, K. B. M. and Porter, R. R.**, The proteolytic activation systems of complement, *Annu. Rev. Biochem.*, 50, 433, 1981.
35. **Reid, K. B. M.**, Proteins involved in the activation and control of the two pathways of human complement, *Biochem. Soc. Trans.*, 11, 1, 1983.
36. **Reid, K. B. M.**, Complete amino acid sequences of the three collagen-like regions present in subcomponent C1q of the first component of human complement, *Biochem. J.*, 179, 367, 1979.
37. **Rosenberry, T. L., Barnett, P., and Mays, C.**, Acetylcholinesterase, *Methods Enzymol.*, 82, 325, 1982.
38. **Anglister, L., Rogozinski, S., and Silman, I.**, Detection of hydroxyproline in preparations of acetylcholinesterase from the electric organ of the electric eel, *FEBS Lett.*, 69, 129, 1976.

39. **Bhattacharyya, S. N. and Lynn, W. S.**, Characterization of the collagen- and non-collagen-like regions present in a glycoprotein isolated from alveoli of patients with alveolar proteinosis, *Biochim. Biophys. Acta*, 625, 343, 1980.

40. **Hagwood, S., Efrati, H., Schilling, J., and Benson, B. J.**, Chemical characterization of lung surfactant apoproteins: amino acid composition, N-terminal sequence and enzymatic digestion, *Biochem. Soc. Trans.*, 13, 1092, 1985.

41. **Floros, J., Steinbrink, R., Jacobs, K., Phelps, D., Kriz, R., Recny, M., Sultzman, L., Jones, S., Taeush, H. W., Frank, H. A., and Fritsch, E. F.**, Isolation and characterization of cDNA clones for the 35-kDa pulmonary surfactant-associated protein, *J. Biol. Chem.*, 261, 9029, 1986.

42. **Possmayer, F.**, A proposed nomenature for pulmonary surfactant-associated proteins, *Am. Rev. Respir. Dis.*, 138, 990, 1988.

43. **Drickamer, K., Dordal, M. S., and Reynolds, L.**, Mannose-binding proteins isolated from rat liver contain carbohydrate-recognition domains linked to collagenous tails. Complete primary structures and homology with pulmonary surfactant apoprotein, *J. Biol. Chem.*, 261, 6878, 1986.

44. **Colley, K. J. and Baenziger, J. U.**, Identification of the post-translational modifications of the core-specific lectin. The core-specific lectin contains hydroxyproline hydroxylysine, and glucosylgalactosylhydroxylysine residues, *J. Biol. Chem.*, 262, 10290, 1987.

45. **Drickamer, K.**, Two distinct classes of carbohydrate-recognition domains in animal lectins, *J. Biol. Chem.*, 263, 9557, 1988.

46. **Kodama, T., Freeman, M., Rohrer, L., Zabrecky, J., Matsudaira, P., and Krieger, M.**, Type I macrophage scavenger receptor contains α-helical and collagen-like coiled coils, *Nature*, 343, 531, 1990.

47. **Rohrer, L., Freeman, M., Kodama, T., Penman, M., and Krieger, M.**, Coiled-coil fibrous domains mediate ligand binding by macrophage scavenger receptor type II, *Nature*, 343, 6258, 1990.

48. **Matsumoto, A., Naito, M., Itakura, H., Ikemoto, S., Asaoka, H., Hayakawa, I., Kanamori, H., Aburatani, H., Takaku, F., Suzuki, H., Kobari, Y., Miyai, T., Takahashi, K., Cohen, E. H., Wydro, R., Housman, D. E., and Kodama, T.**, Human macrophage scavenger receptors: primary structure, expression, and localization in atherosclerotic lesions, *Proc. Natl. Acad. Sci. U.S.A.*, 87, 9133, 1990.

49. **Davis, A. E., III and Lachmann, P. J.**, Bovine conglutinin is a collagen-like protein, *Biochemistry*, 23, 2139, 1984.

50. **Davidson, J. M.**, Elastin: Structure and biology, in *Connective Tissue Disease, Molecular Pathology of the Extracellular Matrix*, Uitto, J. and Perejda, A. J., Eds., Marcel Dekker, New York, 1987, 29.

51. **Rosenbloom, J.**, Elastin: An overview, *Methods Enzymol.*, 144, 172, 1987.

52. **Uitto, J., Hoffmann, H.-P., and Prockop, D. J.**, Synthesis of elastin and procollagen by cells from embryonic aorta. Differences in the role of hydroxyproline and the effects of proline analogs on the secretion of the two proteins, *Arch. Biochem. Biophys.*, 173, 187, 1976.

53. **Butkowski, R. J., Langeveld, J. P. M., Wieslander, J., Hamilton, J., and Hudson, B. G.**, Localization of the Goodpasture epitope to a novel chain of basement membrane collagen, *J. Biol. Chem.*, 262, 7874, 1987.

54. **Sasaguri, M., Ikeda, M., Ideishi, M., and Arakawa, K.**, Identification of [hydroxyproline[3]-lysyl-bradykinin released from human plasma protein by kallikrein, *Biochem. Biophys. Res. Commun.*, 150, 511, 1988.

55. **Maeda, H., Matsumura, Y., and Kato, H.**, Purification and identification of [hydroxyprolyl[3]]bradykinin in ascitic fluid from a patient with gastric cancer, *J. Biol. Chem.*, 263, 16051, 1988.

56. **Gautron, J.-P., Pattou, E., Bauer, K., and Kordon, C.**, (Hydroxyproline[9]) luteinizing hormone-releasing hormone: a novel peptide in mammalian and frog hypothalamus, *Neurochem. Int.*, 18, 221, 1991.

57. **Andrews, P. C., Hawke, D., Shively, J. E., and Dixon, J. E.**, Anglerfish preprosomatostatin II is processed to somatostatin-28 and contains hydroxylysine at residue 23, *J. Biol. Chem.*, 259, 15021, 1984.

58. **Spiess, J. and Noe, B. D.**, Processing of an anglerfish somatostatin precursor to a hydroxylysine-containing somatostatin 28, *Proc. Natl. Adac. Sci. U.S.A.*, 82, 277, 1985.

59. **Adams, E.**, Invertebrate collagens. Marked differences from vertebrate collagens appear in only a few invertebrate groups, *Science*, 202, 591, 1978.

60. **Murray, L. W., Waite, J. H., Tanzer, M. L., and Hauschka, P. V.**, Preparation and characterization of invertebrate collagens, *Methods Enzymol.*, 82, 65, 1982.

61. **Lamport, D. T. and Northcote, D. H.**, Hydroxyproline in primary cell walls of higher plants, *Nature (London)*, 188, 665, 1960.

62. **Chen, J. and Varner, J. E.**, An extracellular matrix protein in plants: characterization of a genomic clone for carrot extensin, *EMBO J.*, 4, 2145, 1985.

63. **Showalter, A. M., Bell, J. N., Cramer, C. L., Bailey, J. A., Varner, J. E., and Lamb, C. J.**, Accumulation of hydroxyproline-rich glycoprotein mRNAs in response to fungal elicitor and infection, *Proc. Natl. Adac. Sci. U.S.A.*, 82, 6551, 1985.

64. **Corbin, D. R., Norbert, S., and Lamb, C. J.**, Differential regulation of a hydroxyproline-rich glycoprotein gene family in wounded and infected plants, *Mol. Cell. Biol.*, 7, 4337, 1987.

65. **van Holst, G.-J., Martin, S. R., Allen, A. K., Ashford, D., Desai, N. N., and Neuberger, A.,** Protein conformation of potato (Solanum tuberosum) lectin determined by circular dichroism, *Biochem. J.,* 233, 731, 1986.

66. **Ashford, D., Desai, N. N., Allen, A. K., and Neuberger, A.,** Structural studies of the carbohydrate moieties of lectins from potato (Solanum tuberosum) tubers and thorn-apple (Datura stramonium) seeds, *Biochem. J.,* 201, 199, 1982.

67. **Roberts, K., Grief, C., Hills, G. J., and Shaw, P. J.,** Cell wall glycoproteins: structure and function, *J. Cell Sci. Suppl.,* 2, 105, 1985.

68. **Schlipfenbacher, R., Wenzl, S., Lottspeich, F., and Sumper, M.,** An extremely hydroxyproline-rich glycoprotein is expressed in inverting *Volvox* embryos, *FEBS Lett.,* 209, 57, 1986.

69. **Ramachandran, G. N.,** Stereochemistry of collagen, *Int. J. Pept. Protein Res.,* 31, 1, 1988.

70. **Gustavson, K. H.,** The function of hydroxyproline in collagens, *Nature (London),* 175, 70, 1955.

71. **Berg, R. A. and Prockop, D. J.,** The thermal transition of a nonhydroxylated form of collagen. Evidence for a role for hydroxyproline in stabilizing the triple-helix of collagen, *Biochem. Biophys. Res. Commun.,* 52, 115, 1973.

72. **Jimenez, S. A., Harsch, M., and Rosenbloom, J.,** Hydroxyproline stabilizes the triple helix of chick tendon collagen, *Biochem. Biophys. Res. Commun.,* 52, 106, 1973.

73. **Rosenbloom, J. and Harsch, M.,** Hydroxyproline content determines the denaturation temperature of chick tendon collagen, *Arch. Biochem. Biophys.,* 158, 478, 1973.

74. **Sakakibara, S., Inouye, K., Shudo, K., Kishida, Y., Kobayashi, Y., and Prockop, D. J.,** Synthesis of (Pro-Hyp-Gly)$_n$ of defined molecular weights. Evidence for the stabilization of collagen triple helix by hydroxyproline, *Biochim. Biophys. Acta,* 303, 198, 1973.

75. **Ward, A. R. and Mason, P.,** Influence of proline hydroxylation upon the thermal stability of collagen fragment α1CB2, *J. Mol. Biol.,* 79, 431, 1973.

76. **Ramachandran, G. N., Bansal, M., and Bhatnagar, R. S.,** A hypothesis on the role of hydroxyproline in stabilizing collagen structure, *Biochim. Biophys. Acta,* 322, 166, 1973.

77. **Traub, W.,** Some stereochemical implications of the molecular conformation of collagen, *Isr. J. Chem.,* 12, 435, 1974.

78. **Suzuki, E., Fraser, R. D. B., and MacRae, T. P.,** Role of hydroxyproline in the stabilization of the collagen molecule via water molecules, *Int. J. Biol. Macromol.,* 2, 54, 1980.

79. **Spiro, R. G.,** The structure of the disaccharide unit of the renal glomerular basement membrane, *J. Biol. Chem.,* 242, 4813, 1967.

80. **Kivirikko, K. I. and Myllylä, R.,** Collagen glycosyltransferases, *Int. Rev. Connect. Tissue Res.,* 8, 23, 1979.

81. **Eyre, D. R., Paz, M. A., and Gallop, P. M.,** Cross-linking in collagen and elastin, *Annu. Rev. Biochem.,* 53, 717, 1984.

82. **Bailey, A. J. and Light, N. D.,** Intermolecular cross-linking in fibrotic collagen, in *Fibrosis,* 114th Ciba Foundation Symp., Pitman, London, 1985, 80.

83. **Eyre, D.,** Collagen cross-linking amino acids, *Methods Enzymol.,* 144, 115, 1987.

84. **Fujimoto, D.,** Evidence for natural existence of pyridinoline cross-link in collagen, *Biochem. Biophys. Res. Commun.,* 93, 948, 1980.

85. **Halme, J., Kivirikko, K. I., and Simons, K.,** Isolation and partial characterization of highly purified protocollagen proline hydroxylase, *Biochim. Biophys. Acta,* 198, 460, 1970.

86. **Pänkäläinen, M., Aro, H., Simons, K., and Kivirikko, K. I.,** Protocollagen proline hydroxylase: molecular weight, subunits and isoelectric point, *Biochim. Biophys. Acta,* 221, 559, 1970.

87. **Rhoads, R. E. and Udenfriend, S.,** Purification and properties of collagen proline hydroxylase from newborn rat skin, *Arch. Biochem. Biophys.,* 139, 329, 1970.

88. **Berg, R. A. and Prockop, D. J.,** Affinity column purification of protocollagen proline hydroxylase from chick embryos and further characterization of the enzyme, *J. Biol. Chem.,* 248, 1175, 1973.

89. **Tuderman, L., Kuutti, E.-R., and Kivirikko, K. I.,** An affinity-column procedure using poly(L-proline) for the purification of prolyl hydroxylase. Purification of the enzyme from chick embryos, *Eur. J. Biochem.,* 52, 9, 1975.

90. **Risteli, J., Tuderman, L., and Kivirikko, K. I.,** Intracellular enzymes of collagen biosynthesis in rat liver as a function of age and in hepatic injury induced by dimethylnitrosamine. Purification of rat prolyl hydroxylase and comparison of changes in prolyl hydroxylase activity with changes in immunoreactive prolyl hydroxylase, *Biochem. J.,* 158, 369, 1976.

91. **Chen-Kiang, S., Cardinale, G. J., and Udenfriend, S.,** Homology between a prolyl hydroxylase subunit and a tissue protein that crossreacts immunologically with the enzyme, *Proc. Natl. Adac. Sci. U.S.A.,* 74, 4420, 1977.

92. **Kao, W. W.-Y. and Berg, R. A.,** Cell density-dependent increase in prolyl hydroxylase activity in cultured L-929 cells requires de novo protein synthesis, *Biochim. Biophys. Acta,* 586, 528, 1979.

93. **Kedersha, N. L. and Berg, R. A.,** An improved method for the purification of vertebrate prolyl hydroxylase by affinity chromatography, *Coll. Rel. Res.,* 1, 345, 1981.

94. **Kuutti, E.-R., Tuderman, L., and Kivirikko, K. I.,** Human prolyl hydroxylase. Purification, partial characterization and preparation of antiserum to the enzyme, *Eur. J. Biochem.,* 57, 181, 1975.

95. **Guzman, N. A., Oronsky, A. L., Suarez, G., Meyerson, L. R., Cutroneo, K. R., Olsen, B. R., and Prockop, D. J.,** Prolyl 4-hydroxylase from human placenta. Simultaneous isolation of immunoglobulin G which binds to *Ascaris* cuticle collagen, *Coll. Rel. Res.,* 2, 365, 1982.

96. **Olsen, B. R., Berg, R. A., Kivirikko, K. I., and Prockop, D. J.,** Structure of protocollagen proline hydroxylase from chick embryos, *Eur. J. Biochem.,* 35, 135, 1973.

97. **Berg, R. A., Kedersha, N. L., and Guzman, N. A.,** Purification and partial characterization of the two nonidentical subunits of prolyl hydroxylase, *J. Biol. Chem.,* 254, 3111, 1979.

98. **Nietfeld, J. J., van der Kraan, J., and Kemp, A.,** Dissociation and reassociation of prolyl 4-hydroxylase subunits after cross-linking of monomers, *Biochim. Biophys. Acta,* 661, 21, 1981.

99. **Koivu, J. and Myllylä, R.,** Protein disulfide isomerase retains procollagen prolyl 4-hydroxylase structure in its native conformation, *Biochemistry,* 25, 5982, 1986.

100. **Guzman, N. A., Berg, R. A., and Prockop, D. J.,** Concanavalin A binds to purified prolyl hydroxylase and partially inhibits its enzymic activity, *Biochem. Biophys. Res. Commun.,* 73, 279, 1976.

101. **Kedersha, N. L., Tkacz, J. S., and Berg, R. A.,** Characterization of the oligosaccharides of prolyl hydroxylase, a microsomal glycoprotein, *Biochemistry,* 24, 5952, 1985.

102. **Kedersha, N. L., Tkacz, J. S., and Berg, R. A.,** Biosynthesis of prolyl hydroxylase: evidence for two separate dolichol-mediated pathways of glycosylation, *Biochemistry,* 24, 5960, 1985.

103. **Berg, R. A., Kao, W. W.-Y., and Kedersha, N. L.,** The assembly of tetrameric prolyl hydroxylase in tendon fibroblasts from newly synthesized α-subunits and from preformed cross-reacting protein, *Biochem. J.,* 189, 491, 1980.

104. **Fujimoto, D. and Prockop, D. J.,** Protocollagen proline hydroxylase from *Ascaris lumbricoides, J. Biol. Chem.,* 244, 205, 1969.

105. **Adams, E. and Lamon, M.,** Prolyl hydroxylase of earthworms, *J. Biol. Chem.,* 252, 7591, 1977.

106. **Leushner, J. R. A. and Pasternak, J.,** Partial purification and characterization of prolyl hydroxylase from the free-living nematode *Panagrellus silusiae, Can. J. Zool.,* 56, 159, 1978.

107. **Marumo, K. and Waite, J. H.,** Prolyl 4-hydroxylase in the foot of the marine mussel *Mytilus edulis* L.: Purification and characterization, *J. Exp. Zool.,* 244, 365, 1987.

108. **Sadava, D. and Chrispeels, M. J.,** Hydroxyproline biosynthesis in plant cells. Peptidyl proline hydroxylase from carrot disks, *Biochim. Biophys. Acta,* 227, 278, 1971.

109. **Tanaka, M., Shibata, H., and Uchida, T.,** A new prolyl hydroxylase acting on poly-L-proline, from suspension cultured cells of *Vinca rosea, Biochim. Biophys. Acta,* 616, 188, 1980.

110. **Tanaka, M., Sato, K., and Uchida, T.,** Plant prolyl hydroxylase recognizes poly(L-proline) II helix, *J. Biol. Chem.,* 256, 11397, 1981.

111. **Cohen, P. B., Schibeci, A., and Fincher, G. B.,** Biosynthesis of arabinogalactan-protein in *Lolium multiforum* (Reygrass) endosperm cells, subcellular distribution of prolyl hydroxylase, *Plant Physiol.,* 72, 754, 1983.

112. **Sauer, A. and Robinson, D. G.,** Intracellular localization of post-translational modifications in the synthesis of hydroxyproline-rich glycoproteins. Peptidyl proline hydroxylation in maize roots, *Planta,* 164, 287, 1985.

113. **Bolwell, G. P., Robbins, M. P., and Dixon, R. A.,** Metabolic changes in elicitor-treated bean cells. Enzymic responses associated with rapid changes in cell wall components, *Eur. J. Biochem.,* 148, 571, 1985.

114. **Blankenstein, P., Lang, W. C., and Robinson, D. G.,** Isolation and characterization of prolyl hydroxylase from *Chlamydomonas reinhardii, Planta,* 169, 238, 1986.

115. **Kaska, D. D., Günzler, V., Kivirikko, K. I., and Myllylä, R.,** Characterization of a low relative molecular-mass prolyl 4-hydroxylase from the green alga *Chlamydomonas reinhardii, Biochem. J.,* 241, 483, 1987.

116. **Kaska, D. D., Myllylä, R., Günzler, V., Gibor, A., and Kivirikko, K. I.,** Prolyl 4-hydroxylase from *Volvox carteri.* A low-Mr enzyme antigenically related to the α subunit of the vertebrate enzyme, *Biochem. J.,* 256, 257, 1988.

117. **Miller, R. L.,** Chromatographic separation of the enzymes required for hydroxylation of lysine and proline residues of protocollagen, *Arch. Biochem. Biophys.,* 147, 339, 1971.

118. **Kivirikko, K. I. and Prockop, D. J.,** Partial purification and characterization of protocollagen lysine hydroxylase from chick embryos, *Biochim. Biophys. Acta,* 258, 366, 1972.

119. **Ryhänen, L.,** Lysyl hydroxylase. Further purification and characterization of the enzyme from chick embryos and chick embryo cartilage, *Biochim. Biophys. Acta,* 438, 71, 1976.

120. **Turpeenniemi, T. M., Puistola, U., Anttinen, H., and Kivirikko, K. I.,** Affinity chromatography of lysyl hydroxylase on concanavalin A-agarose, *Biochim. Biophys. Acta,* 483, 215, 1977.

121. **Turpeenniemi-Hujanen, T. M., Puistola, U., and Kivirikko, K. I.,** Isolation of lysyl hydroxylase, an enzyme of collagen synthesis, from chick embryos as a homogeneous protein, *Biochem. J.,* 189, 247, 1980.

122. **Turpeenniemi-Hujanen, T. M., Puistola, U., and Kivirikko, K. I.,** Human lysyl hydroxylase: Purification to homogeneity, partial characterization and comparison of catalytic properties with those of a mutant enzyme from Ehlers-Danlos syndrome type VI fibroblasts, *Coll. Rel. Res.,* 1, 355, 1981.

123. **Myllylä, R., Pajunen, L., and Kivirikko, K. I.,** Polyclonal and monoclonal antibodies to human lysyl hydroxylase and studies of the molecular heterogeneity of the enzyme, *Biochem. J.,* 253, 489, 1988.

124. **Puistola, U.,** Catalytic properties of lysyl hydroxylase from cells synthesizing genetically different collagen types, *Biochem. J.,* 201, 215, 1982.

125. **Pihlajaniemi, T., Myllylä, R., Alitalo, K., Vaheri, A., and Kivirikko, K. I.,** Post-translational modifications in the biosynthesis of type IV collagen by a human tumor cell line, *Biochemistry,* 20, 7409, 1981.

126. **Puistola, U. and Anttinen, H.,** Evidence for a relative excess of lysyl hydroxylase in chick-embryo tendon and cartilage compared with bone and skin, *Biochim. Biophys. Acta,* 717, 118, 1982.

127. **Turpeenniemi-Hujanen, T. M.,** Immunological characterization of lysyl hydroxylase, an enzyme of collagen synthesis, *Biochem. J.,* 195, 669, 1981.

128. **Royce, P. M. and Barnes, M. J.,** Failure of highly purified lysyl hydroxylase to hydroxylate lysyl residues in the non-helical regions of collagen, *Biochem. J.,* 230, 475, 1985.

129. **Risteli, J., Tryggvason, K., and Kivirikko, K. I.,** Prolyl 3-hydroxylase: partial characterization of the enzyme from rat kidney cortex, *Eur. J. Biochem.,* 73, 485, 1977.

130. **Tryggvason, K., Risteli, J., and Kivirikko, K. I.,** Separation of prolyl 3-hydroxylase and 4-hydroxylase activities and the 4-hydroxyproline requirement for synthesis of 3-hydroxyproline, *Biochem. Biophys. Res. Commun.,* 76, 275, 1977.

131. **Tryggvason, K., Majamaa, K., Risteli, J., and Kivirikko, K. I.,** Prolyl 3-hydroxylase and 4-hydroxylase activities in certain rat and chick-embryo tissues and age-related changes in their activites in the rat, *Biochem. J.,* 183, 303, 1979.

132. **Prockop, D. J. and Juva, K.,** Hydroxylation of proline in particulate fractions from cartilage, *Biochem. Biophys. Res. Commun.,* 18, 54, 1965.

133. **Kivirikko, K. I. and Prockop, D. J.,** Purification and partial characterization of the enzyme for the hydroxylation of proline in protocollagen, *Arch. Biochem. Biophys.,* 118, 611, 1967.

134. **Kivirikko, K. I. and Prockop, D. J.,** Enzymatic hydroxylation of proline and lysine in protocollagen, *Proc. Natl. Acad. Sci. U.S.A.,* 57, 782, 1967.

135. **Hutton, J. J., Jr., Marglin, A., Witkop, B., Kurtz, J., Berger, A., and Udenfriend, S.,** Synthetic polypeptides as substrates and inhibitors of collagen proline hydroxylase, *Arch. Biochem. Biophys.,* 125, 779, 1968.

136. **Kivirikko, K. I., Prockop, D. J., Lorenzi, G. P., and Blout, E. R.,** Oligopeptides with the sequences Ala-Pro-Gly and Gly-Pro-Gly as substrates or inhibitors for protocollagen proline hydroxylase, *J. Biol. Chem.,* 244, 2755, 1969.

137. **Kikuchi, Y., Fujimoto, D., and Tamiya, N.,** The enzymic hydroxylation of protocollagen models, *Biochem. J.,* 115, 569, 1969.

138. **Suzuki, F. and Koyama, E.,** Hydroxylation of proline in collagen model peptide, *Biochim. Biophys. Acta,* 177, 154, 1969.

139. **Rhoads, R. E. and Udenfriend, S.,** Substrate specificity of collagen proline hydroxylase: hydroxylation of a specific proline residue in bradykinin, *Arch. Biochem. Biophys.,* 133, 108, 1969.

140. **McGee, J. O'D., Rhoads, R. E., and Udenfriend, S.,** The substrate recognition site of collagen proline hydroxylase. The hydroxylation of -X-Pro-Gly- sequences in bradykinin analogs and other peptides, *Arch. Biochem. Biophys.,* 144, 343, 1971.

141. **Kivirikko, K. I. and Prockop, D. J.,** Hydroxylation of proline in synthetic polypeptides with purified protocollagen hydroxylase, *J. Biol. Chem.,* 242, 4007, 1967.

142. **Rao, N. V. and Adams, S.,** Partial reaction of prolyl hydroxylase. (Gly-Pro-Ala)$_n$ stimulates α-ketoglutarate decarboxylation without prolyl hydroxylation, *J. Biol. Chem.,* 253, 6327, 1978.

143. **Hutton, J. J., Jr., Kaplan, A., and Udenfriend, S.,** Conversion of the amino acid sequence gly-pro-pro in protein to gly-pro-hyp by collagen proline hydroxylase, *Arch. Biochem. Biophys.,* 121, 384, 1967.

144. **Bhathnagar, R. S. and Rapaka, R. S.,** Synthetic polypeptide models of collagen: Synthesis and applications, in *Biochemistry of Collagen,* Ramachandran, G. N. and Reddi, A. H., Eds., Plenum Press, New York, 1976, 479.

145. **Atreya, P. L. and Ananthanarayanan, V. S.,** Interaction of prolyl 4-hydroxylase with synthetic peptide substrates. A conformational model for collagen proline hydroxylation, *J. Biol. Chem.,* 266, 2852, 1991.

146. **Kivirikko, K. I., Kishida, Y., Sakakibara, S., and Prockop, D. J.,** Hydroxylation of (X-Pro-Gly)$_n$ by protocollagen proline hydroxylase. Effect of chain length, helical conformation and amino acid sequence in the substrate, *Biochim. Biophys. Acta,* 271, 347, 1972.

147. **Okada, K., Kikuchi, Y. K., Kawashiri, Y., and Hiromoto, M.,** Synthesis and enzymic hydroxylation of protocollagen model peptides containing glutamyl or leucyl residues, *FEBS Lett.,* 28, 226, 1972.

148. **Rapaka, R. S., Renugopalakrishnan, V., Urry, D. W., and Bhatnagar, R. S.,** Hydroxylation of proline in polytripeptide models of collagen: stereochemistry of polytripeptide–prolyl hydroxylase interaction, *Biochemistry,* 17, 2892, 1978.

149. **Berg, R. A., Kishida, Y., Sakakibara, S., and Prockop, D. J.,** Hydroxylation of (Pro-Pro-Gly)$_5$ and (Pro-Pro-Gly)$_{10}$ by prolyl hydroxylase. Evidence for an asymmetric active site in the enzyme, *Biochemistry,* 16, 1615, 1977.

150. **Rhoads, R. E. and Udenfriend, S.,** *In vitro* enzymatic hydroxylation of prolyl residues in the α1-CB2 fragments of rat collagen, *J. Biol. Chem.,* 246, 4138, 1971.

151. **Berg, R. A. and Prockop, D. J.,** Purification of [^{14}C]protocollagen and its hydroxylation by prolyl-hydroxylase, *Biochemistry,* 12, 3395, 1973.

152. **Murphy, L. and Rosenbloom, J.,** Evidence that chick tendon procollagen must be denatured to serve as substrate for proline hydroxylase, *Biochem. J.,* 135, 249, 1973.

153. **Brahmachari, S. K. and Ananthanarayanan, V. S.,** β-Turns in nascent procollagen are sites of post-translational enzymatic hydroxylation of proline, *Proc. Natl. Acad. Sci. U.S.A.,* 76, 5119, 1979.

154. **Chopra, R. K. and Ananthanarayanan, V. S.,** Conformational implications of enzymatic proline hydroxylation in collagen, *Proc. Natl. Acad. Sci. U.S.A.,* 79, 7180, 1982.

155. **Bansal, M. and Ananthanarayanan, V. S.,** The role of hydroxyproline in collagen folding: conformational energy calculations on oligopeptides containing proline and hydroxyproline, *Biopolymers,* 27, 299, 1988.

156. **Kivirikko, K. I., Suga, K., Kishida, Y., Sakakibara, S., and Prockop, D. J.,** Asymmetry in the hydroxylation of (Pro-Pro-Gly)$_5$ by protocollagen proline hydroxylase, *Biochem. Biophys. Res. Commun.,* 45, 1591, 1971.

157. **Kivirikko, K. I., Shudo, K., Sakakibara, S., and Prockop, D. J.,** Studies on protocollagen lysine hydroxylase. Hydroxylation of synthetic peptides and the stoichiometric decarboxylation of α-ketoglutarate, *Biochemistry,* 11, 122, 1972.

158. **Ryhänen, L.,** Hydroxylation of lysyl residues in lysine-rich and arginine-rich histones by lysyl hydroxylase in vitro, *Biochim. Biophys. Acta,* 391, 50, 1975.

159. **Kikuchi, Y. and Tamiya, N.,** Solid-phase synthesis of substrate model peptides and their hydroxylation with collagen lysyl hydroxylase, *Bull. Chem. Soc. Jpn.,* 55, 1556, 1982.

160. **Glass, D. B., Dembure, P. P., Priest, J. H., and Elsas, L. J., II,** A [^{3}H]lysine-containing synthetic peptide substrate for human protocollagen lysyl hydroxylase, *Biochim. Biophys. Acta,* 840, 149, 1985.

161. **Ryhänen, L. and Kivirikko, K. I.,** Hydroxylation of lysyl residues in native and denatured protocollagen by protocollagen lysyl hydroxylase *in vitro, Biochim. Biophys. Acta,* 343, 129, 1974.

162. **Kivirikko, K. I., Ryhänen, L., Anttinen, H., Bornstein, P., and Prockop, D. J.,** Further hydroxylation of lysyl residues in collagen by protocollagen lysyl hydroxylase *in vitro, Biochemistry,* 12, 4966, 1973.

163. **Takahashi, K., Kang, Y. K., Nemethy, G., and Scheraga, H. A.,** Low-energy conformations of two lysine-containing tetrapeptides of collagen: Implications for posttranslational lysine hydroxylation, *Biopolymers,* 26, 1781, 1987.

164. **Risteli, J., Tryggvason, K., and Kivirikko, K. I.,** A rapid assay for prolyl 3-hydroxylase activity, *Anal. Biochem.,* 84, 423, 1978.

165. **Myllylä, R., Tuderman, L., and Kivirikko, K. I.,** Mechanism of the prolyl hydroxylase reaction. II. Kinetic analysis of the reaction sequence, *Eur. J. Biochem.,* 80, 349, 1977.

166. **Puistola, U., Turpeenniemi-Hujanen, T. M., Myllylä, R., and Kivirikko, K. I.,** Studies on the lysyl hydroxylase reaction. II. Inhibition kinetics and the reaction mechanism, *Biochim. Biophys. Acta,* 611, 51, 1980.

167. **Hurych, J. and Chvapil, M.,** Influence of chelating agents on the biosynthesis of collagen, *Biochim. Biophys. Acta,* 97, 361, 1965.

168. **Nietfeld, J. J. and Kemp, A.,** Properties of prolyl 4-hydroxylase containing firmly-bound iron, *Biochim. Biophys. Acta,* 613, 349, 1980.

169. **Tuderman, L., Myllylä, R., and Kivirikko, K. I.,** Mechanism of the prolyl hydroxylase reaction. I. Role of co-substrates, *Eur. J. Biochem.,* 80, 341, 1977.

170. **Nietfeld, J. J., de Jong, L., and Kemp, A.,** The influence of 2-oxoglutarate on the activity of prolyl 4-hydroxylase, *Biochim. Biophys. Acta,* 704, 321, 1982.

171. **Günzler, V., Majamaa, K., Hanauske-Abel, H. M., and Kivirikko, K. I.,** Catalytically active ferrous ions are not released from prolyl 4-hydroxylase under turnover conditions, *Biochim. Biophys. Acta,* 873, 38, 1986.

172. **De Jong, L. and Kemp, A.,** Two Fe^{2+} atoms are present in activated prolyl 4-hydroxylase, *Biochim. Biophys. Acta,* 709, 142, 1982.

173. **Hanauske-Abel, H. and Günzler, V.,** A stereochemical concept for the catalytic mechanism of prolyl-hydroxylase. Applicability to classification and design of inhibitors, *J. Theor. Biol.,* 94, 421, 1982.

174. **Hutton, J. J., Tappel, A. L., and Udenfriend, S.,** Cofactor and substrate requirements of collagen proline hydroxylase, *Arch. Biochem. Biophys.,* 118, 231, 1967.

175. **Rhoads, R. E. and Udenfriend, S.,** Decarboxylation of α-ketoglutarate coupled to collagen proline hydroxylase, *Proc. Natl. Acad. Sci. U.S.A.,* 60, 1473, 1968.

176. **Majamaa, K., Hanauske-Abel, H. M., Günzler, V., and Kivirikko, K. I.,** The 2-oxoglutarate binding site of prolyl 4-hydroxylase. Identification of distinct subsites and evidence for 2-oxoglutarate decarboxylation in a ligand reaction at the enzyme-bound ferrous ion, *Eur. J. Biochem.,* 138, 239, 1984.

177. **Majamaa, K., Turpeenniemi-Hujanen, T. M., Latipää, P., Günzler, V., Hanauske-Able, H. M., Hassinen, I. E., and Kivirikko, K. I.,** Differences between collagen hydroxylases and 2-oxoglutarate dehydrogenase in their inhibition by structural analogues of 2-oxoglutarate, *Biochem. J.,* 229, 127, 1985.

178. **Counts, D. F., Cardinale, G. J., and Udenfriend, S.,** Prolyl hydroxylase half reaction: Peptidyl prolyl-independent decarboxylation of α-ketoglutarate, *Proc. Natl. Acad. Sci. U.S.A.,* 75, 2145, 1978.

179. **Myllylä, R., Majamaa, K., Günzler, V., Hanauske-Abel, H. M., and Kivirikko, K. I.,** Ascorbate is consumed stoichiometrically in the uncoupled reactions catalyzed by prolyl 4-hydroxylase and lysyl hydroxylase, *J. Biol. Chem.,* 259, 5403, 1984.

180. **de Jong, L. and Kemp, A.,** Stoichiometry and kinetics of the prolyl 4-hydroxylase partial reaction, *Biochim. Biophys. Acta,* 787, 105, 1984.

181. **Fujimoto, D. and Tamiya, N.,** Incorporation of ^{18}O from air into hydroxyproline by chick embryos, *Biochem. J.,* 84, 333, 1962.

182. **Prockop, D. J., Kaplan, A., and Udenfriend, S.,** Oxygen-18 studies on the conversion of proline to collagen hydroxyproline, *Arch. Biochem. Biophys.,* 101, 499, 1963.

183. **Cardinale, G. J., Rhoads, R. E., and Udenfriend, S.,** Simultaneous incorporation of ^{18}O into succinate and hydroxyproline catalyzed by collagen proline hydroxylase, *Biochem. Biophys. Res. Commun.,* 43, 537, 1971.

184. **Kikuchi, Y., Suzuki, Y., and Tamiya, N.,** The source of oxygen in the reaction catalyzed by collagen lysyl hydroxylase, *Biochem. J.,* 213, 507, 1983.

185. **Siegel, B.,** α-Ketoglutarate dependent dioxygenases: a mechanism for prolyl 4-hydroxylase action, *Bioorg. Chem.,* 8, 219, 1979.

186. **Peterkofsky, B. and Udenfriend, S.,** Enzymic hydroxylation of proline in microsomal polypeptide leading to formation of collagen, *Proc. Natl. Acad. Sci. U.S.A.,* 53, 335, 1965.

187. **Englard, S. and Seifter, S.,** The biochemical functions of ascorbic acid, *Annu. Rev. Nutr.,* 6, 365, 1986.

188. **Myllylä, R., Kuutti-Savolainen, E.-R., and Kivirikko, K. I.,** Role of ascorbate in the prolyl hydroxylase reaction, *Biochem. Biophys. Res. Commun.,* 83, 441, 1978.

189. **Majamaa, K., Günzler, V., Hanauske-Abel, H., Myllylä, R., and Kivirikko, K. I.,** Partial identity of the 2-oxoglutarate and ascorbate binding sites of prolyl 4-hydroxylase, *J. Biol. Chem.,* 261, 7819, 1986.

190. **Mata, J. M., Assad, R., and Peterkofsky, B.,** An intramembranous reductant which participates in the proline hydroxylation reaction with intracisternal prolyl hydroxylase and underhydroxylated procollagen in isolated microsomes from L-929 cells, *Arch. Biochem. Biophys.,* 206, 93, 1981.

191. **Chauhan, U., Assad, R., and Peterkofsky, B.,** Cysteinyl-cysteine and the microsomal protein from which it is derived act as reducing cofactor for prolyl hydroxylase, *Biochem. Biophys. Res. Commun.,* 131, 277, 1985.

192. **Nietfeld, J. J. and Kemp, A.,** The function of ascorbate with respect to prolyl 4-hydroxylase activity, *Biochim. Biophys. Acta,* 657, 159, 1981.

193. **de Jong, L., Albracht, S. P. J., and Kemp, A.,** Prolyl 4-hydroxylase activity in relation to the oxidation state of enzyme-bound iron. The role of ascorbate in peptidyl proline hydroxylation, *Biochim. Biophys. Acta,* 704, 326, 1982.

194. **Puistola, U., Turpeenniemi-Hujanen, T., Myllylä, R., and Klvirikko, K. I.,** Studies on the lysyl hydroxylase reaction. II. Inhibition kinetics and the reaction mechanism, *Biochim. Biophys. Acta,* 611, 51, 1980.

195. **Günzler, V., Brocks, D., Henke, S., Myllylä, R., Geiger, R., and Kivirikko, K. I.,** Syncatalytic inactivation of prolyl 4-hydroxylase by synthetic peptides containing the unphysiologic amino acid 5-oxaproline, *J. Biol. Chem.,* 263, 19498, 1988.

196. **Myllylä, R., Schubotz, L. M., Weser, U., and Kivirikko, K. I.,** Involvement of superoxide in the prolyl and lysyl hydroxylase reactions, *Biochem. Biophys. Res. Commun.,* 89, 98, 1979.

197. **de Waal, A., de Jong, L., Hartog, A. F., and Kemp, A.,** Photoaffinity labeling of the peptide binding sites of prolyl 4-hydroxylase with *N*-(4-azido-2-nitrophenyl)glycyl-(Pro-Pro-Gly)₅, *Biochemistry,* 24, 6493, 1985.

198. **de Waal, A., Hartog, A. F., and de Jong, L.,** Localization and quantification of the 2-oxoglutarate binding sites of prolyl 4-hydroxylase by photoaffinity labeling with tritiated 5-azidopyridine-2-carboxylic acid, *Biochim. Biophys. Acta,* 953, 20, 1988.

199. **de Waal, A. and de Jong, L.,** Processive action of the two peptide binding sites of prolyl 4-hydroxylase in the hydroxylation of procollagen, *Biochemistry,* 27, 150, 1988.

200. **de Waal, A.,** Structure and function of prolyl 4-hydroxylase, thesis, Laboratory of Biochemistry, B.C.P. Jansen Institute, University of Amsterdam, 1988.

201. **de Jong, L., van der Kraan, I., and de Waal, A.,** The kinetics of the hydroxylation of procollagen by prolyl 4-hydroxylase. Proposal for a processive mechanism of binding of the dimeric hydroxylating enzyme in relation to the high k_{cat}/K_m ratio and a conformational requirement for hydroxylation of -X-Pro-Gly-sequences, *Biochim. Biophys, Acta,* 1079, 103, 1991.

202. **Kivirikko, K. I. and Majamaa, K.,** Synthesis of collagen: chemical regulation of post-translational events, in *Fibrosis,* 114th Ciba Foundation Symp., Pitman, London, 1985, 34.

203. **Kivirikko, K. I. and Savolainen, E.-R.,** Hepatic collagen metabolism and its modification by drugs, in *Liver Drugs: From Experimental Pharmacology to Therapeutic Application,* Testa, B. and Perrissound, D., Eds., CRC Press, Boca Raton, 1988, 193.

204. **McGee, O. D., Jimenez, M. H., Felix, A. M., Cardinale, G. J., and Udenfriend, S.,** Inhibition of prolyl hydroxylase activity by bradykinin analogs containing a prolyl-like residue, *Arch. Biochem. Biophys.,* 154, 482, 1973.

205. **Prockop, D. J. and Kivirikko, K. I.,** Effect of polymer size on the inhibition of protocollagen proline hydroxylase by polyproline. II, *J. Biol. Chem.,* 244, 4388, 1969.

206. **Anttinen, H., Puistola, U., Pihlajaniemi, T., and Kivirikko, K. I.,** Differences between proline and lysine hydroxylations in their inhibition by zinc or by ascorbate deficiency during collagen synthesis in various cell types, *Biochim. Biophys. Acta,* 674, 336, 1981.

207. **Anttinen, H., Ryhänen, L., and Oikarinen, A.,** Effects of divalent cations on collagen biosynthesis in isolated chick embryo tendon cells, *Biochem. Biophys. Acta,* 609, 321, 1980.

208. **Anttinen, H., Ryhänen, L., Puistola, U., Arranto, A., and Oikarinen, A.,** Decrease in liver collagen accumulation in carbon tetrachloride-injured and normal growing rats upon administration of zinc, *Gastroenterology,* 86, 532, 1984.

209. **Cunliffe, C. J. and Franklin, T. J.,** Inhibition of prolyl 4-hydroxylase by hydroxyanthraquinones, *Biochem. J.,* 239, 311, 1986.

210. **Tschank, G., Brocks, D. G., Engelbart, K., Mohr, J., Günzler, V., and Hanauske-Abel, H. M.,** Inhibition of prolyl hydroxylation and procollagen processing in chick embryo calvaria by a derivative of pyridine-2,4-dicarboxylate: characterization of the diethyl ester as a suitable proinhibitor, *Biochem. J.,* 275, 469, 1991.

211. **Bickel, M., Baader, E., Brocks, D. G., Engelbart, K., Günzler, V., Schmidts, H. L., and Vogel, G. H.,** Beneficial effects of inhibitors of prolyl 4-hydroxylase in CCl_4-induced fibrosis of the liver in rats, *J. Hepatol.,* in press.

212. **Tschank, G., Raghunath, M., Günzler, V., and Hanauske-Abel, H. M.,** Pyridine-dicarboxylates, the first mechanism-derived inhibitors for prolyl 4-hydroxylase, selectively suppress cellular hydroxyproline biosynthesis. Reduction of interstitial collagen and C1q secretion *in vivo, Biochem. J.,* 248, 625, 1987.

213. **Majamaa, K., Sasaki, T., and Uitto, J.,** Inhibition of prolyl hydroxylation during collagen biosynthesis in human skin fibroblast cultures by ethyl 3,4-dihydroxybenzoate, *J. Invest. Dermatol.,* 89, 405, 1987.

214. **Sasaki, T., Majamaa, K., and Uitto, J.,** Reduction of collagen production in keloid fibroblast cultures by ethyl-3,4-dihydroxybenzoate, *J. Biol. Chem.,* 262, 9397, 1987.

215. **Ishimaru, T., Kanamaru, T., Takahashi, T., Ohta, K., and Okazaki, H.,** Inhibition of prolyl hydroxylase activity and collagen biosynthesis by the anthraquinone glycoside, P-1894B, an inhibitor produced by *Streptomyces albogriseolus, Biochem. Pharmacol.,* 31, 915, 1982.

216. **Ishimaru, T., Kanamaru, T., Takahashi, T., and Okazaki, H.,** Inhibition of prolyl hydroxylase activity and collagen biosynthesis by fibrostatin C, a novel inhibitor produced by *Streptomyces catenulae* subsp. *griseospora* no. 23924, *J. Antibiotics,* 41, 1668, 1988.

217. **Günzler, V., Hanauske-Abel, H. M., Myllylä, R., Mohr, J., and Kivirikko, K. I.,** Time-dependent inactivation of chick embryo prolyl 4-hydroxylase by coumalic acid. Evidence for a syncatalytic mechanism, *Biochem. J.,* 242, 163, 1987.

218. **Günzler, V., Hanauske-Abel, H. M., Myllylä, R., Kaska, D., Hanauske, A., and Kivirikko, K. I.,** Syncatalytic inactivation of prolyl 4-hydroxylase by anthracyclines, *Biochem. J.,* 251, 365, 1988.

219. **Sasaki, T., Holeyfield, C., and Uitto, J.,** Doxorubicin-induced inhibition of prolyl hydroxylation during collagen biosynthesis in human skin fibroblast cultures, *J. Clin. Invest.,* 80, 1735, 1987.

220. **Karvonen, K., Ala-Kokko, L., Pihlajaniemi, T., Helaakoski, T., Günzler, V., Henke, S., Kivirikko, K. I., and Savolainen, E.-R.,** Specific inactivation of prolyl 4-hydroxylase and inhibition of collagen synthesis by oxaproline-containing peptides in cultured human skin fibroblasts, *J. Biol. Chem.,* 265, 8415, 1990.

221. **Myllylä, R., Helaakoski, T., Günzler, V., Tripier, D., Henke, S., and Kivirikko, K. I.,** Labeling of prolyl 4-hydroxylase catalytic sites by a suicide inactivator, a 5-oxaproline-containing peptide, manuscript in preparation.

222. **de Waal, A., Hartog, A. F., and de Jong, L.,** Protoaffinity labeling of the 2-oxoglutarate binding sites of prolyl 4-hydroxylase with 5-azidopyridine-2-carboxylic acid, *Biochim. Biophys. Acta,* 912, 151, 1987.

223. **Höyhtyä, M., Myllylä, R., Piuva, J., Kivirikko, K. I., and Tryggvason, K.,** Monoclonal antibodies to human prolyl 4-hydroxylase, *Eur. J. Biochem.,* 141, 477, 1984.

224. **Bai, Y., Muragaki, Y., Obata, K.-I., Iwata, K., and Ooshima, A.,** Immunological properties of monoclonal antibodies to human and rat prolyl 4-hydroxylase, *J. Biochem.,* 99, 1563, 1986.

225. **Helaakoski, T., Vuori, K., Myllylä, R., Kivirikko, K. I., and Pihlajaniemi, T.,** Molecular cloning of the α-subunit of human prolyl 4-hydroxylase: the complete cDNA-derived amino acid sequence and evidence for alternative splicing of RNA transcripts, *Proc. Natl. Adac. Sci. U.S.A.,* 86, 4392, 1989.

226. **Kivirikko, K. I., Helaakoski, T., Tasanen, K., Vuori, K., Myllylä, R., Parkkonen, T., and Pihlajaniemi, T.,** Molecular biology of prolyl 4-hydroxylase, *Ann. N.Y. Acad. Sci.,* 580, 132, 1990.

227. **Helaakoski, T., Vuori, K., Parkkonen, T., Myllylä, R., Chow, L. T., Kivirikko, K. I., and Pihlajaniemi, T.,** Characterization of cDNA and genomic clones for the α subunit of human prolyl 4-hydroxylase, *Ann. N.Y. Acad. Sci.,* 580, 473, 1990.

228. **Pajunen, L., Jones, T. A., Helaakoski, T., Pihlajaniemi, T., Solomon, E., Sheer, D., and Kivirikko, K. I.,** Assignment of the gene coding for the α-subunit of prolyl 4-hydroxylase to human chromosome region 10q21.3-23.1, *Am. J. Hum. Genet.,* 45, 829, 1989.

229. **Pihlajaniemi, T., Helaakoski, T., Tasanen, K., Myllylä, R., Huhtala, M.-L., Koivu, J., and Kivirikko, K. I.,** Molecular cloning of the β-subunit of human prolyl 4-hydroxylase. This subunit and protein disulphide isomerase are products of the same gene, *EMBO J.,* 6, 643, 1987.

230. **Edman, J. C., Ellis, L., Blacher, R. W., Roth, R. A., and Rutter, W. J.,** Sequence of protein disulphide isomerase and implications of its relationship to thioredoxin, *Nature (London),* 317, 267, 1985.

231. **Anfinsen, C. B. and Scheraga, H. A.,** Experimental and theoretical aspects of protein folding, *Adv. Protein Chem.,* 29, 205, 1975.

232. **Freedman, R. B., Hawkins, H. C., Murant, S. J., and Reid, L.,** Protein disulphide-isomerase: a homologue of thioredoxin implicated in the biosynthesis of secretory proteins, *Biochem. Soc. Trans.,* 16, 96, 1988.

233. **Freedman, R. B.,** Protein disulfide isomerase: multiple roles in the modification of nascent secretory proteins, *Cell,* 57, 1069, 1989.

234. **Koivu, J. and Myllylä, R.,** Interchain disulfide bond formation in type I and II procollagen. Evidence for a protein disulfide-isomerase catalyzing bond formation, *J. Biol. Chem.,* 262, 6159, 1987.

235. **Koivu, J., Myllylä, R., Helaakoski, T., Pihlajaniemi, T., Tasanen, K., and Kivirikko, K. I.,** A single polypeptide acts both as the β subunit of prolyl 4-hydroxylase and as a protein disulfide-isomerase, *J. Biol. Chem.,* 262, 6447, 1987.

236. **Myllylä, R., Kaska, D. D., and Kivirikko, K. I.,** The catalytic mechanism of the hydroxylation reaction of peptidyl proline and lysine does not require protein disulfide isomerase activity, *Biochem. J.,* 263, 609, 1989.

237. **Cheng, S-y., Gong, Q-h., Parkison, C., Robinson, E. A., Appella, E., Merlino, G. T., and Pastan, I.,** The nucleotide sequence of a human cellular thyroid hormone binding protein present in endoplasmic reticulum, *J. Biol. Chem.,* 262, 11221, 1987.

238. **Yamauchi, K., Yamamoto, T., Hayashi, H., Koya, S., Takikawa, H., Toyoshima, K., and Horiuchi, R.,** Sequence of membrane-associated thyroid hormone binding protein from bovine liver: its identity with protein disulphide isomerase, *Biochem. Biophys. Res. Commun.,* 146, 1485, 1987.

239. **Hasumara, S., Kitagawa, S., Lovelace, E., Willingham, M. C., Pastan, I., and Cheng, S.,** Characterization of a membrane-associated 3,3′,5-triido L-thyronine binding protein by use of monoclonal antibodies, *Biochemistry,* 25, 7881, 1986.

240. **Geetha-Habid, M., Noiva, R., Kaplan, H. A., and Lennarz, W. J.,** Glycosylation site binding protein, a component of oligosaccharyl transferase, is highly similar to three other 57 kd luminal proteins of the ER, *Cell,* 54, 1053, 1988.

241. **Bulleid, N. J. and Freedman, R. B.,** Cotranslational glycosylation of proteins in systems depleted of protein disulphide isomerase, *EMBO J.,* 9, 3527, 1990.

242. **Noiva, R., Kaplan, H. A., and Lennarz, W. J.,** Glycosylation site-binding protein is not required for N-linked glycoprotein synthesis, *Proc. Natl. Acad. Sci. U.S.A.,* 88, 1986, 1991.

243. **Wetterau, J. R., Combs, K. A., Spinner, S. N., and Joiner, B. J.,** Protein disulfide isomerase is a component of the microsomal triglyceride transfer protein complex, *J. Biol. Chem.,* 265, 9800, 1990.

244. **Wetterau, J. R., Aggerbeck, L. P., Laplaud, P. M., and McLean, L. R.,** Structural properties of the microsomal triglyceride-transfer protein complex, *Biochemistry,* 30, 4406, 1991.

245. **Parkkonen, T., Kivirikko, K. I., and Pihlajaniemi, T.,** Molecular cloning of a multifunctional chicken protein acting as the prolyl 4-hydroxylase β-subunit, protein disulphide isomerase and a cellular thyroid binding protein. Comparison of cDNA-deduced amino acid sequences with those in other species, *Biochem. J.,* 256, 1005, 1988.

246. **Kao, W. W.-Y., Nakazawa, M., Aida, T., Everson, W. V., Kao, C. W.-C., Seyer, J. M., and Hughes, S. H.,** Isolation of cDNA and genomic clones of β-subunit of chicken prolyl 4-hydroxylase, *Connect. Tissue Res.,* 18, 157, 1988.

247. **Bennett, C. F., Balcarek, J. M., Varrichio, A., and Crooke, S. T.,** Molecular cloning and complete amino acid sequence of form-I phosphoinositide-specific phospholipase C, *Nature (London),* 334, 268, 1988.

248. **Holmgren, A.,** Thioredoxin, *Annu. Rev. Biochem.,* 54, 237, 1985.

249. **Gleason, F. K., Whittaker, M. M., Holmgren, A., and Jörnvall, H.,** The primary structure of thioredoxin from the filamentous *Cyanobacterium anabaena* sp. 7119, *J. Biol. Chem.,* 260, 9567, 1985.

250. **Lim, C.-J., Geraghty, D., and Fuchs, J. A.,** Cloning and nucleotide sequence of the frxA gene of Escherichia coli K-12, *J. Bacteriol.,* 163, 311, 1985.

251. **Maeda, K., Tsugita, A., Dalzoppo, D., Vilbois, F., and Schurmann, P.,** Further characterization and amino acid sequence of m-type thioredoxins from spinach chloroplasts, *Eur. J. Biochem.,* 154, 197, 1986.

252. **Jones, S. W. and Luk, K.-C.,** Isolation of a chicken thioredoxin cDNA clone. Thioredoxin mRNA is differentially expressed in normal and Rous sarcoma virus-transformed chicken embryo fibroblasts, *J. Biol. Chem.,* 263, 9607, 1988.

253. **Wollman, E. E., d'Auriol, L., Rimsky, L., Shaw, A., Jacquet, J.-P., Wingfield, P., Graber, P., Dessarps, F., Robin, P., Galiver, F., Bertoglio, J., and Fradelizi, D.,** Cloning and expression of a cDNA for human thioredoxin, *J. Biol. Chem.,* 263, 15506, 1988.

254. **Mazzarella, R. A., Srinivasan, M., Haugejorden, S. M., and Green, M.,** ERp72, an abundant luminal endoplasmic reticulum protein, contains three copies of the active site sequences of protein disulfide isomerase, *J. Biol. Chem.,* 265, 1094, 1990.

255. **Tsibris, J. C. M., Hunt, L. T., Ballejo, G., Barker, W. C., Toney, L. J., and Spellacy, W. N.,** Selective inhibition of protein disulfide isomerase by estrogens, *J. Biol. Chem.,* 264, 13967, 1989.

256. **Munro, S. and Pelham, H. R. B.,** A C terminal signal prevents secretion of luminal ER proteins, *Cell,* 48, 899, 1987.

257. **Pelham, H. R. B.,** Evidence that luminal ER proteins are sorted from secreted proteins in a post-ER compartment, *EMBO J.,* 7, 913, 1988.

258. **Fliegel, L., Newton, E., Burns, K., and Michalak, M.,** Molecular cloning of cDNA encoding a 55-kDa multifunctional thyroid hormone binding protein of skeletal muscle sarcoplasmic reticulum, *J. Biol. Chem.,* 265, 15496, 1990.

259. **Scherens, B., Dubois, E., and Messenguy, F.,** Determination of the sequence of the yeast YCLB13 gene localized on chromosome III. Homology with the protein disulfide isomerase (PDI gene product) of other organisms, *Yeast,* 7, 185, 1991.

260. **LaMantia, M., Miura, T., Tachikawa, H., Kaplan, H. A., Lennarz, W. J., and Mizunaga, T.,** Glycosylation site binding protein and protein disulfide isomerase are identical and essential for cell viability in yeast, *Proc. Natl. Acad. Sci. U.S.A.,* 88, 4453, 1991.

261. **Pajunen, L., Myllylä, R., Helaakoski, T., Pihlajaniemi, T., Tasanen, K., Höyhtyä, M., Tryggvason, K., Solomon, E., and Kivirikko, K. I.,** Assignment of the gene coding for both the β-subunit of prolyl 4-hydroxylase and the enzyme protein disulfide isomerase to human chromosome region 17p11→qter, *Cytogenet. Cell Genet.,* 47, 37, 1988.

262. **Popescu, N. C., Cheng, S-y., and Pastan, I.,** Chromosomal localization of the gene for a human thyroid hormone-binding protein, *Am. J. Hum. Genet.,* 42, 560, 1988.

263. **Pajunen, L., Jones, T. A., Goddard, A., Sheer, D., Solomon, E., Pihlajaniemi, T., and Kivirikko, K. I.,** Regional assignment of the human gene coding for a multifunctional polypeptide (P4HB) acting as the β-subunit of prolyl 4-hydroxylase and the enzyme protein disulfide isomerase to 17q25, *Cytogenet. Cell Genet.,* 56, 165, 1991.

264. **Tasanen, K., Parkkonen, T., Chow, L. T., Kivirikko, K. I., and Pihlajaniemi, T.,** Characterization of the human gene for a polypeptide that acts both as the β-subunit of prolyl 4-hydroxylase and as protein disulfide isomerase, *J. Biol. Chem.,* 263, 16218, 1988.

265. **Nakazawa, M., Aida, T., Everson, W. V., Gonda, M. A., Hughes, S. H., and Kao, W. W.-Y.,** Structure of the gene encoding the beta-subunit of chicken prolyl 4-hydroxylase, *Gene,* 71, 451, 1988.

266. **Myllylä, R., Pihlajaniemi, T., Pajunen, L., Turpeenniemi-Hujanen, T., and Kivirikko, K. I.,** Molecular cloning of chick lysyl hydroxylase. Little homology in primary structure to the two types of subunit of prolyl 4-hydroxylase, *J. Biol. Chem.,* 266, 2805, 1991.

267. **Hautala, T., Byers, M. G., Eddy, R. L., Shows, T. B., Kivirikko, K. I., and Myllylä, R.,** Cloning of human lysyl hydroxylase. Complete cDNA-derived amino acid sequence and assignment of the gene to the p36.2-36.3 region of chromosome 1, submitted for publication.

268. **Olsen, B. R., Berg, R. A., Kishida, Y., and Prockop, D. J.,** Collagen synthesis: localization of prolyl hydroxylase in tendon cells detected with ferritin-labeled antibodies, *Science,* 182, 825, 1973.

269. **Cutroneo, K. R., Guzman, N. A., and Sharawy, M. M.,** Evidence for a subcellular vesicular site of collagen prolyl hydroxylation, *J. Biol. Chem.,* 249, 5989, 1974.

270. **Peterkofsky, B. and Assad, R.,** Localization of chick embryo limb bone microsomal lysyl hydroxylase at intracisternal and intramembrane sites, *J. Biol. Chem.,* 254, 4714, 1979.

271. **Tschank, G., Hanauske-Abel, H. M., and Peterkofsky, B.,** The effectiveness of inhibitors of soluble prolyl hydroxylase against the enzyme in the cisternae of isolated bone microsomes, *Arch. Biochem. Biophys.,* 261, 312, 1988.

272. **Peterkofsky, B., Tschank, G., and Luedke, C.,** Iron-dependent uptake of ascorbate into isolated microsomes, *Arch. Biochem. Biophys.,* 254, 282, 1987.

273. **Colley, K. J. and Baenziger, J. U.,** Biosynthesis and secretion of rat core-specific lectin. Relationship of post-translational modification and assembly to attainment of carbohydrate binding activity, *J. Biol. Chem.,* 262, 3415, 1987.

274. **Bolwell, G. P. and Dixon, R. A.,** Membrane-bound hydroxylases in elicitor-treated bean cells. Rapid induction of the synthesis of prolyl hydroxylase and a putative cytochrome P-450, *Eur. J. Biochem.,* 159, 163, 1986.

275. **Hollister, D. W., Byers, P. H., and Holbrook, K. A.,** Genetic disorders of collagen metabolism, *Adv. Hum. Genet.,* 12, 1, 1982.

276. **Krane, S. M.,** Genetic and acquired disorders of collagen deposition, in *Extracellular Matrix Biochemistry,* Piez, K. A. and Reddi, A. H., Eds., Elsevier, New York, 1984, 413.

277. **Pinnell, S. R., Krane, S. M., and Kenzora, J. E.,** A heritable disorder of connective tissue. Hydroxylysine deficient collagen disease, *N. Engl. J. Med.,* 286, 1013, 1972.

278. **Krane, S. M., Pinnell, S. R., and Erbe, R. W.,** Lysyl-protocollagen hydroxylase deficiency in fibroblasts from siblings with hydroxylysine-deficient collagen, *Proc. Natl. Adac. Sci. U.S.A.,* 69, 2899, 1972.

279. **Sussman, M. D., Lichtenstein, J. R., Nigra, T. P., Martin, G. R., and McKusick, V. A.,** Hydroxylysine deficient skin collagen in a patient with a form of Ehlers-Danlos syndrome, *J. Bone Jt. Surg.,* 56A, 1228, 1974.

280. **Elsas, L. J., Miller, R. L., and Pinnell, S. R.,** Inherited human lysyl hydroxylase deficiency: ascorbate response, *J. Pediatr.,* 92, 378, 1978.

281. **Ihme, A., Risteli, L., Krieg, T., Risteli, J., Feldmann, U., Kruse, K., and Müller, P. K.,** Biochemical characterization of variants of the Ehlers-Danlos syndrome type VI, *Eur. J. Clin. Invest.,* 13, 357, 1983.

282. **Judisch, F. G., Waziri, M., and Krachmer, J. H.,** Ocular Ehlers-Danlos syndrome with normal lysyl hydroxylase activity, *Arch. Ophthalmol.,* 94, 1483, 1976.

283. **Krane, S. M.,** Genetic diseases of collagen, in *Gene Families of Collagen and Other Proteins,* Prockop, D. J. and Champe, P. C., Eds., Elsevier, New York, 1980, 57.

284. **Krane, S. M.,** Hydroxylysine-deficient collagen disease: a form of Ehlers-Danlos syndrome type VI, in *American Academy of Orthopaedic Surgeons Symposium on Heritable Disorders of Connective Tissue,* Akeson, W. H., Bornstein, P., and Glimcher, M. J., Eds., Mosby, Saint Louis, 1982, 61.

285. **Miller, R. L., Elsas, L. J., and Priest, R. E.,** Ascorbate action on normal and mutant human lysyl hydroxylases from cultured dermal fibroblasts, *J. Invest. Dermatol.,* 72, 241, 1979.

286. **Risteli, L., Risteli, J., Ihme, A., Krieg, T., and Müller, P. K.,** Preferential hydroxylation of type IV collagen by lysyl hydroxylase from Ehlers-Danlos syndrome type VI fibroblasts, *Biochem. Biophys. Res. Commun.,* 96, 1778, 1980.

287. **Tajima, S., Murad, S., and Pinnell, S. R.,** A comparison of lysyl hydroxylation in various types of collagen from type VI Ehlers-Danlos syndrome fibroblasts, *Coll. Rel. Res.,* 3, 511, 1983.

288. **Ihme, A., Krieg, T., Nerlich, A., Feldmann, V., Rauterberg, J., Glanville, R. W., Edel, G., and Müller, P. K.,** Ehlers-Danlos syndrome type VI: collagen type specificity of defective lysyl hydroxylation in various tissues, *J. Invest. Dermatol.,* 83, 161, 1984.

289. **Myllylä, R., Alitalo, K., Vaheri, A., and Kivirikko, K. I.,** Regulation of collagen post-translational modification in transformed human and chick-embryo cells, *Biochem. J.,* 196, 683, 1981.

290. **Goldberg, B. and Green, H.,** The synthesis of collagen and protocollagen hydroxylase by fibroblastic and nonfibroblastic cell lines, *Proc. Natl. Adac. Sci. U.S.A.,* 59, 1110, 1968.

291. **Chen-Kiang, S., Cardinale, G. J., and Udenfriend, S.,** Expression of collagen biosynthetic activities in lymphocytic cells, *Proc. Natl. Acad. Sci. U.S.A.,* 75, 1379, 1978.

292. **Myllylä, R. and Seppä, H.,** Studies on enzymes of collagen biosynthesis and the synthesis of hydroxyproline in macrophages and mast cells, *Biochem. J.,* 182, 311, 1979.

293. **Breul, S. D., Bradley, K. H., Hance, A. J., Schafer, M. P., Berg, R. A., and Crystal, R. G.,** Control of collagen production by human diploid lung fibroblasts, *J. Biol. Chem.,* 255, 5250, 1980.

294. **Berg, R. A., Schwartz, M. L., and Crystal, R. G.,** Regulation of the production of secretory proteins: intracellular degradation of newly synthesized "defective" collagen, *Proc. Natl. Adac. Sci. U.S.A.,* 77, 4746, 1980.

295. **Majamaa, K., Kuutti-Savolainen, E.-R., Tuderman, L., and Kivirikko, K. I.,** Turnover of prolyl hydroxylase tetramers and the monomer-size protein in chick-embryo cartilaginous bone and lung *in vivo,* *Biochem. J.,* 178, 313, 1979.

296. **Majamaa, K. and Oikarinen, J.,** Labeling of prolyl hydroxylase tetrameric subunits in freshly isolated chick-embryo tendon cells and in certain chick-embryo tissues *in vivo, Biochem. J.,* 204, 737, 1982.

297. **Helaakoski, T., Pajunen, L., Kivirikko, K. I., and Pihlajaniemi, T.,** Increases in mRNA concentrations of the α- and β-subunits of prolyl 4-hydroxylase accompany increased gene expression of type IV collagen during differentiation of mouse F9 cells, *J. Biol. Chem.,* 265, 11413, 1990.

298. **Bassuk, J. A. and Berg, R. A.,** Correlation of the steady-state RNA levels among the α-subunit of prolyl 4-hydroxylase and the α1 and α2 chains of type I collagen during growth of chicken embryo tendon fibroblasts, *Biochem. Biophys. Res. Commun.,* 1174, 169, 1991.

299. **Yeowell, H. N., Murad, S., and Pinnell, S. R.,** Hydralazine differentially increases mRNAs for the α and β subunits of prolyl 4-hydroxylase whereas it decreases proα1(I) collagen mRNAs in human skin fibroblasts, *Arch. Biochem. Biophys.,* 289, 399, 1991.

300. **Risteli, L., Risteli, J., Salo, L., and Kivirikko, K. I.,** Intracellular enzymes of collagen biosynthesis in 3T6 fibroblasts and chick-embryo tendon and cartilage cells, *Eur. J. Biochem.,* 97, 297, 1979.

301. **Majamaa, K., Myllylä, R., Alitalo, K., and Vaheri, A.,** Regulation of proline 3-hydroxylation and prolyl 3-hydroxylase and 4-hydroxylase activities in transformed cells, *Biochem. J.,* 206, 499, 1982.

302. **Oyamada, I., Bird, T. A., and Peterkofsky, R.,** Decreased extracellular matrix production in scurvy involves a humoral factor other than ascorbate, *Biochem. Biophys. Res. Commun.,* 152, 1490, 1988.

303. **Uitto, J. and Prockop, D. J.,** Synthesis and secretion of under-hydroxylated procollagen at various temperatures by cells subject to temporary anoxia, *Biochem. Biophys. Res. Commun.,* 60, 414, 1974.

304. **Oikarinen, A., Anttinen, H., and Kivirikko, K. I.,** Further studies on the effect of the collagen triple-helix formation on the hydroxylation of lysine and the glycosylations of hydroxylysine in chick-embryo tendon and cartilage cells, *Biochem. J.,* 166, 357, 1977.

305. **Oikarinen, A., Anttinen, H., and Kivirikko, K. I.,** Effect of L-azetidine-2-carboxylic acid on glycosylations of collagen in chick-embryo tendon cells, *Biochem. J.,* 160, 639, 1976.

306. **Majamaa, K.,** Effect of prevention of procollagen triple-helix formation on proline 3-hydroxylation in freshly isolated chick-embryo tendon cells, *Biochem. J.,* 196, 203, 1981.

307. **Anttinen, H. and Hulkko, A.,** Regulation of the glycosylations of collagen hydroxylysine in chick embryo tendon and cartilage cells, *Biochim. Biophys. Acta,* 632, 417, 1980.

308. **Bolwell, G. P., Robbins, M. P., and Dixon, R. A.,** Metabolic changes in elicitor-treated bean cells. Enzymic responses associated with rapid changes in cell wall components, *Eur. J. Biochem.,* 148, 571, 1985.

Chapter 2

ADVANCED GLYCOSYLATION ENDPRODUCTS

Richard Bucala, Helen Vlassara, and Anthony Cerami

TABLE OF CONTENTS

I. INTRODUCTION

The formation of brown color during the heating or long-term storage of proteins is a phenomenon that has been recognized by food chemists for many years. In 1912, Louis Camille Maillard recorded his observations of heating sugars with amino acids.[1] He found that yellow-brown pigments formed at a rate which varied with the composition of the amino acid-sugar mixtures. The loss of protein-derived lysine residues was observed by Patton and Hill and led them to propose that a direct reaction occurred between the reducing sugars and the free (ϵ) amino group of lysine[2]. In 1949, Mohammed et al. described the progressive loss of free amino groups in serum albumin after incubation with glucose at physiological temperature and pH.[3] Over the years these effects were called nonenzymatic browning or the Maillard reaction, and became increasingly important as a mechanism to explain the decreased nutritional availability of food which occurs with prolonged storage.[4,5]

At the present time, the chemical pathway which leads from sugar-amino acid interaction to brown pigmentation is only partially understood. Maillard products can be described best as a class of heterogenous sugar-amino acid adducts which form through progressive dehydration, rearrangement, and cyclization reactions. They are characterized by their yellow-brown color, fluorescence, and ability to participate in cross-linking and polymerization reactions. Many studies of Maillard chemistry have focused on its degradative effect on food; however the Maillard reaction also leads to the generation of aromatic, flavorful compounds which enhance the taste of such diverse products as toasted bread, caramel, syrups, and brewed beverages.[4,5]

A role for Maillard chemistry in living organisms has been appreciated only recently. Initial studies of this process *in vivo* focused on the biochemical abnormalities which occur in diabetes mellitus, an illness characterized by high circulating levels of glucose. More recent investigations have implicated nonenzymatic browning in aging and in the normal, physiological turnover of proteins.[6]

The Maillard reaction proceeds from the early products of nonenzymatic glycosylation; that is from the Amadori products which result from the addition of simple, reducing sugars to protein amino groups. Since Maillard products are the long-term consequence of nonenzymatic glycosylation, they have come to be called Advanced Glycosylation Endproducts or AGEs.[7] The following discussion will review our current understanding of advanced glycosylation biochemistry and address the diverse biological effects of this process.

II. NONENZYMATIC GLYCOSYLATION

A. REACTION BETWEEN GLUCOSE AND PROTEIN

Studies of the nonenzymatic glycosylation process began with the structural elucidation of the minor electrophoretic variants of human hemoglobin. In 1968, Rahbar described the presence of elevated amounts of a minor hemoglobin species in the blood of diabetic patients.[8] This species, called hemoglobin A_{1c}, was shown to have a small molecule blocking the α-amino groups of the β-chains.[9] The blocking group had the properties of a Schiff base or other borohydride-reducible moiety,[9,10] and appeared identical to the reaction product of glucose and valylhistidine.[11] Ultimately, proton magnetic resonance spectroscopy demonstrated unequivocally the presence of a glucose Amadori product attached to the N-terminal valine.[12] Formation of hemoglobin A_{1c} was shown to occur as a posttranslational modification whose rate of formation is dependent on the ambient blood glucose concentration.[13,14]

Figure 1 illustrates the reaction scheme between an amino group and glucose as a model reducing sugar. This addition reaction occurs without enzymes. In the first step, an α-amino group of the N-terminal amino acid or the ϵ-amino group of lysine reacts with the sugar-derived carbonyl to form a reversible Schiff base.[15,16] It is likely that the Schiff base adducts

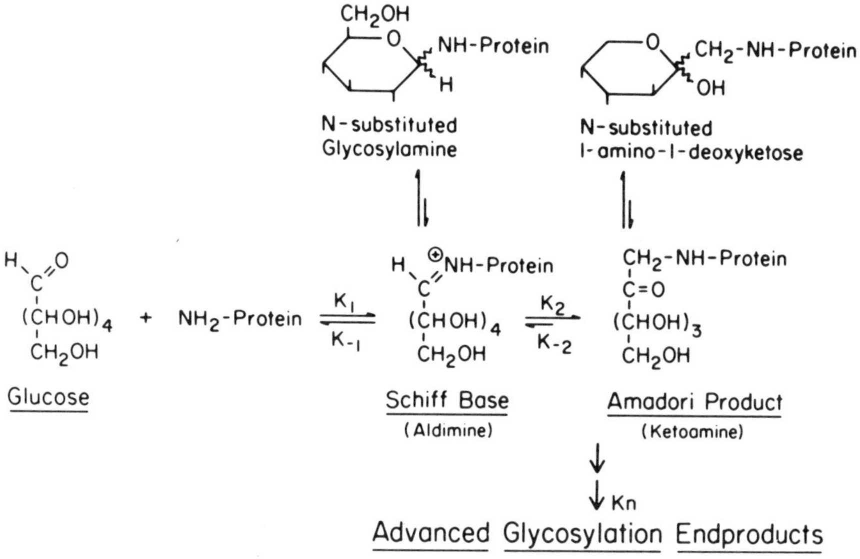

FIGURE 1. Formation of protein bound glycosylation products. Equilibrium levels of the reversible Schiff base and Amadori products are reached within hours and weeks, respectively. Advanced glycosylation products accumulate on long-lived proteins and are irreversibly attached to proteins. (From Brownlee, M., Vlassara, H., and Cerami, A., in *Diabetic Complications,* Crabbe, M. J. C., Ed., Churchill Livingstone, New York, 1987, 94. With permission.)

exist both in an open-chain aldimine form, and a chemically more stable glycosylamine ring form. The Schiff base adduct is freely reversible; the rate of formation is equal approximately to the rate of dissociation. Schiff base products form rapidly and equilibrium for these labile adducts occurs in a matter of a few hours. Over this short time frame, the steady-state level of glucose Schiff bases is determined only by the ambient glucose concentration.[16]

Schiff base adducts of glucose and protein undergo intramolecular rearrangement to form a more stable adduct called the Amadori product (Figure 2.).[17] This product is also chemically reversible, but its equilibrium is reached over a period of 28 days. Experimentally determined rate constants for the formation and dissociation of the Amadori product are $14.2 \times 10^{-6} \text{ s}^{-1}$ and $1.7 \times 10^{-6} \text{ s}^{-1}$ respectively. This yields a calculated equilibrium constant of (K_{eq}) of 8.4.[18] The absolute level of the Amadori product *in vivo* reflects only the ambient glucose concentration and the half-life of the protein substrate. This phenomenon has been exploited in the use of hemoglobin A_{1c} as a clinical marker for longer term (weeks) glucose control (Figure 3).

To date over 25 proteins have been evaluated for their ability to form Amadori products either *in vivo* or *in vitro* (Table 1).[19-44] In addition to hemoglobin, the glycosylation of a variety of proteins has been shown to be more extensive in diabetes. It is important to note that after the period of time required to reach equilibrium with free glucose, the absolute amount of Amadori product attached to protein does not increase with time. This kinetic constraint has been confirmed *in vivo* by measuring Amadori products from diabetic tissues exposed to similar hyperglycemia for varying periods of time. The level of Amadori products is consistently observed to be two- to threefold higher than in nondiabetic proteins; regardless of whether the proteins are obtained after 18 weeks or after many years of hyperglycemia.[45,46] A necessary precondition is that the protein substrate must have a survival time *in vivo* longer than the period required to reach equilibrium for the Amadori product (3 to 4 weeks).

A number of variables affect the extent of glycosylation which can be achieved *in vitro*. These include temperature, incubation time, pH (only uncharged amino groups serve as sites for nucleophilic attack by the sugar carbonyl), and substrate concentration. The first three

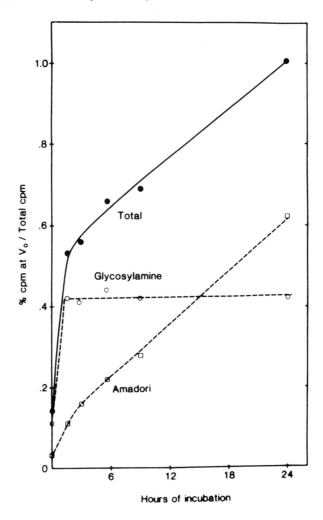

FIGURE 2.   Kinetics of glycosylamine (Schiff base) and Amadori product formation. (From Baynes, J. W., Thorpe, S. R., and Murashaw, M. W., in *Methods in Enzymology, Posttranslational Modifications,* Vol. 103, Wold, F., and Moldave, K., Eds., Academic Press, New York, 1984, 88. With permission.)

variables are fixed *in vivo*, and the preceding arguments make clear how glucose concentration (as in diabetes) and protein half-life are important determinants for the extent of glycosylation of a given protein. A number of structural studies have made it clear that the reactivity of a given amino group is critically dependent on the microenvironment within the protein. Since only unprotonated amines can react to form Schiff bases, the pK_a affects the rate of the initial condensation reaction. This accounts in part for the preferential modification of α-amino groups. In the case of human hemoglobin, the order of prevalence of glycosylation *in vivo* has been determined to be β-Val-1, β-Lys-66, α-Lys-61, β-Lys-17, and α-Val-1.[47] Glycosylation *in vitro* produces a different distribution. In neither case does reactivity correspond to the known pK_a of the amine — with the exception of β-Val-1, which has the lowest pK_a of the series. The first step of nonenzymatic glycosylation, Schiff base formation, does not appear to be rate limiting for the glycosylation of the side chain residues in hemoglobin.

Human serum albumin contains 59 ϵ-amino lysine residues. At least ten lysines are glycosylated *in vivo*, with the most clearly identified residues being Lys-525, Lys-439,

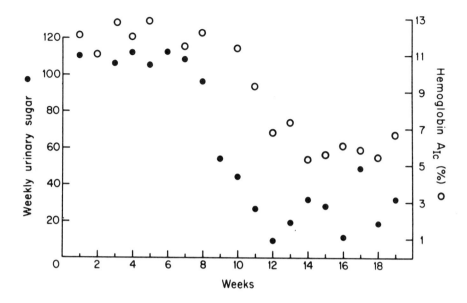

FIGURE 3. Temporal relationship between weekly urinary glucose excretion and hemoglobin A_{lc}. (From Koenig, R. J., Peterson, C. M., Jones, R. L., Saudek, C., Lehrman, M., and Cerami, A., *N. Engl. J. Med.*, 295, 417, 1976. With permission.)

TABLE 1
Proteins Evaluated for Nonenzymatic Glycosylation[19-44]

Extracellular matrix proteins	Membrane proteins
Collagen	Erythrocytes
Myelin	Platelets
Basement membrane	Endothelial cells
Fibronectin	
Lens capsule	Plasma proteins
Fibrin	Albumin
	Immunoglobulins
Cellular proteins	Apolipoproteins AI, AII, B, CI, E
Hemoglobin A	Haptoglobin
Lens crystallins	Ferritin
Tubulin	Transferrin
Cathepsin B	α-1 Anti-trypsin
Ribonuclease	Complement C3
β-NAc-D-glucosaminidase	Plasminogen
Calmodulin	Plasminogen activator
Superoxide dismutase	Fibrinogen
	Antithrombin III

Lys-199, and Lys-281.[48,49] Recent mass spectroscopy and gas chromatography studies also have identified a low level of nonenzymatic glycosylation at the N-terminal aspartic acid.[50] Glucitol-Lys-525 accounts for approximately one third of the overall glycosylation. Glycosylation at Lys-199 was identified initially on the basis of inhibition by aspirin acetylation, which specifically modifies this site.[138] Despite its low pK_a however, Lys-199 is a minor site for glycosylation *in vivo*. The enhanced reactivity of this site *in vitro* may be due to the removal, during albumin purification, of the fatty acid which normally binds at this site. Three of the albumin lysine sites are located adjacent to basic amino acids: Lys-525 is in a Lys-Lys sequence, Lys-439 is in a Lys-His sequence, and Lys-534 is in a Lys-His-Lys sequence. Interestingly, it has been noted that there are similar basic residues in the glycosylation sites of hemoglobin: the N-terminal valine is adjacent to a histidine and close to a lysine, and two other glycosylation sites, β-Lys-66, and α-Lys-61, are located in a Lys-Lys sequence. Iberg and Fluckiger have proposed that in all these cases the adjacent protons participate in local acid-base catalysis to accelerate the Amadori rearrangement.[49]

In normal human serum, about 6 to 15% of albumin exists as a glycosylated fraction. This contrasts with hemoglobin where about 4% is present normally as the A_{1c} fraction. In addition, it has been calculated that there is about one glucosyl lysine per five hemoglobin tetramers.[51,52]

Model studies of ribonuclease have been helpful in revealing the role of accessory ligands and buffer ions in the glycosylation process. Incubation of ribonuclease with glucose in phosphate buffer leads to an inactivation of enzyme activity.[53] This occurs as a result of the glycosylation of a lysine near the catalytic site. Similar incubation in the presence of cationic buffers leads to preferential glycosylation at alternative lysines and much less enzyme inactivation. Because Schiff base formation does not appear to be affected, the authors postulate that inorganic phosphate catalyzes the subsequent stabilization step, the Amadori rearrangement. Interestingly, the glycosylation of the N-terminal valine of hemoglobin also is enhanced by phosphate buffer, as well as by 2,3,diphosphoglycerate. It appears that phosphate moieties may play a role in the differences observed during the *in vitro* versus the *in vivo* glycosylation of hemoglobin.[54]

Nonenzymatic glycosylation is often considered to be a universal phenomenon in the sense that any primary amino group has the potential to participate in Schiff base formation. Yet, the bulk of current evidence suggests that there is great variation in the susceptibility of a given amino group to glycosylation. Changes in protein function which may result from high sugar levels must take into account the inherent variability of a given protein to nonenzymatic glycosylation.

B. FUNCTIONAL CONSEQUENCES OF NONENZYMATIC GLYCOSYLATION

Hemoglobin A_{1c} has a higher affinity for oxygen than hemoglobin A.[55] This is believed to result from the inhibition of 2,3-diphosphoglycerate binding by α-glucosyl-valine-1. It was hypothesized that this effect might underlie the end organ damage in diabetes, via diminished oxygen unloading in the microcirculation. The oxygen affinity of diabetic blood however, is only marginally higher by 2 mm of Hg.

The modification of hemoglobin S by the triose glyceraldehyde inhibits gelation and red cell sickling.[56] This is analogous to the carbamylation reaction of hemoglobin with cyanate described by Cerami and Manning.[57] Both reactions have served as valuable models for the development of pharmacological strategies toward the therapy of sickle-cell anemia.

Nonenzymatic glycosylation of albumin leads to significant changes in the transport properties of this protein. Fluorescence emission studies suggest that glycosylation of lysine-525 induces a profound conformational change. Affinity for the ligand bilirubin is decreased twofold, and for long chain fatty acids twentyfold.[58] Interestingly, glycosylated serum albumin is avidly taken up by endothelial cells, and direct injection of glycosylated serum proteins induces glomerular basement membrane thickening.[59,60] Both effects may be relevant to the histopathological changes observed in diabetes.

Erythrocyte membrane proteins are progressively glycosylated over the course of the red cell life, and the degree of glycosylation increases in the diabetic state.[22,25] Increased glycosylation may play a role in the decreased erythrocyte deformability observed in diabetes, and recent *in vitro* studies of membrane fluidity support this hypothesis.[61]

Low-density lipoprotein, the lipid transport protein which plays an important role in the initiation of atherogenesis, has been shown to be glycosylated both *in vitro* and *in vivo*.[28,30] Glycosylated LDL is less readily taken up and degraded by tissue fibroblasts and macrophages.[62-64] This may contribute to the premature atherosclerosis observed in diabetes. Interestingly glycosylated proteins may be antigenic in some diabetic individuals. Witztum et al. have described patient sera which contain antibodies specific for a glucitol-lysine modification.[65] These antibodies recognize the Amadori product after it has been chemically reduced *in vitro*. Native, unreduced adducts do not compete for antibody recognition. Precisely how the reactive epitope arises *in vivo* is unclear, however a role for these antibodies in the progression of diabetic vascular and renal disease has been postulated.

Nonenzymatic glycosylation affects proteins of the clotting cascade and in diabetes this modification may interfere with vascular wall homeostasis. Glycosylation renders fibrin less susceptible to plasmin digestion.[33] Similarly, the thrombin inhibiting activity of antithrombin III is inhibited by nonenzymatic glycosylation. Evidence suggests that this effect is the result of a reduction in the affinity of antithrombin III for heparin.[36] Platelet membrane proteins also display increased glycosylation in diabetes.[41] This may play a role in the functional platelet defects of diabetes; however as yet there is no *in vivo* data to support this hypothesis.

Lens crystallins, nerve myelin, and matrix collagen turn over extremely slowly and have been the focus of much investigation into nonenzymatic glycosylation *in vivo*. All of these proteins lie within insulin-independent tissues and would be expected to be extensively modified by glucose in patients with diabetes. Interestingly all these tissues: the lens, the vascular wall collagen, and the nerve sheath feature prominently in the clinical sequelae of long-standing diabetes mellitus. In the mid-1970s, the prediction that these proteins in particular display excessive glycosylation *in vivo* formed an important tenet of the theory that diabetic complications resulted from the accumulation of glucose-protein modification.[67,68]

Early studies of the lens crystallins demonstrated that [^{14}C]glucose or [^{14}C]glucose-6-phosphate accumulated in acid-precipitable lens protein in a time-dependent fashion.[68] Furthermore, the protein solutions gradually became opalescent as a function of progressive glycosylation.[19] The increase in light scattering was found to be due to the formation of high-molecular-weight aggregates, which were derived in part from protein-bound Amadori products. Since the N-terminal amino acids in most crystallins are acylated, these glucose derived cross-links are probably restricted to side-chain lysine residues. The remaining covalent cross-links were the result of disulfide-bond interchange, induced by conformational changes within the protein. That glycosylation of lens crystallins produces conformational changes directly has been confirmed by recent studies of circular dichroism and tryptophan fluorescence.[69] Studies of nonenzymatic glycosylation and cataractogenesis have also been extended into both experimental and human diabetic cataracts. A 5- to 10-fold increase in lysine modification has been detected in the cataractous lenses of galactosemic and diabetic rats. Human lenses also have been observed to accumulate lysine Amadori products with time.[70,71]

Perhaps the most significant finding to emerge from lens protein studies has been that with prolonged incubation of lens proteins and sugars, protein-bound pigmented moieties form which possess the spectroscopic properties of pigments isolated from aged lenses.[72] These observations led directly to the hypothesis that the Maillard reaction, and Maillard products, occur *in vivo*.

Collagen and peripheral nerve myelin are glycosylated *in vivo*, and accumulate Amadori products at an accelerated rate in diabetic patients.[34,73] This leads to functional and structural changes which may have direct bearing on the pathophysiology of diabetic complications. The bulk of evidence suggests that these complications may be due primarily to the products of advanced glycosylation (discussed below).

In summary, a number of adverse functional consequences of nonenzymatic glycosylation and Amadori product formation have been described. However, given the widespread distribution of Amadori products and the fact that levels of Amadori products never increase more than 2.5 to 3 times normal, most investigators would concur that Amadori products per se, have not yet been found to contribute *directly* to diabetic pathology. An additional difficulty is that in many cases it is not possible to assess the extent to which products of advanced glycosylation, i.e., the subsequent reactions of the Amadori product, contribute to the adverse effects which have been elucidated.

III. ADVANCED GLYCOSYLATION

Within the time frame of most cellular and plasma protein half-lives, the Amadori product is stable and does not undergo further significant reactions. Over a longer time period (weeks to years) the Amadori product undergoes additional chemical transformations. These reactions incude fragmentation, oxidation, dehydration, and condensation with additional, reactive amino groups. All of these reactions fall within the Maillard pathway and the resultant products have come to be termed advanced glycosylation endproducts (AGEs). These sugar-derived moieties have in common a brown pigmentation, a characteristic fluorescence spectrum, and an ability to participate in protein-protein cross-linking. Maillard products have been implicated in a number of pathological sequelae of both diabetes and normal aging.

A. EARLY STUDIES OF ADVANCED GLYCOSYLATION *IN VIVO*

An important distinction between Amadori products and advanced glycosylation endproducts is that while Amadori products are in equilibrium with glucose, AGEs are irreversibly bound to protein. Thus, on long-lived proteins, the quantity of bound AGE increases proportionally to the concentration at equilibrium of the Amadori product. Studies of lens crystallins glycosylation were the first to implicate a role for the Maillard process *in vivo*. The lens of the eye is composed predominantly of long-lived crystallins proteins which become progressively yellow-brown with age. These pigmented changes lead to the brunescent cataract, and its formation is accelerated in patients with diabetes. Previous work demonstrated that these yellow-brown fluorescent pigments could act as covalent cross-links between lens proteins.[74] Figure 4 illustrates the characteristic fluorescence excitation spectra of lens crystallins incubated with glucose as well as the spectra of freshly isolated young and aged human lenses. The control sample displays only the fluorescence of tryptophan residues (280 nm). Three additional excitation maxima at 360, 400, and 470 nm are found in the lens proteins incubated with hexoses. Similar fluorescent excitation maxima can be observed to have accumulated in the aged lenses. These data were the first to suggest that some of the pigments formed in the human lenses during aging could be the result of advanced glycosylation products which form *in vivo*. The progressive glycosylation of lens crystallins *in vitro* was found to lead to the formation of high-molecular-weight aggregates which displayed an increase in the formation of disulfide and nondisulfide cross-links. These macromolecular aggregates produced light scattering and opalescence, and thus mimicked the progressive opacification of cataractous lenses in diabetes.[72]

Connective tissue collagen is rich in lysine. Not surprisingly, after hemoglobin, collagen was the first protein found to have glycosylated ε-amino groups.[75,76] Collagen turns over very slowly in mature individuals and like crystallins and hemoglobin, it is exposed to ambient blood glucose levels. Early work by Schnider and Kohn had demonstrated an age-related decrease in the solubility and digestibility of connective tissue collagen. Collagen becomes more cross-linked with age and associated with this cross-linking is an increase in the amount of glucose attached to the collagen fibrils. This effect was more pronounced in individuals with diabetes mellitus.[24,77,78] Figure 5 illustrates the result of a study where samples of human dura collagen, which exhibited an increase in glycosylation and a decrease in solubility, were acid-hydrolyzed and analyzed for the presence of AGE associated fluorescence.[38] The fluorescence intensity of the collagen-derived material was plotted versus the age of the connective tissue specimen. AGE associated fluorescence was found to increase with the age of the individual. In subjects with diabetes however, the collagen associated fluorescence was greater than expected for their age. This is consistent with the prediction

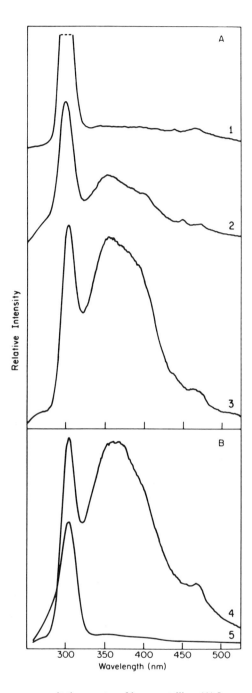

FIGURE 4. Fluorescence excitation spectra of lens crystallins: (A) Lens crystallins incubated (1) without sugar, (2) with 5 m*M* glucose, and (3) with 5 m*M* glucose-6-phosphate. (B) Lens protein from (upper) senile cataractous lens, and (lower) 20 year old normal lens. (From Monnier, V. M. and Cerami, A., *Science,* 211, 491, 1981. Copyright 1981 by the AAAS. With permission.)

that Amadori products are AGE precursors, and that their increased presence in diabetics results in increased accumulation of irreversibly bound, fluorescent AGEs.

Direct experimental evidence for the cross-linking potential of post-Amadori products emerged from elegant *in vitro* studies utilizing ribonuclease as a model protein.[79] Glucose incorporation into protein was observed to plateau after two days (Figure 6). This corre-

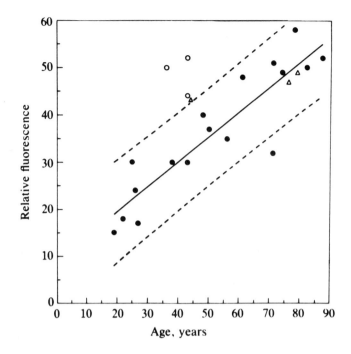

FIGURE 5. Relative fluorescence of advanced glycosylation endproducts in dura collagen from individuals of different ages (measured as fluorescence at 440 nm upon excitation at 370 nm per mg hydroxyproline). The broken line indicates the 95% confidence limits. ● = normal subjects, ○ = patients with juvenile diabetes, △ = patients with adult onset diabetes. (From Monnier, V. M., Kohn, R. R., and Cerami, A., *Proc. Natl. Acad. Sci. U.S.A.*, 81, 583, 1984. With permission.)

sponded to the loss of primary amino groups and free lysine. Gradually, with continued incubation time covalently linked dimers and trimers of ribonuclease appeared (Figure 7). Additional experiments showed that glycosylated ribonuclease could covalently "trap" unglycosylated, radiolabeled ribonuclease. When free glucose was removed from the incubation mixtures, glycosylated ribonuclease continued to polymerize, and this polymerization could be inhibited by the addition of free lysine. The observation that excessively glycosylated protein continues to polymerize in the absence of free glucose has profound pathophysiological implications for the progression of diabetic complications. Theoretically, it can be argued that once proteins have been "primed" with excess Amadori products during periods of severe hyperglycemia, restoration of the strictest glucose control will not prevent the inevitable consequences of advanced glycosylation.

B. STRUCTURAL STUDIES OF ADVANCED GLYCOSYLATION ENDPRODUCTS

Elucidation of the synthetic pathway leading from nonenzymatic glycosylation to AGE formation has been hampered by the slowness of AGE formation, the heterogeneity of the structural intermediates, and the instability of these products to both acid and alkaline hydrolysis. In 1984, the first synthetic AGE: furoyl-furanyl imidazole (FFI), was isolated from an *in vitro* preparation of polylysine and glucose (Figure 8).[80] The structure of this yellow-brown fluorescent compound appeared to result from the condensation of two Amadori products and immediately suggested the cross-linking mechanism of this particular AGE. Subsequent studies utilizing ^{15}N enriched NH_4Cl suggested that the greatest portion of FFI forms during chemical isolation: acid hydrolysis, followed by basification and organic

mol HL + HL'/mol RNase A ($\bullet$—$\bullet$)
mol GLUCOSE/mol RNase A (○—○)

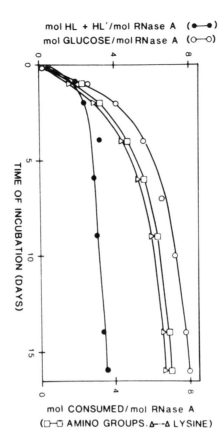

TIME OF INCUBATION (DAYS)

mol CONSUMED/mol RNase A
(□—□ AMINO GROUPS, △--△ LYSINE)

FIGURE 6.   Modification of ribonuclease by glucose *in vitro*. At indicated times the ribonu-
clease-glucose incubation was assayed for ^{3}H-glucose incorporation (by acid precipitation), loss
of primary amino groups, decrease in lysine content, and yield of hexitol-lysine (HL + HL').
(From Eble, A. S., Thorpe, S. R., and Baynes, J. W., *J. Biol. Chem.*, 258, 9406, 1983. With
permission.)

extraction.[81,82] Nevertheless FFI has been indispensable in the development of immunoassays
for *in vivo* AGE products,[83] and in the elucidation of a receptor mediated pathway for AGE
degradation.[84] It would appear that structures very similar to FFI form *in vivo* as a result of
prolonged glucose modification.

Amadori products can undergo oxidative degradation, producing reactive deoxygluco-
sones which can reattach to protein amino groups.[85] "Pyrraline" and other glucose-derived
pyrroles appear to form via this mechanism,[86] and a recent immunochemical study has
detected these adducts on diabetic proteins.[87] Additional cross-linking structures which have
been described include the AFGPs (1-alkyl-2-formyl-3,4-diglycosyl pyrroles).[88] These compounds were isolated from synthetic mixtures which utilized sulfite to inhibit the more
terminal stages of advanced glycosylation. In the food and pharmaceutical industry, sulfite
has been a widely used agent to control the adverse effects of the Maillard process.[89] It was
reasoned that by inhibiting the terminal stages of advanced glycosylation, the reaction
products might be enriched for an intermediate which could be readily isolated and identified.
Figure 9 shows the proposed pathway for formation of glucose-derived AFGP (G-AFGP)
from the reaction of an Amadori product with 3-deoxyglucosone.

Recently, Sell and Monnier have isolated and identified an acid-resistant fluorescent
molecule from pooled human dura collagen.[90] This molecule "pentosidine" has the unexpected structure of a ribose or five-carbon sugar as the cross-linking moiety. Although ribose

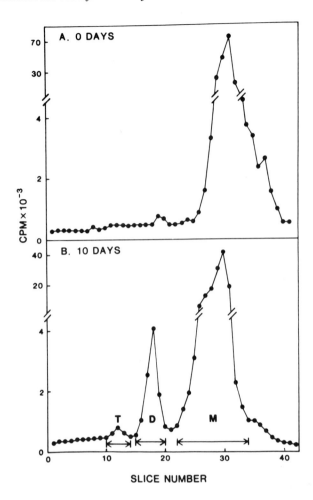

FIGURE 7.   Radioactivity profile of sodium dodecyl sulphate polyacrylamide gel electrophoresis of ^{125}I-ribonuclease incubated with glucose for 0 and 10 days. Ribonuclease monomer, dimer, and trimer regions are indicted by M, D, and T, respectively. The low-molecular weight region is on the right and the high-molecular weight region is on the left. (From Eble, A. S., Thorpe, S. R., and Baynes, J. W., *J. Biol. Chem.*, 258, 9406, 1983. With permission.)

glycosylates proteins readily (and at a faster rate than glucose) the total pentose levels in the human plasma have been estimated to be only 44 μM. Pentosidine was found in a variety of human tissues and its levels on collagen correlate linearly with human age.

An additional pathway for Amadori product degradation leads to the formation of protein bound carboxymethyl-lysine and erythronic acid.[91] Carboxymethyl-lysine was found in human lens crystallins, and at higher levels than detectable Amadori products. Both erythronic acid and carboxymethyl-lysine are present in human urine. These findings suggest that some oxidative degradation of Amadori products occurs *in vivo*. This may limit the total amount of Amadori products available to participate in protein-protein cross-linking.

IV. REACTIONS BETWEEN PROTEINS AND OTHER SUGARS

Glucose is the predominant extracellular sugar and most studies have emphasized its role in initiating nonenzymatic glycosylation. It is important to realize however, that the chemistry of advanced glycosylation can be generalized to other aldoses and ketoses (Table 2.). Pentoses, trioses, and phosphorylated sugars such as glucose-6-phosphate and glycer-

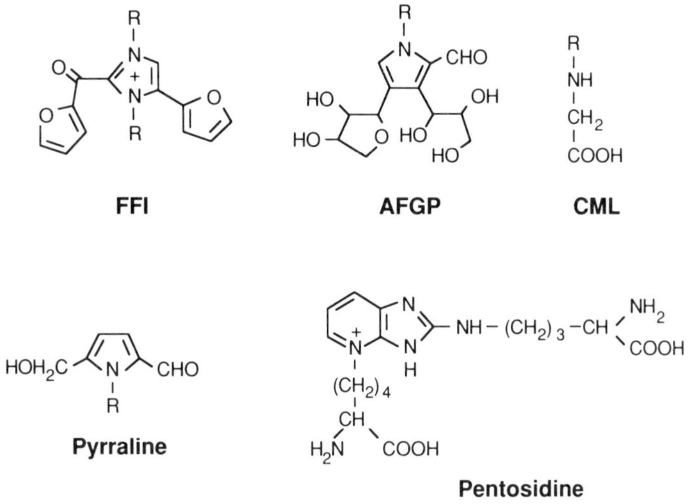

FIGURE 8.   Structures of advanced glycosylation endproducts. The presumed lysine-derived amine nitrogens are shown unsubstituted. FFI, pyrraline, and AFGP are synthetic products. FFI and pentosidine were isolated after acid hydrolysis. CML: carboxymethyl-lysine.

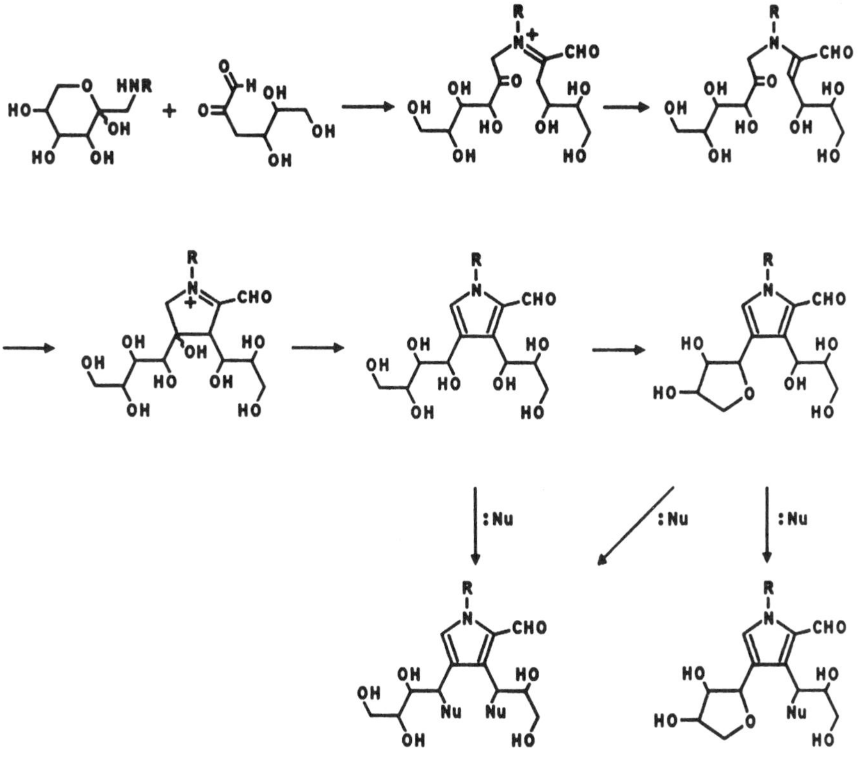

FIGURE 9.   Proposed pathway of formation of glucose-derived AFGP from the Amadori product of amino-hexanoic acid and 3-deoxyglucosone. Adduct formation with additional nucleophiles (Nu) such as amines or sulfite is also shown. (From Farmar, J., Ulrich, P., and Cerami, A., *J. Org. Chem.*, 53, 2346, 1988. Copyright 1988 American Chemical Society. With permission.)

TABLE 2
Approximate Rates of Reaction of Selected Reducing Sugars with Protein/Amino Acids

	Percent carbonyl[93,95]	Rate of Schiff base formation[93] K_1 (10^{-3} mM^{-1} h[1])	Overall browning rate[a6,94]
Hexoses			
Glucose	0.002	0.6	1
Galactose	0.02	2.8	4.6
Mannose	0.005	3.2	3
Fructose	0.7	4.5	10
Hexose Phosphates			
Glucose-6-PO$_4$	<0.4		18
Fructose-6-PO$_4$	4-5		10
Pentoses			
Xylose	0.02	2.9	7
Ribose	0.05	10.0	129

Note: Compiled from References 6, 93-95. Reaction conditions are not identical between studies and numerical values should serve as a relative guide only.

[a] Relative to glucose.

aldehyde-3-phosphate are much more reactive than glucose toward protein amino groups.[92,93] The rate of nonenzymatic glycosylation is in part a function of the anomerization rate and equilibrium toward the reactive carbonyl form.[94,95] Thus glucose-6-phosphate is 20 times more reactive than glucose and frequently is used to accelerate the advanced glycosylation process experimentally. Fructose produces protein-bound fluorophores and cross-links at ten times the rate of glucose. It has been postulated that fructose may play a pathogenic role in tissues rich in polyol dehydrogenase, such as the ocular lens and the peripheral nerve.[96,97]

V. BIOLOGICAL CONSEQUENCES OF ADVANCED GLYCOSYLATION

A. COLLAGEN AND EXTRACELLULAR MATRIX

One consequence of biological aging which is accelerated in diabetes is an increase in the cross-linking of collagen. This cross-linking leads to decreased solubility and susceptibility to enzymatic digestion,[24,77] while imparting increased rigidity to collagen-containing tissue.[98-101] Model studies of rat tail collagen have supported the hypothesis that advanced glycosylation contributes significantly to the physico chemical changes of aged collagen. In one series of studies, the breaking time of tendon fibers was measured after incubation with various reducing sugars.[102] It had been observed previously that tendons from older rodents, which had a higher degree of collagen cross-linking, are resistant to breakage in 7 M urea at 45°.[103,104] Incubation of tail collagen with reducing sugars was found to lead to nonenzymatic glycosylation, and an increase in absorbance, fluorescence, and breaking time; mimicking the changes observed in aged collagen *in vivo* (Figure 10A, B).

In more recent studies, the presence of covalent, intermolecular bonds was demonstrated by an increased hydrothermal isometric tension at 90°C.[105,106] Under these conditions the noncovalent and aldimine bonds of collagen are disrupted. The solubility of glycosylated collagen in 0.5 M acetic acid, cyanogen bromide, and sodium dodecyl sulphate was markedly reduced. Peptide mapping of cyanogen-bromide digested collagen demonstrated significant amounts of high molecular weight, cross-linked material in the advanced glycosylated samples. The glucose-derived cross-links were found to occur throughout the length of the collagen molecule, in distinction to the cross-links normally produced by the enzyme lysyl

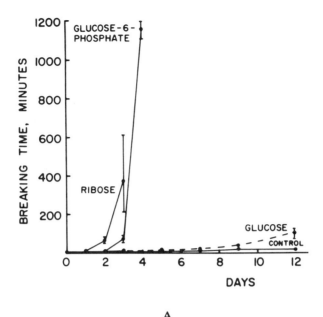

A

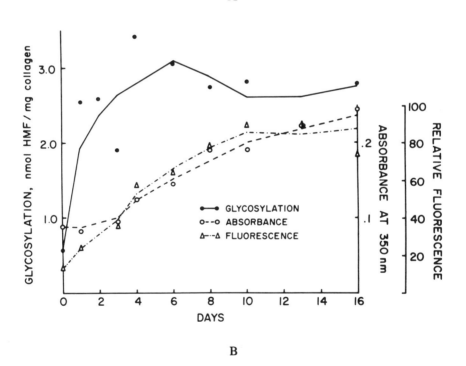

B

FIGURE 10. (A) Breaking time of rat tail tendon fibers in 7 M urea at 45° after incubation with 0.1 M sugars for the indicated times. (B) Time course for nonenzymatic glycosylation (measured by hydroxymethyl furfural) and the amounts of fluorescent and pigmented compounds in digests of rat tail tendon incubated with 0.1 M glucose-6-phosphate. Absorbance was measured at 350 nm. Relative fluorescence was measured at 440 nm upon excitation at 370 nm. (From Kohn, R. R., Cerami, A., and Monnier, V. M., *Diabetes,* 33, 57, 1954. Copyright by the American Diabetes Association. With permission.)

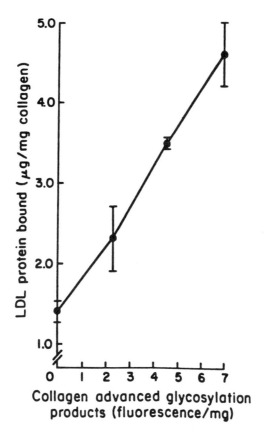

FIGURE 11.   Covalent binding of radiolabeled LDL by immobilized collagen as a function of collagen advanced glycosylation. (From Brownlee, M., Vlassara, H., and Cerami, A., *Diabetes,* 34, 938, 1985. Copyright by the American Diabetes Association. With permission.)

oxidase, which occur at discrete sites at the amino- and carboxyl-terminal ends of the molecule. Reduction in protease digestion has been observed with diabetic collagen *in vivo*, and may play a role in the basement membrane thickening that occurs progressively with time in diabetics.

Diabetic vasculature also is characterized by an early and a progressive deposition of a variety of plasma proteins. An accelerated rate of atherosclerosis occurs which is independent of other cardiovascular risk factors. Abnormalities in both platelet adhesion and lipoprotein metabolism occur in diabetes and may contribute to this process, but evidence suggests that advanced glycosylation may be pivotal in initiating atherogenesis. The deposition and accumulation of the serum lipoprotein LDL (low-density lipoprotein) has been linked directly to the formation of vessel wall atheroma.[107] In analogy to the studies of ribonuclease glycosylation, where it was demonstrated that glycosylated ribonuclease can cross-link to unmodified ribonuclease, glycosylated collagen can serve as a cross-linking matrix to trap a variety of serum proteins, including immunoglobulin and low-density lipoprotein (LDL). As shown in Figure 11, the degree of covalent trapping was found to increase proportionately with the extent of advanced glycosylation. Similarly, other plasma proteins such as albumin and immunoglobulin will bind covalently to glycosylated collagen.[108] Experimentally, albumin and anti-albumin immunoglobulin which was bound to collagen *in vitro* retained its ability to form anti-antibody complexes when the corresponding free antigen or antibody was added. These observations suggest a mechansim to explain the linear deposition of immunoglobulin and albumin seen by immunofluorescence in diabetic basement membrane

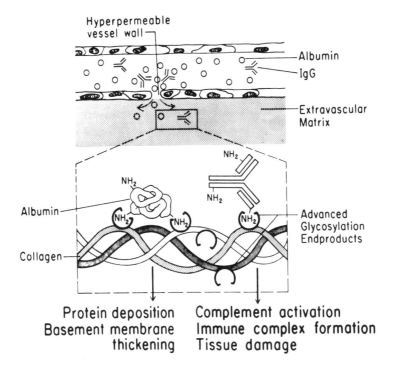

FIGURE 12.   Scheme for the covalent trapping of plasma proteins by advanced glycosylation endproducts on vascular wall collagen. (From Brownlee, M., Vlassara, H., and Cerami, A., *Ann. Intern. Med.*, 101, 527, 1984. With permission.)

(Figure 12).[109] Once these short lived proteins become trapped onto basement membrane, additional advanced glycosylation products will form on these proteins, which then can serve as additional reactive sites for the covalent attachment of additional proteins. Furthermore, immunoglobulin trapped via this mechanism could initiate complement activation and ensuing complement-mediated damage. In glomeruli, abundant poly-C9 deposition has been noted to occur.[110]

Direct evidence for a correlation between the severity of diabetic complication and AGE formation has been reported by Monnier et al.[111] Collagen-linked fluorescence was measured in 66 skin biopsy specimens obtained from a group of diabetic and control subjects. The mean age-adjusted fluorescence values was found to be twofold higher in the diabetic subjects than in the control group. In the diabetic group, a significant correlation was observed between the level of fluorescence and objective measures of retinopathy, arterial stiffness, and joint stiffness.

B. MYELIN AND PERIPHERAL NERVE

The peripheral neuropathy that results from chronic diabetes is associated with characteristic morphological changes which include segmental demyelination and axonal degeneration. Investigations into the potential role of excessive glycosylation in neuropathy were stimulated by the observation of Cerami and colleagues that diabetic nerve lesions resembled those produced by the carbamylating and anti-sickling agent cyanate.[112] Peripheral nerve myelin indeed was observed to be more heavily glycosylated in diabetic versus normal individuals,[73] with a principal target protein being the high-molecular-weight PO protein.[34] An additional histopathological feature of diabetic neuropathy is the infiltration of mononuclear phagocytes within the nerve which appear to be taking up and degrading the myelin sheath.[113] This latter feature led Vlassara et al. to test the hypothesis that the products of

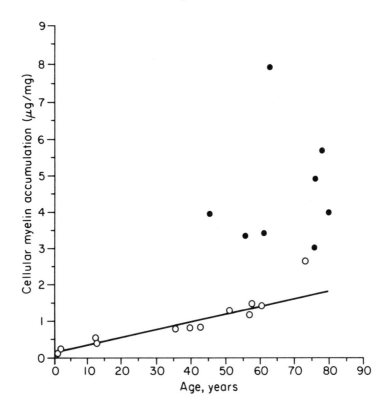

FIGURE 13.   Intracellular accumulation of ^{125}I-myelin from normal (○) and diabetic
(●) individuals by macrophages as a function of myelin age. (From Vlassara, H.,
Brownlee, M., Monogue, K., Pasagian, A., Dinarello, C., and Cerami, A., in *Monokines and Other Non-Lymphocytic Cytokines*, Alan R. Liss, New York, 1985, 145.
With permission.)

advanced glycosylation might serve as recognition signals for the uptake and degradation
of glycosylated myelin. Studies of AGE-myelin uptake in cultured macrophages supported
this concept (Figure 13). It was observed that radiolabeled diabetic myelin was much more
susceptible to macrophage uptake, and that this increased uptake could be mimicked by
glycosylating normal myelin *in vitro*.[114] Competition studies utilizing both synthetic AGEs
and a variety of AGE-modified proteins demonstrated that uptake was mediated via a specific
AGE-ligand receptor.[115] This early study led directly to the conception of a macrophage
scavenger system that exists for the selective removal of the products of advanced glycosylation (discussed below).

C. VASCULAR OCCLUSION AND MATRIX REMODELING

The identification of a macrophage scavenger receptor for AGE uptake and degradation
opened new avenues of investigation into the role of connective tissue homeostasis in aging
and diabetes. Collagen and basement membrane, although very long-lived macromolecules,
are degraded and replaced at a slow rate throughout adult life.[116] The monocyte-derived
macrophage traditionally has been considered to play an important role in this process by
removing altered macromolecules and initiating steps leading to tissue remodeling. In the
case of atherosclerosis, for example, the efficiency of LDL removal and degradation is
believed to influence the rate at which the atherosclerotic plaque develops.[117,118] Scavenger
receptors for acetylated, maleaylated and formylated proteins have been described.[119-122]
Since these ligands do not appear to form naturally, the precise biological role for these
various ligand-receptors interactions has yet to be identified.

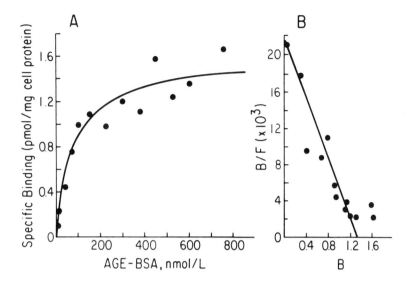

FIGURE 14. (A) Specific binding of ^{125}I-AGE-bovine serum albumin to mouse peritoneal macrophages. (B) Scatchard analysis of binding data. Axis units are Bound = pmol/mg cell protein, Bound/Free = pmol/nM. (From Vlassara, H., Brownlee, M., and Cerami, A., *Diabetes*, 34, 553, 1985. Copyright by the American Diabetes Association. With permission.)

Detailed studies of the macrophage AGE receptor have revealed that it has an AGE K_a of $1.7 \times 10^7 \, M^{-1}$ and that there are approximately 1.5×10^5 receptors per cell (Figure 14).[113] Affinity labeling of receptor derived from the monocyte line RAW 264.7 revealed the presence of at least one subunit of molecular weight 90 kDa.[123] Competition experiments suggest that the macrophage AGE receptor is distinct from previously described scavenger receptors and mannose/fucose ligand receptors.[124] Elucidation of the respective specificities and biological roles of the various scavenger receptors ultimately must await the molecular cloning of these proteins.

Thus, it appears that the macrophage can recognize AGE ligands which form *in vivo* and mediate their uptake and degradation. The net accumulation of AGE-proteins in tissue matrix must reflect a balance between glucose-accelerated accumulation and macrophage-mediated removal.

Receptor mediated uptake of AGE-proteins initiates a sequence of cytokine-mediated processes which promote tissue remodeling. The macrophage releases tumor-necrosis factor (TNF) and interleukin-1 (IL-1) in response to receptor occupancy by an AGE.[125] These cytokines recruit additional cells and processes that are involved in vessel wall homeostasis. Released monokines stimulate mesenchymal cells to release collagenase and other proteases. Matrix proteins and proteoglycans are degraded and simultaneously, new collagen is synthesized by fibroblasts.[126,127] In the vessel wall, interleukin-1 can enhance fibroblast, smooth muscle cell, and endothelial cell proliferation.[128,129] Tumor-necrosis factor is able to induce thrombosis-promoting effects on vascular endothelial cells. These include induction of a tissue factor-like procoagulant, suppression of the anti-coagulant protein C pathway, and synthesis of a plasminogen activator inhibitor.[130] These effects would then promote the release of platelet derived growth factor from aggregating platelets and from endothelial cells.[131] In patients with long-standing diabetes, these growth promoting effects would further stimulate cellular proliferation and matrix production and contribute to the enhanced basement membrane thickening and atherogenesis characteristic of the diabetic vessel wall.

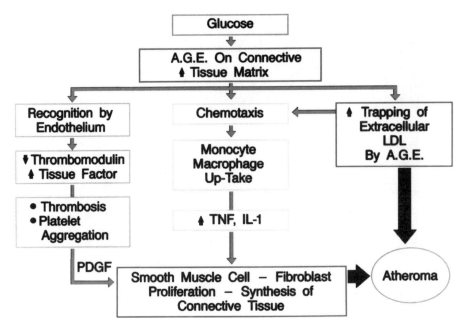

FIGURE 15.   Model relating the potential mechanisms by which advanced glycosylation contributes to the vasculopathy of diabetes mellitus. Abbreviations: AGE, advanced glycosylation endproducts; IL-1, interleukin-1; LDL, low-density lipoprotein; PDGF, platelet-derived growth factor; TNF, tumor necrosis factor.

D. ENDOTHELIAL CELL FUNCTION AND VASCULAR PERMEABILITY

Perhaps the earliest vascular lesion attributed to chronic hyperglycemia is an increased endothelial permeability to serum components. In an early study, glycosylated albumin was found to extravasate preferentially from the venous microcirculation of normal hamsters.[132] Recently, it has been determined that in addition to the macrophage, the endothelial cell also interacts with advanced glycosylation products in a ligand-specific manner.[133] When albumin modified with advanced glycosylation products was added to bovine endothelial cells in culture, an increase in cell permeability was observed. This was measured by the passage of an inert macromolecular tracer, inulin, across the monolayer. Furthermore, endothelial receptor occupancy by AGE-modified albumin led to a down-regulation of the anticoagulant cofactor thrombomodulin, and induction of the synthesis and cell-surface expression of the procoagulant cofactor tissue factor. Thus, one of the earliest observed events in diabetic vascular disease, increased vascular permeability may also be attributed, at least in part by the cellular response of endothelial cells to advanced glycosylation endproducts (Figure 15).

Additional adverse effects on endothelial cell function may result from the direct effects of AGEs on nucleic acid function. Culture of endothelial cells in high glucose leads to nucleic acid breakage and the induction of DNA repair synthesis.[134] In model systems, the interaction of either glucose or an amine catalyzed glucose "reactive intermediate" leads to DNA damage, mutation, and gene transposition.[135-139] The full potential of glucose-induced genotoxicity has yet to be evaluated, however its role in adversely affecting DNA function in the insulin independent cells found in the retina, kidney, and endothelium may be very significant.

VI. PHARMACOLOGICAL INHIBITION

The possibility of interfering with the adverse consequences of glucose-protein interaction was considered during investigations of the effects of glucose on lens crystallins confor-

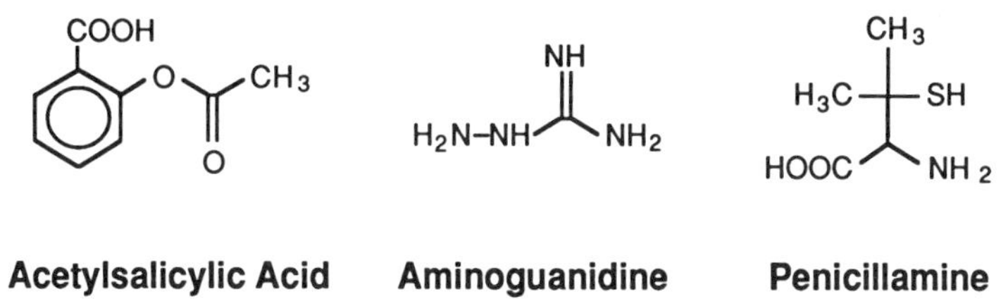

Acetylsalicylic Acid   **Aminoguanidine**   **Penicillamine**

FIGURE 16.   Structures of inhibitors of both the early and advanced glycosylation process.

mation. Acetylsalicylic acid (aspirin) is a mild acetylating agent whose primary pharmacologic effect is anti-inflammatory (Figure 16). This occurs through the inhibition of prostaglandin synthetase and is mediated by the acetylation of a specific serine hydroxyl within the enzyme.[140,141] Aspirin also acetylates primary amino groups and in serum albumin specifically modifies the lys-199 side chain, preventing nonenzymatic glycosylation at this site.[142] Treatment of lens crystallins with aspirin *in vitro* has been shown to protect against the formation of the high-molecular weight aggregates induced by glycosylation.[143] Similar protective effects of aspirin acetylation have been observed in lens proteins which have been opacified with other protein modification agents such as cyanate[144] and corticosteroids.[145] In a recent study, the progression of cataract was slowed markedly by administering aspirin to rats which had streptozotocin-induced diabetes.[146] Retrospective studies of cataract incidence in human populations have found evidence to support an ameliorative effect of chronic aspirin use.[147,148]

Amadori product formation represents an important branch-point in glycosylation chemistry because progression to fluorescent, cross-linking moieties requires the covalent attachment of an additional glucose or protein-derived group. This realization suggested a possible strategy for intervening in the glycosylation process before progression to cross-linking occurred. Aminoguanidine, a small hydrazine-like compound, can react with Amadori products and inhibit further reaction with protein amino groups. Aminoguanidine does not prevent Amadori product formation, instead it prevents further dehydration and rearrangement which cause protein-protein cross-linking to occur.

Rats which were made diabetic and treated with aminoguanidine were found to have significantly less collagen cross-linking in the vascular wall, as measured by protease digestibility.[106] Similarly, covalent trapping of lipoprotein was reduced. Other studies have demonstrated decreased protein trapping in the basement membrane of the glomerulus. This work suggests that aminoguanidine markedly inhibits the glucose mediated cross-linking that occurs in experimental diabetes.[149]

Aminoguanidine displays little toxicity (the LD_{50} in the rat is 1.5 g per kg) and clinical trials evaluating its efficacy in the progression of diabetic complications are currently underway in the United States.

D-Penicillamine (D-β,β-dimethylcysteine) is a heavy metal chelator and anti-rheumatic agent. *In vitro* studies have demonstrated that it inhibits both glucose incorporation (Amadori product) and advanced glycosylation, as assessed by measurements of protein cross-linking and fluorescence.[96] Penicillamine is widely used in patients with rheumatoid arthritis. It would be of interest to examine retrospectively whether chronic therapy with this drug selectively spares any of the complications of diabetes or believed to be mediated by advanced glycosylation.

The potential role of advanced glycosylation inhibitors such as aminoguanidine in medical applications is significant. Experimentally, the use of aminoguanidine will provide insight into the relative contribution of advanced glycosylation in the evolution of various diabetic or aging sequelae.

VII. FUTURE DIRECTIONS

Despite difficulties inherent in the structural elucidation of advanced glycosylation end-products *in vivo*, significant progress has been achieved in the synthesis of model compounds and in the characterization of *in vivo* cross-links. The identification of additional structures will assist in working out the precise chemical pathway which leads from Amadori precursors to the terminal stages of advanced glycosylation. Studies of the metabolism of these products both in normal aging and in hyperglycemic states will be of general biological interest. Diabetes mellitus has served as a useful *in vivo* model in assessing the damaging effects of advanced glycosylation. In an analogous fashion, studies of patients with end stage renal disese may be revealing in working out the normal degradative and excretory pathways of AGEs. Toward this end, new approaches and strategies toward the detection of AGEs and their precursors will need to be developed. In addition to sensitive chemical and spectroscopic methods, assays that rely on either enzymatic or biological amplification will be required. One recent approach utilizes specific polyclonal or monoclonal antibodies and ELISA technology.[83,87,150] Under conditions of carefully defined cross-reactivity and sensitivity, this strategy offers much promise toward both detecting and quantitating AGEs *in vivo*. Extension of these studies *in situ* with tissue immunohistochemistry will do much to extend our understanding of the distribution and metabolic fate of advanced glycosylation endproducts.

Historically, since the earliest work on hemoglobin A_{1c}, studies of the nonenzymatic glycosylation process have focussed on the role of these protein modifications in tissue pathology. Given the universal nature of these reactions however, it is not difficult to imagine that advanced glycosylation chemistry might have been exploited by living systems in normal cellular physiology. Recent studies on the specific role of genetic transposition induced by AGEs are intriguing in this regard. However, elucidation of a normal or "positive" role for advanced glycosylation must await further studies of both the biochemistry and cell biology of this process.

REFERENCES

1. **Maillard, L. C.,** Action des acides amines sur les sucres; formation des melanoidines par voie methodique, *C. R. Acad. Sci.,* 154, 66, 1912.
2. **Patton, A. R. and Hill, E. G.,** Inactivation of nutrients by heating with glucose, *Science,* 107, 68, 1948.
3. **Mohammad, A., Olcott, H. S., and Fraenkel-Conrat, H.,** The "browning" reaction of proteins with glucose, *Arch. Biochem.,* 24, 157, 1949.
4. **Reynolds, T. M.,** Chemistry of nonenzymatic browning. II, *Adv. Food Res.,* 14, 167, 1965.
5. **Eriksson, C., Ed.,** *Maillard Reactions in Food. Prog. Food and Nutr. Sci.,* 1981, 5.
6. **Baynes, J. W. and Monnier, V. M., Eds.,** *The Maillard Reaction in Aging, Diabetes, and Nutrition, Prog. Clin. Biol. Res.,* 1989, 304.
7. **Brownlee, M., Vlassara, H., and Cerami, A.,** Nonenzymatic glycosylation and the pathogenesis of diabetic complications, *Ann. Intern. Med.,* 101, 527, 1984.
8. **Rahbar, S.,** An abnormal hemoglobin in red cells of diabetes, *Clin. Chim. Acta,* 22, 296, 1968.
9. **Holmquist, W. R. and Schoeder, W. A.,** A new N-terminal blocking group involving a Schiff base in Hemoglobin A_{1c}, *Biochemistry,* 5, 2489, 1966.
10. **Bookchin, R. M. and Gallop, P. M.,** Structure of Hemoglobin A_{1c}; nature of the N-terminal α-chain blocking group, *Biochem. Biophys. Res. Commun.,* 32, 86, 1968.
11. **Dixon, H. B. F.,** A reaction of glucose with peptides, *Biochem. J.,* 129, 203, 1972.
12. **Koenig, R. J., Blobstein, S. H., and Cerami, A.,** Structure of carbohydrate of hemoglobin A_{1c}, *J. Biol. Chem.,* 252, 2992, 1977.
13. **Koenig, R. J., Peterson, C. M., Jones, R. L., Saudek, C., Lehrman, M., and Cerami, A.,** Correlation of glucose regulation and hemoglobin A_{1c} in diabetes mellitus, *N. Engl. J. Med.,* 295, 417, 1976.
14. **Jovanovic, L. and Peterson, C. M.,** The clinical utility of glycohemoglobins, *Am. J. Med.,* 70, 331, 1981.

15. **Gottschalk, A.**, Interaction between reducing sugars and amino acids under neutral and acid conditions, in *The glycoproteins*, Gottschalk, A., Ed., Elsevier, New York, 1972, 141.

16. **Baynes, J. W., Thorpe, S. R., and Murashaw, M. W.**, Nonenzymatic glycosylation of lysine residues in albumin, in *Methods in Enzymology, Posttranslational Modifications*, Vol. 103, Wold, F. and Moldave, K., Eds., Academic Press, New York, 1984, 88.

17. **Higgins, P. J. and Bunn, H. F.**, Kinetic analysis of the nonenzymatic glycosylation of hemoglobin, *J. Biol. Chem.*, 256, 5204, 1981.

18. **Mortensen, H. B. and Christophersen, C.**, Glucosylation of haemoglobin A in red blood cells studies *in vitro*. Kinetics of the formation and dissociation of haemoglobin A_{1c}, *Clin. Chim. Acta*, 134, 317, 1983.

19. **Monnier, V. M., Stevens, V. J., and Cerami, A.**, Nonenzymatic glycosylation, sulfhydryl oxidation, and aggregation of lens proteins in experimental sugar cataracts, *J. Exp. Med.*, 150, 1098, 1979.

20. **Day, J. R., Thorpe, S. R., and Baynes, J. W.**, Nonenzymatically glycosylated albumin: *in vitro* preparation and isolation from normal human serum, *J. Biol. Chem.*, 254, 595, 1979.

21. **Rosenberg, H., Modrak, J. B., Hassing, J. M., Al-Turk, W. A., and Stohs, S. J.**, Glycosylated collagen, *Biochem. Biophys. Res. Commun.*, 91, 498, 1979.

22. **Miller, J. A., Gravallese, E., and Bunn, H. F.**, Nonenzymatic glycosylation of erythrocyte membrane proteins: relevance to diabetes, *J. Clin. Invest.*, 65, 896, 1980.

23. **Cohen, M. P., Urdanivia, E., Surma, M., and Wu, V. Y.**, Increased glycosylation of glomerular basement membrane in diabetes, *Biochem. Biophys. Res. Commun.*, 95, 765, 1980.

24. **Schnider, S. L. and Kohn, R. R.**, Glucosylation of human collagen in aging and diabetes mellitus, *J. Clin. Invest.*, 66, 1179, 1980.

25. **Schleicher, E., Scheller, L., and Wieland, O. H.**, Quantitation of lysine-bound glucose of normal and diabetic erythrocyte membranes by HPLC analysis of furosine (ϵ-N(L-furoylmethyl)-L-lysine), *Biochem. Biophys. Res. Commun.*, 99, 1011, 1981.

26. **Zaman, Z. and Verwilghen, R. L.**, Non-enzymatic glycosylation of horse spleen and rat liver ferritins, *Biochem. Biophys. Acta*, 669, 120, 1981.

27. **McVerry, B. A., Thorpe, S., and Joe, F.**, Non-enzymatic glycosylation of fibrinogen, *Hemostasis*, 10, 261, 1981.

28. **Schleicher, E., Deufel, T., and Wieland, O. H.**, Non-enzymatic glycosylation of human serum lipoproteins: elevated epsilon-lysine glycosylated low density lipoprotein in diabetic patents, FEBS Lett., 129, 1, 1981.

29. **Coradello, H., Pollack, A., Pugnano, et al.**, Nonenzymatic glycosylation of cathepsin B: possible influence on conversion of proinsulin to insulin, *IRCS Med. Sci.*, 9, 766, 1981.

30. **Witztum, J. L., Fisher, E. M., and Peitro, T.**, Nonenzymatic glycosylation of high-density lipoprotein accelerates its catabolism in guinea pigs, *Diabetes*, 31, 1029, 1982.

31. **Williams, S. K., Howarth, N. L., Devenny, J. J., and Bitensky, M. W.**, Structural and functional consequences of increased tubulin glycosylation in diabetes mellitus, *Proc. Natl. Acad. Sci. U.S.A.*, 79, 6546, 1982.

32. **Dolhofer, R., Siess, E. A., and Wieland, O. H.**, Inactivation of bovine kidney beta-N-acetyl-D-glucosaminidase by nonenzymatic glycosylation, *Hoppe Seylers Z. Physiol. Chem.*, 363, 1427, 1982.

33. **Brownlee, M., Vlassara, H., and Cerami, A.**, Nonenzymatic glucosylation reduces the susceptibility of fibrin to degradation by plasmin, *Diabetes*, 32, 680, 1983.

34. **Vlassara, H., Brownlee, M., and Cerami, A.**, Excessive glycosylation of peripheral and central nervous system myelin components in diabetic rats, *Diabetes*, 32, 670, 1983.

35. **Geiger, M. and Binder, B. R.**, Plasminogen activation in diabetes mellitus, *J. Biol. Chem.*, 259, 2976, 1984.

36. **Brownlee, M., Vlassara, H., and Cerami, A.**, Inhibition of heparin-catalyzed human antithrombin III activity by nonenzymatic glycosylation: possible role in fibrin deposition in diabetes, *Diabetes*, 33, 532, 1984.

37. **Cohen, M. P. and Ku, L.**, Inhibition of fibronectin binding to matrix components by nonenzymatic glycosylation, *Diabetes*, 33, 970, 1984.

38. **Monnier, V. M., Kohn, R. R., and Cerami, A.**, Accelerated age-related browning of human collagen in diabetes mellitus, *Proc. Natl. Acad. Sci. U.S.A.*, 81, 583, 1984.

39. **Curtiss, L. K. and Witztum, J. L.**, Plasma apolipoproteins AI, AIII, B, CI, and E are glucosylated in hyperglycemic diabetic subjects, *Diabetes*, 34, 452, 1985.

40. **Tarsio, J. F., Wigness, B., and Rhode, T. D.**, Nonenzymatic glycosylation of fibronectin and alterations in the molecular association of cell matrix and basement membrane components in diabetes mellitus, *Diabetes*, 34, 477, 1985.

41. **Sampietro, T., Lenzi, S., Cecchetti, P., Giampietro, O., Crushelli, L., and Navalesi, R.**, Nonenzymatic glycation of human platelet membrane proteins *in vitro* and *in vivo*, *Clin. Chem.*, 32, 1328, 1986.

42. **Austin, G. E., Mullins, R. H., and Morin, L. G.**, Non-enzymic glycation of individual plasma proteins in normoglycemic and hyperglycemic patients, *Clin. Chem.*, 33, 2220, 1987.

43. **Kowluru, R. A., Heidorn, D. B., Edmondson, S. P., Bitensky, M. W., Kowluru, A., Downer, N. W., Whaley, T. W., and Trewhella, J.,** Glycation of calmodulin: chemistry and structural and functional consequences, *Biochemistry,* 28, 2220, 1989.

44. **Arai, K., Maguchi, S., Fuji, S., Ishibashi, H., Oikawa, K., and Taniguchi, N.,** Glycation and inactivation of human Cu-Zn-superoxide dismutase, *J. Biol. Chem.,* 262, 16969, 1989.

45. **Yue, D. K., McLennan, S., and Turtle, J. R.,** Non-enzymatic tissue glycosylation of tissue protein in diabetes in the rat, *Diabetologia,* 24, 377, 1983.

46. **Vogt, B. W., Schleicher, E. D., and Wieland, O. H.,** Epsilon-amino-lysine-bound glucose in human tissues obtained at autopsy: increase in diabetes mellitus, *Diabetes,* 31, 1123, 1982.

47. **Shapiro, R., McManus, M. J., Zalut, C., and Bunn, H. F.,** Sites of nonenzymatic glycosylation of human hemoglobin A, *J. Biol. Chem.,* 255, 3120, 1980.

48. **Garlick, R. L. and Mazur, J. S.,** The principal site of nonenzymatic glycosylation of human serum albumin *in vivo, J. Biol. Chem.,* 258, 6142, 1983.

49. **Iberg, N. and Fluckiger, R.,** Nonenzymatic glycosylation of albumin *in vivo, J. Biol. Chem.,* 261, 13542, 1986.

50. **Robb, D. A., Olufemi, O. S., Williams, D. A., and Midgley, J. M.,** Identification of glycation at the N-terminus of albumin by gas chromatography-mass spectroscopy, *Biochem. J.,* 261, 871, 1989.

51. **Bunn, H. F., Shapiro, R., McManus, M., Garrick, L., McDonald, M. J., Gallop, P. M., and Gabbay, K. H.,** Heterogeneity of human hemoglobin A$_0$ due to non-enzymatic glycosylation, *J. Biol. Chem.,* 254, 3892, 1979.

52. **Guthrow, C. E., Morris, M. A., Day, J. F., Thorpe, S. R., and Baynes, J. W.,** Enhanced nonenzymatic glucosylation of human serum albumin in diabetes mellitus, *Proc. Natl. Acad. Sci. U.S.A.,* 76, 4258, 1979.

53. **Watkins, N. G., Neglia-Fisher, C. I., Dyer, D. G., Thorpe, S., and Baynes, J. W.,** Effect of phosphate on the kinetics and specificity of glycation of protein, *J. Biol. Chem.,* 262, 7207, 1987.

54. **Haney, D. N. and Bunn, H. F.,** Glycosylation of hemoglobin *in vitro*: affinity labeling of hemoglobin by glucose-6-phosphate, *Proc. Natl. Acad. Sci. U.S.A.,* 73, 3534, 1976.

55. **McDonald, M. J., Bleichman, M., Bunn, H. F., and Noble, R. W.,** Functional properties of the glycosylated minor components of human adult hemoglobin, *J. Biol. Chem.,* 254, 702, 1979.

56. **Nigen, A. M. and Manning, J. M.,** Inhibition of erythrocyte sickling *in vitro* by DL-glyceraldehyde, *Proc. Natl. Acad. Sci. U.S.A.,* 74, 367, 1977.

57. **Cerami, A. and Manning, J.,** Potassium cyanate as an inhibitor of the sickling of erythrocytes *in vitro, Proc. Natl. Acad. Sci. U.S.A.,* 68, 1180, 1971.

58. **Shaklai, N., Garlick, R. L., and Bunn, H. F.,** Nonenzymatic glycosylation of human serum albumin alters its conformation and function, *J. Biol. Chem.,* 259, 3812, 1984.

59. **Williams, S. K., Devenny, J. J., and Bitensky, M. W.,** Micropinocytic ingestion of glycosylated albumin by isolated microvessels: possible role in the pathogenesis of diabetic microangiopathy, *Proc. Natl. Acad. Sci. U.S.A.,* 78, 2393, 1981.

60. **McVerry, B. A., Fisher, C., Hopp, A., and Huehns, E. R.,** Production of pseudodiabetic renal glomerular changes in mice after repeated injections of glucosylated proteins, *Lancet,* 1, 8171, 738, 1980.

61. **Bryszewska, M. and Szosland, K.,** Association between glycation of erythrocyte membrane proteins and membrane fluidity, *Clin. Biochem.,* 21, 49, 1988.

62. **Gonen, B., Baenziger, J., Schonfeld, G., Jacobson, D., and Farrar, P.,** Nonenzymatic glycosylation of low density lipoprotein *in vitro*. Effects on cell interactive properties, *Diabetes,* 30, 875, 1981.

63. **Sasaki, J. and Cottam, G. L.,** Glycosylation of LDL decreases its ability to interact with high-affinity receptors of human fibroblasts *in vitro* and decreases its clearance from rabbit plasma *in vivo, Biochim. Biophys. Acta,* 713, 199, 1982.

64. **Lorenzi, M., Cagliero, E., Henriksen, T., Witztum, J. L., and Sampietro, T.,** Interaction of human endothelial cells with elevated glucose concentrations and native and glycosylated low density lipoprotein, *Diabetologia,* 26, 218, 1984.

65. **Witztum, J. L., Steinbrecher, U. P., Kesaniemi, Y. A., and Fisher, M.,** Autoantibodies to glucosylated proteins in the plasma of patients with diabetes mellitus, *Proc. Natl. Acad. Sci. U.S.A.,* 81, 3204, 1984.

66. **Witztum, J. L., Steinbrecher, U. P., Fisher, M., and Kesaniemi, A.,** Nonenzymatic glycosylation of homologous low-density lipoprotein and albumin renders them immunogenic in the guinea pig, *Proc. Natl. Acad. Sci. U.S.A.,* 80, 2757, 1983.

67. **Koenig, R. J. and Cerami, A.,** Synthesis of hemoglobin A$_{Ic}$ in normal and diabetic mice: Potential model of basement membrane thickening, *Proc. Natl. Acad. Sci. U.S.A.,* 72, 3687, 1975.

68. **Stevens, V. J., Rouzer, C. A., Monnier, V. M., and Cerami, A.,** Diabetic cataract formation: potential role of glycosylation of lens crystallins, *Proc. Natl. Acad. Sci. U.S.A.,* 75, 2918, 1978.

69. **Beswick, H. T. and Harding, J. J.,** Conformational changes induced in lens α- and γ-crystallins by modification with glucose-6-phosphate; implications for cataract, *Biochem. J.,* 246, 761, 1987.

70. **Pande, A., Garner, W. H., and Spector, A.,** Glycosylation of human lens protein and cataractogenesis, *Biochem. Biophys. Res. Commun.,* 89, 1260, 1979.

71. **Ansari, N. H., Awasthi, Y. L., and Srivastiva, S. K.,** Role of glycosylation in protein disulfide formation and cataractogenesis, *Exp. Eye Res.,* 31, 9, 1980.

72. **Monnier, V. M. and Cerami, A.,** Nonenzymatic browning *in vivo*: possible process for aging of long-lived proteins, *Science,* 211, 491, 1981.

73. **Vlassara, H., Brownlee, M., and Cerami, A.,** Nonenzymatic glycosylation of peripheral nerve protein in diabetes mellitus, *Proc. Natl. Acad. Sci. U.S.A.,* 78, 5190, 1981.

74. **Van Heyningen, R.,** Fluorescent glucosides in the human lens, *Nature,* 230, 393, 1971.

75. **Tanzer, M. L., Fairweather, R., and Gallop, P. M.,** Collagen cross-links: isolation of N-ε-hexosyl hydroxylysine from borohydride-reduced calf-skin insoluble collagen, *Arch. Biochem. Biophys.,* 151, 137, 1972.

76. **Robins, S. P. and Bailey, A. J.,** Age-related changes in collagen: the identification of reducible lysine-carbohydrate condensation products, *Biochem. Biophys. Res. Commun.,* 48, 76, 1972.

77. **Schnider, S. L. and Kohn, R. R.,** Effects of age and diabetes mellitus on the solubility and nonenzymatic glycosylation of human skin collagen, *J. Clin. Invest.,* 67, 1630, 1981.

78. **Schnider, S. L. and Kohn, R. R.,** Effects of age and diabetes mellitus on the solubility of collagen from human skin, tracheal cartilage and dura mater, *Exp. Gerontol.,* 17, 185, 1982.

79. **Eble, A. S., Thorpe, S. R., and Baynes, J. W.,** Nonenzymatic glucosylation and glucose-dependent cross-linking of proteins, *J. Biol. Chem.,* 258, 9406, 1983.

80. **Pongor, S., Ulrich, P. C., Bencsath, F. A., and Cerami, A.,** Aging of proteins: isolation and identification of a fluorescent chromophore from the reaction of polypeptides with glucose, *Proc. Natl. Acad. Sci. U.S.A.,* 81, 2684, 1984.

81. **Njoroge, F. G., Fernandes, A. A., and Monnier, V.,** Mechanism of formation of the putative advanced glycosylation endproduct and protein cross-link 2-(2-furoyl)-4(5)-(2-furanyl)-1H-imidazole, *J. Biol. Chem.,* 263, 10646, 1988.

82. **Horiuchi, S., Shiga, M., Araki, N., Takata, K., Saitoh, M., and Morino, Y.,** Evidence against *in vivo* presence of 2-(2-furoyl)-4(5)-(2-furanyl)-1H-imidazole, a major fluorescent advanced glycosylation end-product generated by nonenzymatic glycosylation, *J. Biol. Chem.,* 263, 18821, 1988.

83. **Chang, J. C. F., Bucala, R., Ulrich, P. C., and Cerami, A.,** Detection of an advanced glycosylation product bound to protein *in situ, J. Biol. Chem.,* 260, 7970, 1985.

84. **Vlassara, H., Brownlee, M., and Cerami, A.,** Novel macrophage receptor for glucose-modified protein is distinct from previously described scavenger receptors, *J. Exp. Med.,* 164, 1301, 1986.

85. **Kato, H., Fumitaka, H., Shin, D. B., Oimom, M., and Baba, S.,** 3-Deoxyglucosone, an intermediate product of the Maillard reaction, in *The Maillard Reaction in Aging, Diabetes, and Nutrition,* Baynes, J. W. and Monnier, V. M., Eds., Alan R. Liss, New York, 1989, 69.

86. **Njoroge, F. G., Sayre, L. M., and Monnier, V. M.,** Detection of D-glucose derived pyrrole compounds during Maillard reaction under physiological conditions, *Carbohydr. Res.,* 167, 211, 1987.

87. **Hayase, F., Nagaraj, R. H., Miyata, S., Njoroge, F. G., and Monnier, V. M.,** Aging of proteins: immunological detection of a glucose-derived pyrrole formed during Maillard reaction *in vivo, J. Biol. Chem.,* 264, 3758, 1989.

88. **Farmar, J., Ulrich, P., and Cerami, A.,** Novel pyrroles from sulfite-inhibited Maillard reactions: insight into the mechanism of inhibition, *J. Org. Chem.,* 53, 2346, 1988.

89. **McWeeney, D. J.,** The chemical behavior of food additives, *Proc. Nutr. Sci.,* 38, 129, 1979.

90. **Sell, D. R. and Monnier, V. M.,** Structure elucidation of a senescence cross-link from human extracellular matrix, *J. Biol. Chem.,* 264, 21597, 1989.

91. **Ahmed, M. U., Thorpe, S. R., and Baynes, J. W.,** Identification of carboxymethyllysine as a degradation product of fructosyllysine in glycated protein, *J. Biol. Chem.,* 261, 4889, 1988.

92. **Stevens, V. J., Vlassara, H., Abati, A., and Cerami, A.,** Nonenzymatic glycosylation of hemoglobin, *J. Biol. Chem.,* 252, 2998, 1977.

93. **Bunn, H. F. and Higgins, P. J.,** Reaction of monosaccharides with proteins: possible evolutionary significance, *Science,* 213, 222, 1981.

94. **Schwimmer, S. and Olcott, H. S.,** Reaction between glycine and hexose phosphates, *J. Am. Chem. Soc.,* 75, 4855, 1953.

95. **Means, G. E. and Chang, M. K.,** Nonenzymatic glycosylation of proteins, structure and function changes, *Diabetes,* 31, (Suppl. 3), 1, 1982.

96. **McPherson, J. D., Shilton, B. H., and Walton, D. J.,** Role of fructose in glycation and cross-linking of proteins, *Biochemistry,* 27, 1901, 1988.

97. **Suarez, G., Rajaram, R., Oronsky, A. L., and Gawinowicz, M. A.,** Nonenzymatic glycation of bovine serum albumin by fructose (fructation), comparison with the Maillard reaction initiated by glucose, *J. Biol. Chem.,* 264, 3674, 1989.

98. **Grgic, A., Rosenbloom, A. L., Weber, F. T., and Giordano, B.,** Joint contracture in childhood diabetes, *N. Engl. J. Med.,* 292, 372, 1975.

99. **Hamlin, C. R., Kohn, R. R., and Luschin, J. H.,** Apparent accelerated aging of human collagen in diabetes mellitus, *Diabetes,* 24, 902, 1975.

100. **Pillsbury, H. C., Hung, W., Kyle, M. C., and Freis, E. D.,** Arterial pulse wave and velocity and systolic time intervals in diabetic children, *Am. Heart J.,* 87, 783, 1974.

101. **Schuyler, M. R., Niewoehner, D. E., Inkley, S. R., and Kohn, R. R.,** Abnormal lung elasticity in juvenile diabetes mellitus, *Am. Rev. Respir. Dis.,* 113, 37, 1976.

102. **Kohn, R. R., Cerami, A., and Monnier, V. M.,** Collagen aging *in vitro* by nonenzymatic glycosylation and browning, *Diabetes,* 33, 57, 1984.

103. **Bochantin, J. and Mays, L. L.,** Age-dependence of collagen tail fiber breaking strength in Sprague-Dawley and Fisher 344 rats, *Exp. Gerontol.,* 16, 101, 1981.

104. **Harrison, D. E. and Archer, J. R.,** Measurement of changes in mouse tail collagen with age: temperature dependence and procedural details, *Exp. Gerontol.,* 13, 75, 1978.

105. **Kent, M. J. C., Light, N. D., and Bailey, A. J.,** Evidence for glucose-mediated covalent cross-linking of collagen after glycosylation *in vitro, Biochem. J.,* 225, 745, 1985.

106. **Brownlee, M., Vlassara, H., Kooney, T., Ulrich, P., and Cerami, A.,** Aminoguanidine prevents diabetes-induced arterial wall protein cross-linking, *Science,* 232, 1629, 1986.

107. **Brownlee, M., Vlassara, H., and Cerami, A.,** Nonenzymatic glycosylation products on collagen covalently trap low-density lipoprotein, *Diabetes,* 34, 938, 1985.

108. **Brownlee, M., Pongor, S., and Cerami, A.,** Covalent attachment of soluble proteins by nonenzymatically glycosylated collagen; role in the *in situ* formation of immune complexes, *J. Exp. Med.,* 158, 1739, 1983.

109. **Miller, K. and Michael, A. F.,** Immunopathology of renal extracellular membranes in diabetes: specificity of tubular basement-membrane immunofluorescence, *Diabetes,* 25, 701, 1976.

110. **Falk, R. J., Dalmasso, A. P., and Kim, Y.,** Neoantigen of the polymerized ninth component of complement: characterization of a monoclonal antibody and immunohistochemical localization in renal disease, *J. Clin. Invest.,* 72, 560, 1983.

111. **Monnier, V. M., Vishwanath, V., Frank, K. E., Elmets, C. A., Dauchot, P., and Kohn, R. R.,** Relation between complications of Type I diabetes mellitus and collagen-linked fluorescence, *N. Engl. J. Med.,* 314, 403, 1986.

112. **Peterson, C. M., Tsairis, P., Ohnishi, A., Lu, Y. S., Grady, R., Cerami, A., and Dyck, P. J.,** Sodium cyanate induced polyneuropathy in patients with sickle-cell disease, *Ann. Intern. Med.,* 81, 152, 1974.

113. **Thomas, P. and Eliasson, S.,** Diabetic neuropathy, in *Peripheral Neuropathy,* Vol. 2, Dyck, P. J., Thomas, P. K., and Lambert, E. H., Eds., W. B. Saunders, Philadelphia, 1975, 956.

114. **Vlassara, H., Brownlee, M., and Cerami, A.,** Recognition and uptake of human diabetic peripheral nerve myelin by macrophages, *Diabetes,* 34, 553, 1985.

115. **Vlassara, H., Brownlee, M., and Cerami, A.,** High-affinity receptor-mediated uptake and degradation of glucose-modified proteins: a potential mechanism for the removal of senescent macromolecules, *Proc. Natl. Acad. Sci. U.S.A.,* 82, 5588, 1985.

116. **Pinnell, S. P.,** Disorders of collagen, in *The Metabolic Basis of Inherited Disease,* Stanbury, J. B., Wyngaarden, J. B., and Fredrickson, D. S., Eds., McGraw-Hill, New York, 1978, 1366.

117. **Brown, M. S. and Goldstein, J. L.,** Lipoprotein metabolism in the macrophage: implications for cholesterol deposition in atherosclerosis, *Annu. Rev. Biochem.,* 52, 223, 1983.

118. **Steinberg, D.,** Lipoproteins and atherosclerosis—a look back and a look ahead, *Atherosclerosis,* 3, 283, 1983.

119. **Goldstein, J. L., Ho, Y. K., Basu, S. K., and Brown, M. S.,** Binding site on macrophages that mediates uptake and degradation of acetylated low-density lipoprotein, producing massive cholesterol deposition, *Proc. Natl. Acad. Sci. U.S.A.,* 76, 333, 1979.

120. **Horiuchi, S., Takata, K., Maeda, H., and Morino, Y.,** Scavenger function of sinusoidal liver cells, *J. Biol. Chem.,* 259, 53, 1985.

121. **Horiuchi, S., Takata, K., and Morino, Y.,** Purification of a receptor for formaldehyde-treated serum albumin from rat liver, *J. Biol. Chem.,* 260, 482, 1985.

122. **Takata, K., Horiuchi, S., Araki, N., Shiga, M., Saitoh, M., and Morino, Y.,** Endocytic uptake of nonenzymatically glycosylated proteins is mediated by scavenger receptor for aldehyde-modified proteins, *J. Biol. Chem.,* 263, 14819, 1988.

123. **Radoff, S., Vlassara, H., and Cerami, A.,** Characterization of a solubilized cell surface binding protein on macrophages specific for proteins modified nonenzymatically by advanced glycosylation endproducts, *Arch. Biochem. Biophys.,* 263, 418, 1988.

124. **Vlassara, H., Brownlee, M., and Cerami, A.,** Novel macrophage receptor for glucose-modified proteins is distinct from previously described scavenger receptors, *J. Exp. Med.,* 164, 1301, 1986.

125. **Vlassara, H., Brownlee, M., Manogue, K., Dinarello, C., and Cerami, A.,** Cachectin/TNF and IL-1 induced by glucose-modified proteins: role in normal tissue remodeling, *Science,* 240, 1546, 1988.

126. **Dinarello, C.,** Biology of interleukin 1, *FASEB J.,* 2, 108, 1988.

127. **Hansch, G. M., Torbohm, I., Kempis, J., and Rother, K.,** Modulation of collagen synthesis in glomerular epithelial cells by interleukin 1 and supernatants of mesangial cells, *Kidney Int.,* 33, 317A, 1988.

128. **Lovett, D. H., Ryan, J. L., and Sterzel, R. B.,** Stimulation of rat mesangial proliferation by macrophage interleukin 1, *J. Immunol.,* 136, 3700, 1983.

129. **Libby, P., Warner, S. J. C., and Friedman, G. B.,** Interleukin 1: a mitogen for human vascular smooth muscle cells that induces the release of inhibitory prostanoids, *J. Clin. Invest.,* 81, 487, 1988.

130. **Nawroth, P. P., Handley, D. A., Esmon, C. T., and Stern, D. M.,** Interleukin 1 induces endothelial cell pro-coagulant while suppressing cell-surface anticoagulant activity, *Proc. Natl. Acad. Sci. U.S.A.,* 83, 3460, 1986.

131. **Le, J. and Vilcek, J.,** Tumor necrosis factor and interleukin 1: cytokines with multiple overlapping activities, *Lab. Invest.,* 56, 234, 1987.

132. **Sampietro, T., Bertuglia, S., Colantuoni, A., Bionda, A., Lenzi, S., and Donato, L.,** Increased permeability of hamster microcirculation to glycosylated albumin, *Lancet II,* 994, 1987.

133. **Esposito, C. Gerlach, H., Brett, J., Stern, D., and Vlassara, H.,** Endothelial receptor-mediated binding of glucose-modified albumin is associated with increased monolayer permeability and modulation of cell surface coagulant properties, *J. Exp. Med.,* 170, 1387, 1989.

134. **Lorenzi, M., Montisano, D. F., and Toledo, S.,** High glucose and DNA damage in endothelial cells, *J. Clin. Invest.,* 77, 322, 1986.

135. **Bucala, R., Model, P., and Cerami, A.,** Modification of DNA by reducing sugars: A possible mechanism for nucleic acid aging and age-related dysfunction in gene expression, *Proc. Natl. Acad. Sci. U.S.A.,* 81, 105, 1984.

136. **Bucala, R., Model, P., Russel, M., and Cerami, A.,** Modification of DNA by glucose-6-phosphate induces DNA rearrangements in an E. coli plasmid, *Proc. Natl. Acad. Sci. U.S.A.,* 82, 8439, 1985.

137. **Lee, A. T. and Cerami, A.,** The formation of reactive intermediate(s) of glucose-6-phosphate and lysine capable of rapidly reacting with DNA, *Mutat. Res.,* 179, 151, 1987.

138. **Lee, A. T. and Cerami, A.,** Elevated glucose-6-phosphate levels are associated with plasmid mutations *in vivo, Proc. Natl. Acad. Sci. U.S.A.,* 84, 8311, 1987.

139. **Lee, A. T. and Cerami, A.,** Induction of $\gamma\delta$ transposition in response to elevated glucose-6-phosphate levels, *Mutation Res,,* 249, 125, 1991.

140. **Roth, G. J., Stanford, N., and Majerus, P. W.,** Acetylation of prostaglandin synthase by aspirin, *Proc. Natl. Acad. Sci. U.S.A.,* 72, 3073, 1975.

141. **Roth, G. J., Machuga, E. T., and Ozol, J.,** Isolation and covalent structure of the aspirin-modified, active site region of prostaglandin synthetase, *Biochemistry,* 22, 4672, 1983.

142. **Walker, J. E.,** Lysine residue 199 of human serum albumin is modified by acetylsalicylic acid, *FEBS Lett.,* 66, 173, 1976.

143. **Huby, R. and Harding, J. J.,** Non-enzymic glycosylation (glycation) of lens proteins by galactose and protection by aspirin and reduced glutathione, *Exp. Eye Res.,* 47, 53, 1988.

144. **Crompton, M., Rixon, K. C., and Harding, J. J.,** Aspirin prevents carbamylation of soluble lens proteins and prevents cyanate-induced phase separation opacities *in vitro*: a possible mechanism by which aspirin could prevent cataract, *Exp. Eye Res.,* 40, 297, 1985.

145. **Bucala, R., Manabe, S., Urban, R. C., and Cerami, A.,** Nonenzymatic modification of lens crystallins by prednisolone induces sulfhydryl oxidation and aggregate formation: *in vitro* and *in vivo* studies, *Exp. Eye Res.,* 41, 353, 1985.

146. **Swamy, M. S. and Abraham, E. C.,** Inhibiton of lens crystallin glycation and high molecular weight aggregate formation by aspirin *in vitro* and *in vivo, Invest. Ophthalmol. Vis. Sci.,* 30, 1120, 1989.

147. **Cotlier, E. and Sharma, Y. R.,** Aspirin and senile cataracts in rheumatoid arthritis, *Lancet,* 1, 338, 1981.

148. **Cotlier, E.,** Senile cataracts: evidence for acceleration by diabetes and deceleration by salicylate, *Can. J. Ophthalmol.,* 16, 113, 1981.

149. **Brownlee, M., Vlassara, H., and Cerami, A.,** Aminoguanidine prevents hyperglycemia-induced defect in binding of heparin by matrix molecules, *Diabetes,* 36, 85A, 1987.

150. **Nakayama, H., Taneda, S., Kuwajima, S., Aoki, S., Kuroda, Y., Misawa, K., and Nakagawa, S.,** Production and characterization of antibodies to advanced glycation products on proteins, *Biochem. Biophys. Res. Commun.,* 162, 740, 1989.

Chapter 3

RABBIT MUSCLE GLYCOGEN PHOSPHORYLASE *b*. THE STRUCTURAL BASIS OF ACTIVATION AND CATALYSIS

Nikos G. Oikonomakos, K. Ravi Acharya, and Louise N. Johnson

TABLE OF CONTENTS

I. PROLOGUE

Glycogen phosphorylase (GP) is the best known paradigm of an enzyme that exemplifies two fundamental mechanisms of control in catalysis: (1) allosteric activation or inhibition, where enzyme activity is regulated by the action of specific metabolites on at least two conformations R and T existing in equilibrium[186] and (2) phosphorylation, the most widely occurring reversible posttranslational modification used in the regulation of the activity of many enzymes.[32,150] It was the first enzyme to be shown to undergo allosteric activation in the presence of AMP,[35,36] to exist in oligomeric form,[160] and to be covalently interconvertible (by phosphorylation-dephosphorylation of one specific serine residue) between two forms *b* and *a*.[59,62,81,149] The physical understanding of these properties has provided a challenge whose solution should be of relevance to other control systems. The application of the X-ray diffraction techniques on GP has played a dominant role. The crystallographic studies were aimed at providing a detailed stereochemical explanation of the fascinating control properties of the enzyme. The crystal structure of T-state (less active) GP*b*[1,223,269] and T-state GP*a*[241,244] have been solved at 1.9 Å and 2.1 Å resolution respectively and a comparison of the two structures[244] has shown the conformational changes that result from phosphorylation of Ser-14 and create an allosteric site with a higher affinity for AMP.[243] In the T-state structure lattice forces prevent communication of these changes to the catalytic site, which is over 30 Å from the Ser-14 and AMP allosteric site. The interactions made, the conformational changes that occur, and the mode of binding of a great number of substrates and T-state and R-state inhibitors[7,114,117,174,175,178,200,210,242,274] at the catalytic site of T-state GP have been investigated by crystallographic methods. A major achievement was the solution of the tetrameric R-state structure of GP*b* and GP*a*.[8,9] These crystal structure determinations which showed how quaternary structural changes between T-state and R-state GP result in tertiary structural changes at the ligand binding sites permitted a detailed description of the T to R allosteric transition. GP is completely dependent on the coenzyme pyridoxal 5'-phosphate (PLP) for activity[5] and the role of 5'-phosphate group in catalysis has been shown to be quite different from that in conventional vitamin B_6 dependent enzymes.[70,101,124,170] The basis for understanding the catalytic mechanism has come from extensive kinetic studies with natural, synthetic substrates and relevant ligands,[142-144,206,249] with a plethora of cofactor analogues and even model compounds to mimic the normal enzymic reaction,[26,28,70,83,101,170] NMR observations, comparative studies, and mutagenesis experiments[28,208,224] coupled with detailed knowledge of the environment of the natural and modified cofactors in the protein molecule[199,203] and a series of studies on catalysis in the crystal[89,125,178] carried out with T-state GPb. The structural and functional properties of GP have been the subject of several excellent reviews.[43,66,67,83,101,124,169,196] This review is mostly concerned with the contribution of X-ray diffraction studies to our understanding of the stereochemical basis of the enzyme control and function.

II. INTRODUCTION TO THE ENZYME PROPERTIES

GP, 1,4-α-D-glucan:orthophosphate α-glucosyltransferase (EC 2.4.1.1), catalyzes the first step in the intracellular catabolism of glycogen in muscle, liver and brain (see Newgard et al.,[196] for a recent review of the physiological role of the three isozymes). Glycogen, the principal storage form of carbohydrate, consists of polysaccharide chains of α-D-glucose units linked by α-1,4-glycosidic bonds with α-1,6 branch points occurring at approximately every 10 to 12 sugar units. As electron microscopy of striated muscle thin sections has revealed[268] most of the glycogen occurs in particles of average diameter with 400 Å. The enzyme utilizes phosphate (P_i) to cleave a glucose unit, resulting in the energy-conserving formation of glucose-1-phosphate (glucose-1-P) (Figure 1). The phosphorylytic cleavage of the α-1,4-glycosidic linkage occurs at the nonreducing ends of the glycogen chains and GP digests to within four residues of a branch point. The reverse reaction, lengthening of the polysaccharide, is taking place by addition of the glucosyl moiety of glucose-1-P to the glycogen chain with concomitant production of P_i. Although the equilibrium constant, K_{eq}, is 0.3, the enzyme *in vivo* functions in the direction of glycogen degradation because the concentration of P_i exceeds that of glucose-1-P.[83,124,169] Glycogen synthesis takes place mostly by means of the uridine diphosphoglucose (UDP-Glc) pathway, i.e., a separate enzyme system involving glycogen synthetase.[171] GP contains 1 mol per mole of PLP[5] linked to the enzyme by a Schiff base.[61] In skeletal muscle GP is present in such high concentrations so that there is more PLP associated with phosphorylase than all the vitamin B_6 dependent enzymes together.[101]

GP, in resting muscle, exists in the inactive b form but can be activated by the cooperative binding of AMP and some analogues (as IMP) and inhibited by ATP, ADP, glucose-6-P, UDP-Glc, purines (as caffeine) and D-glucose. In addition, *in vitro*, activation of the enzyme can be produced by high concentrations of substrate anions and several anions high on the Hofmeister series[20,50] and certain organic solvents.[47,256] These effectors are also able to stimulate the AMP (or IMP)-dependent activity of GPb as do protamine, spermine and other polyamines, fluoride, divalent cations such as Ca^{2+} and Mg^{2+}.[47,83,148,190,227,237,256,257,265] There is some evidence that the stimulation of IMP-dependent activity of GPb by spermine may have a physiological role.[159] The list also includes some less thoroughly characterized effectors such as phenothiazines, polycarboxylates, sulphated polysaccharides, polymyxins, prostaglandins, aliphatic amines and sodium cholate.[151-153,237-240]

The main pathway for physiological activation or inhibition involves conversion of the GPb to the GPa form by phosphorylation of Ser-14 by a specific kinase (which is also regulated by means of extracellular and hormonal signals) or conversion of GPa to GPb by GPa phosphatase-catalyzed hydrolysis of the Ser-14-P bond.[29-31] Conversion of GPb to GPa results in an enzyme which is active in the absence of AMP. GPa exhibits an increased affinity for AMP (by about 100-fold) but a decreased affinity for ATP, ADP and glucose-6-P.[43,66,83,124,169] During the phosphorylation and dephosphorylation reactions a phospho-dephospho hybrid ab enzyme is formed which exhibits intermediate properties.[43,261]

The oligomeric nature of GP was first described by Madsen and Cori[160] who determined the subunit structure of both forms showing that, *in vitro*, GPb was a dimer of presumably two identical subunits and GPa was tetramer. Under physiological conditions, GP is asso-

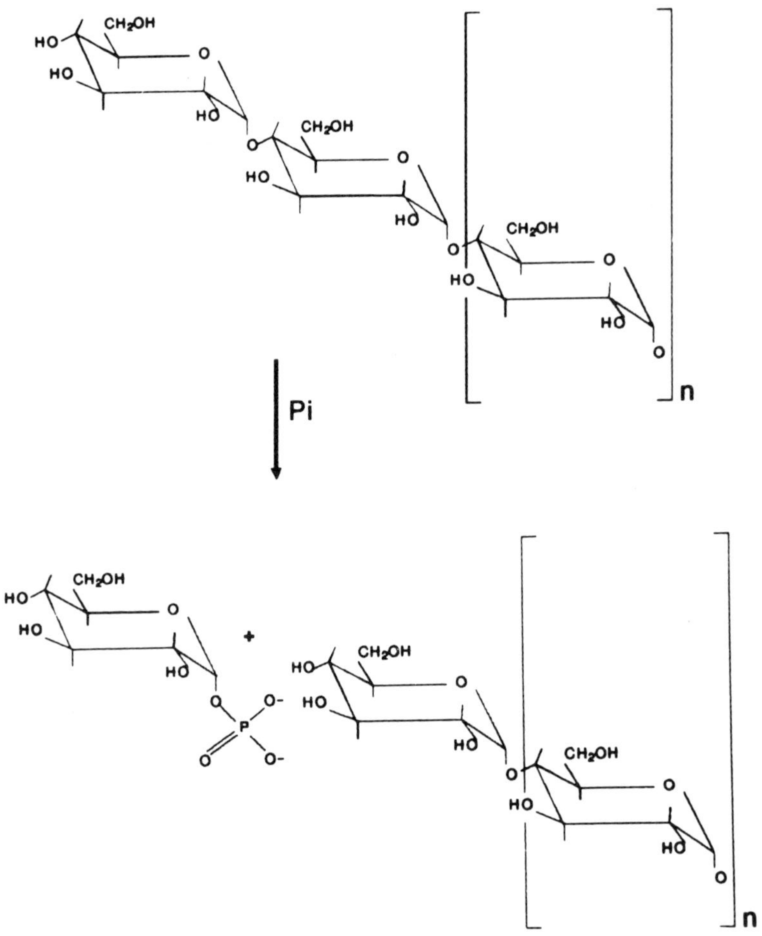

FIGURE 1.   The phosphorylytic reaction catalyzed by glycogen phosphorylase.

ciated with glycogen particles in the cell.[124] Binding to glycogen dissociates the tetrameric form to dimers.[182,263] Consequently, the catalytically active molecular species of GP is the dimer. The tetrameric species exhibits a lower affinity for glycogen and a lower specific activity.[109] In contrast, the monomeric subunit of GP seems to be in a catalytically inactive conformation.[56]

Cooperativity is the characteristic property of GP*b*. It requires that, in an enzyme that has more than one binding site for its substrate or allosteric effector, binding of, e.g., one molecule to the enzyme alters the affinity with which others are bound. In the presence of substrate, the AMP saturation curve is sigmoid rather than hyperbolic,[99,163] i.e., enzyme catalytic activity can undergo great alterations within a narrow range of AMP concentrations. Since the nucleotide (allosteric-αλλos = other, στερεos = indicating three-dimensional quality or solidity-ligand) does not participate in the reaction, this sigmoidal behaviour implies interactions between AMP binding sites, that is, the binding of the first molecule at one site induces changes in the affinity or/and catalytic efficiency at the other, usually remote, binding site. The inhibition of GP by glucose first described by Cori and Cori[37] and Cori et al.[38] represents another example of manifestation of cooperativity. In the presence of glucose (Figure 2) the rate-substrate concentration curve is also sigmoid and the enzyme does not obey the simple Michaelis-Menten equation. These cooperative effects closely comparable to the classical haem-haem interaction in haemoglobin can be expressed by

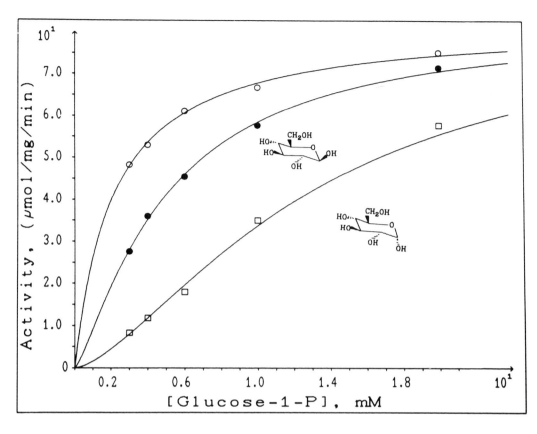

FIGURE 2.   Activity of phosphorylase *b* at 30°C and pH 6.8, at saturating concentrations of AMP (1 m*M*) and glycogen (1%), as a function of substrate concentration in the absence or presence of 10 m*M* of α-D-glucose or 10 m*M* β-D-glucose.

applying the linear tranformation of Hill equation.[104] Although this mathematical treatment does not attempt to provide an explanation of the phenomenon in molecular terms it provides however a useful method of analysing some allosteric effects,[100,163,213] where the value of the Hill coefficient (n) can be considered as an expression of the degree of cooperativity. In terms of conformational changes two models have been postulated. In the symmetry or concerted model,[186] the sigmoid AMP-binding curve is interpreted by postulating that the amount of active enzyme in solution is usually increasing in the narrow region of nucleotide concentrations. In this model the cooperative enzyme is envisaged as (pre-) existing in at least two conformations, designated T and R, in equilibrium. A major assumption of the model is that the association between subunits may be such as to confer an element of symmetry on the molecule and that this element of symmetry is conserved in the transition between the two states.[186]The T state exhibits low affinity for substrates, allosteric activators and R-state inhibitors and a high affinity for T-state inhibitors. The R state is stabilized by, and exhibits high affinity for substrates, activators and R-state inhibitors and low affinity for T-state inhibitors. In the sequential model[147] which was based on earlier proposals for the induced fit theory of enzyme action,[146] it is supposed that the binding of an allosteric ligand alters the conformation of a subunit, e.g., from T to R, and this alteration affects the stability of the other subunit because of the change in the subunit-subunit interface. In the case where this change could make it easier for the second subunit to change its conformation to R, the binding of the second ligand molecule would occur more easily. Both the symmetry and the sequential model are based on subunit interactions and the two models are limiting cases of a more general model which allows for the occurrence of pre-existing equilibria as well as ligand-induced conformational changes.[103] GP is one of the most widely studied examples of allosteric enzymes. Both GP*a* and GP*b* exhibit at least two major quaternary structural states. The active (R) state is induced by AMP, substrates or certain substrate analogues such as glucose-1,2-P and UDP-Glc and the inactive (T) state is stabilized by the binding of ATP, glucose-6-P, glucose and caffeine.[2,3,100,108,129,130,132,133,136,168,181,263,264,271,272] Positive homotropic effects (interactions between identical ligands in the terminology of Monod et al.[186]) and positive and negative heterotropic (interactions between different ligands) effects have been observed (reviewed by Graves and Wang,[83] Busby and Radda,[22] Dombradi,[43] Madsen,[169] Johnson et al.[123]). Most of these ligand effects can be discussed in terms of the two-state symmetry model, where T and R states should be considered as two extremes of a conformational spectrum.[66,78,83,123,169]

Glucose, an allosteric competitive inhibitor of the enzyme with respect to the substrate glucose-1-P (or P_i) controls the activity of GP*a* by increasing the cooperativity between substrate binding sites and stabilizing the inactive dimeric T-state conformation[100] which is a better substrate for GP phosphatase thus promoting the conversion of GP*a* to GP*b*[4,41,42,172] (reviewed by Dombradi,[43] and Madsen,[169]). D-glucose is a less prominent physiological inhibitor of muscle GP, since in muscle cells there is virtually no free glucose.[138,171,196] However, in the liver, which is almost freely permeable to glucose, this direct effect of glucose to control the activity of GP may be an important physiological mechanism in the regulation of glycogen synthesis/degradation in response to fluctuating glucose levels.[102,135,166,169,171,185,245,246,280] In earlier studies β-D-glucose was shown to be noninhibitory[37] but subsequent kinetic studies[248] have shown that α-D-glucose exhibits a threefold lower K_i than β-D-glucose for GP*b* (with respect to glucose-1-P) suggesting a preference for, but not exclusive binding of, the α-anomer. More recent results[17,53] supported the α-stereospecificity of GP*a* by showing that α-D-glucose was more effective than the β-D-glucose in inhibiting GP*a* activity as well as in the dephosphorylation of GP*a* by GP phosphatase.

The GP reaction mechanism is not fully understood. The kinetic mechanism is rapid-equilibrium random bi bi.[51,52] GP can bind either glycogen or phosphate and the resulting ternary complex, enzyme-glycogen-phosphate, is then transformed to enzyme-glycogen (with one less glucose unit)-glucose-1-P. The kinetic mechanism is not altered by the interconversion between GP*b* and GP*a*, nor by changing the normal Michaelis-Menten kinetics to the allosteric type in the presence of ATP or glucose. In this type of kinetic mechanism, the apparent Michaelis constants are equivalent to the corresponding apparent dissociation constants.[169] The breaking of the glycosidic a-Cl-Ol bond and the retainment of the α-configuration in Cl[34] suggest a chemical mechanism involving a double displacement, analogous to that proposed for the "retaining" glycosyl-transferring enzymes such as lysozyme or β-galactosidase.[15,16,233] Such a mechanism would require an enzyme nucleophilic group (as the side-chain carboxylate group of an aspartate or glutamate residue) which attacks from the opposite side of the pyranose ring to the leaving group.[234] The lack of conclusive evidence for an enzymic nucleophile required by this mechanism has, however, led to a concerted mechanism in accordance with internal nucleophilic substitution ($S_N i$).[144] In this mechanism the first step is the same as the first step of the $S_N 1$ mechanism but in the second step part of the leaving group attacks from the front and this results in retention of configuration. Extensive studies have established an obligatory role for the cofactor 5'-phosphate group in catalysis and the ionization state of the 5'-phosphate group is sensitive to the activation state of the enzyme.[101,124,170,212] Current proposals for the GP reaction mechanism suggest that upon the formation of the ternary complex the cofactor 5'-phosphate functions either as an acid-base catalyst serving to promote, through a direct hydrogen bonding interaction, protonation of the substrate phosphate or with the phosphorus atom as an electrophile serving to effect a partial withdrawal of electrons from the substrate glucose-1-P, leading to a pseudo-pyrophosphate intermediate.[101,127,144,208,251,272,273] Both proposed mechanisms depend on the correct orientation of the substrate phosphate and sugar groups with respect to the 5'-phosphate group of the cofactor. A key question concerns the groups on the protein involved in the stabilization of the phosphate-phosphate interaction.

III. ENZYME STRUCTURE

A. CRYSTALLIZATION

The GP molecule is conformationally flexible; therefore, when attempting to crystallize any conformation of the enzyme, one might attempt to "freeze" all the molecules in this state conformation or at least to attain the minimum degree of structural heterogeneity.[180] Assuming that the simplest case of the concerted model for allosteric transitions[186] is applicable to GP, the enzyme in solution is envisaged as existing in two extreme conformations in equilibrium, T and R, which differ in their activity. The allosteric constant L (the equilibrium constant for the transition between the R and T states, $L = [T_o]/[R_o]$) for GPb is at least 3300 and for GPa is between 3 to 13 (evidence in Madsen[169]). In solution GPb has a low activity and for appreciable catalysis an R-state activator (e.g., AMP) must be present. GPa is active and does not require such an activator, although addition of AMP can produce a 10 to 20% increase in activity. AMP activated GPb and unliganded GPa display comparable catalytic activities; hence even at saturating concentrations of AMP GPb possesses only 7 to 25% of the enzyme molecules in the R state. Concomitant with conversion to the R state, rabbit muscle GP dimers associate to tetramers as a function of concentration, pH, ionic strength and temperature.[43] Thus, GPa exists in a tetrameric state even at relatively low concentrations, while GPb at fairly high concentrations, when activated, associates to form tetramers. Tetramerization appears to be a unique property of the skeletal muscle enzyme since liver, potato and E. coli phosphorylase dimers do not associate. In solution GPa tetramers have only 12 to 33% (at 20 to 35°C) of the full GP activity, but may be dissociated by glycogen (or oligosaccharides) and T-state inhibitors (e.g., glucose) to give fully active or fully inactive dimers, respectively.[100,109,182,263,264,271] Micro-crystals of rabbit muscle GPb first described by Fischer and Krebs[60] were obtained at low temperature from 1 mM AMP and 10 mM $Mg(CH_3CO_2)_2$. Under the conditions of crystallization, dimerization of the protein molecule is observed.[137] These microcrystals are very soluble at room temperature and this property is now routinely used in the purification of the enzyme.[63] In contrast, GPa forms micro-crystalline material readily in the absence of effectors but the microcrystals are relatively insoluble (except in the presence of high salt concentrations). Macrocrystals of R-state rabbit muscle GPa and GPb (space group P2₁, a = 119.0 Å, b = 190.0 Å, c = 88.2 Å, β = 109.35° with one tetramer of 390,000 per asymmetric unit), first described by Matthews,[177] Madsen et al.,[165] and Fasold et al.[54] were obtained in the presence of 1.0 M ammonium sulphate and optionally AMP and Mg^{2+}. A favourable T-state tetragonal macrocrystal of rabbit muscle GPb, with suitable unit cell parameters, was discovered in Oxford.[49] The enzyme can be crystallized at a low ionic strength or with no difficulty in the presence of IMP and Mg^{2+} or IMP and spermine,[115,198] in space group P4₃2₁2 with unit cell dimensions a = b = 128.5 Å and c = 116.3 Å. The asymmetric unit contains one GP monomer and the two monomers (subunits) of the physiologically active dimer are related by the crystallographic twofold axis at $z = \frac{1}{2}$. Similarly, GPa was crystallized in the presence of glucose in the same tetragonal space group as GPb.[65] For both enzyme crystals, the conformation is very close to the T state. While crystals of the fully active dimeric AMP-activated GPb have not been prepared for the rabbit enzyme, small crystals of the R state of enzyme have been obtained in the presence of either AMP, maltotriose and polyethylene glycol or organic solvents[221] but they have relatively large unit cells and proved to be extremely radiation sensitive. Crystals of GPb in the presence of ammonium sulphate and AMP (or IMP) have been recently produced.[201,283] The GP*b*-AMP crystals belong to a monoclinic space group P2$_1$ with unit cell dimensions a = 119 Å, b = 190 Å, c = 176 Å, β = 110° (with two tetrameric molecules per asymmetric unit) and a structural analysis of the co-crystallized complex is in progress. Hence, in the case of rabbit muscle GP, where an equilibrium among various conformations exists, crystallization can select only one unique conformation.[90,119,131] Crystals for conformation R have been also prepared for muscle GP from shark[66] and pig[198,201,202] and for bovine liver GP*a*[112] but no detailed analyses are available. T-state yeast *(S. cerevisiae)* GP has been recently reported to crystallize in a tetragonal form (space group P422, a = b = 161.12 Å, c = 175.49 Å with a dimer in the asymmetric unit.[113] Some representative crystals of rabbit muscle GP are shown in Figure 3.

B. CRYSTAL STRUCTURE OF THE T-STATE CONFORMATION
1. GP*b*

The X-ray crystal structure of T-state GP*b* (+ IMP) has been solved[223,269] and refined at 1.9 Å to a crystallographic R-factor (defined as $\Sigma \mid [F_o]-[F_c] \mid /\Sigma \ [F_o]$, where $[F_o]$ and $[F_c]$ are the observed and calculated structure factor amplitudes, respectively) of 0.185 with data collected using synchrotron radiation.[1] The knowledge of the amino acid sequence[253] has helped in solving the enzyme structure, while the determination of the entire nucleotide sequence of the rabbit muscle GP cDNA[194] has revealed only a few corrections. The enzyme contains 842 amino acid residues and one pyridoxal 5'-phosphate giving a subunit mol wt of approximately 97,500. The structure is relatively compact with a radius of about 30 Å. It is an α/β protein comprising 53% α-helices and 19% β-sheets (for residues 10 to 842). The polypeptide chain can be divided in two domains: domain 1 or regulatory domain (residues 10 to 484) and domain 2 or catalytic domain (residues 485 to 842) (Figure 4). The two subunits of the functionally active dimer are related by a crystallographic two fold symmetry axis. The subunit-subunit contacts are not extensive. The main interactions involve the cap (residues 36 to 45) and the tower helices α7 (residues 262 to 278). The cap inter-actions provide a link to the C-terminal tail of the other subunit (His 36' to Asp 838), van der Waals interactions to the α2 helix (residues 47 to 78), hydrogen bonds from the main chain oxygens of Leu 39' and Val 40' to Arg 193 (β7, residues 191 to 193) and an ionic link between Lys 41' and Glu 195. The tower helices run antiparallel to one another with an angle of about 20° between them. Residues from the top of the tower are in van der Waals contact with side chains in the 280s loop (residues 280 to 292) of the other subunit. The tower is also linked to the cap interface by contacts from the tower residues to those in the loop β6'/β7'. The N terminal tail in GP*b* is in an extended β-stranded conformation ending at Gly 20 where α1 helix (residues 20 to 38) begins. The tail is positioned in a rather broad depression on the surface of the protein and is in an anionic environment (there are 11 acidic residues and only 7 basic residues within 15 Å of the Ser 14 Cα. Until recently the N terminal tail from the residue 1 to residue 18 could not be located in the electron density map and was assumed to be mobile. However, during refinement of the GP*b*-glucose-1-P complex,[174] additional electron density was observed at the N terminus for residues 10 to 18. The tail has been implicated in the control properties of the enzyme and it is thought to mediate both homotropic and heterotropic interactions between subunits.[43,66,124,169] This latest electron density map also showed that the carboxylate of Pro 842, the last residue at the C terminus appears to interact with Asn 44', a residue which is important in subunit-subunit interactions.

2. GP*a*

The X-ray crystal structure of GP*a* (+ glucose) is known at 2.1 Å resolution[243] and the current R factor is 0.16.[244] Although GP*a* differs from GP*b* in that it is almost fully active in the absence of AMP, it strongly binds nucleotide in a noncooperative mode and is not significantly inhibited by glucose-6-P (which is a powerful inhibitor of GP*b*),[66,158,181] this enzyme form adopts, however, a similar conformation in the tetragonal crystal with significant differences only in the subunit interface near the phosphorylation site. The comparison of the T-state crystal structure of GP*b* and GP*a* has revealed the conformational changes that represent the first step in the activation of the enzyme by phosphorylation. Overall GP*b* and GP*a* are very similar (rms differences in Cα positions = 0.6 Å). A major difference in conformation between the two structures consists in the burial and ordering of the N terminal (residues 5 to 16) which carries the Ser 14 (P) site and the exposure and disorder of the C terminal tail (residues 838 to 842) on the *b* to *a* transition. Upon conversion to GP*a*, the N terminal tail folds into a distorted 3_{10} helix and the Ser 14 (P) makes contact to two arginine residues at the subunit interface, Arg 69 from its own subunit and Arg 43′ from the other subunit. The interactions that Ser 14 (P) makes with Arg 69, Arg 43′ and Val 15 and Gln 72 trigger the formation of inter-subunit contacts involving residues remote from the Ser (P) site. The residues Ile 13 and Val 15 located on either side of Ser 14 (P) are completely buried in pockets lined with nonpolar groups from side chains of the α2 helix. Arg 16 in GP*a* occupies part of the position of Glu 839 in GP*b*. The position occupied by Arg 43′ in GP*b* is taken by the N terminal tail Arg 10 in GP*a*. Rotation of Arg 43′ creates a hydrophobic pocket which accommodates Ile 13. These transitions affect His 36, the salt bridge between His 36′ and Asp 838 being ruptured, and the α2 helix and serve to create a high affinity binding site for AMP in GP*a*. In GP*b* the α2 helix is partially unwound at residues 65 and 66 resulting in a transition of about 0.8 Å with respect to GP*a*. Thus Gln 72 and Tyr 75 in this helix adopt different conformations in GP*b* and GP*a* and are able to make more favorable interactions with AMP in GP*a*. Likewise the cap residues Asp 42′ and Asn 44′, shifted slightly in response to the movement of Arg 43′, are also able to interact more strongly with the nucleotides in GP*a*. Movements of the α2 helix results in a different conformation of a second Gln 71. In GP*b* Gln 71 makes important interactions with the inhibitor glucose-6-P, while in GP*a* the different conformation of this residue blocks the site, thus providing a steric explanation for the inability of glucose-6-P to inhibit GP*a*. In general the phosphorylation of Ser 14 promotes close intersubunit contacts which increase the affinity of the enzyme for AMP but inhibit glucose-6-P binding.[244]

3. Binding Sites

GP tetragonal crystals are highly solvated. They contain a considerable amount of liquid solvent forming a three-dimensional network of channels. Thus, the packing of the molecules in the crystals results in large channels between the protein molecules with diameters approximately 20 Å that run the whole length of the unique axis of the crystal (Plate 1*). These channels allow small molecules to diffuse into preformed crystals and to reach their corresponding binding sites.[123] Various crystallographic studies have led to identification of five metabolite-binding sites in GP (Figure 4).

1. The catalytic site, identified by the binding of the substrates glucose-1-P[117,269] and P_i[79] lies as a deep crevice between domain 1 and domain 2 at the centre of the molecule and incorporates the essential cofactor pyridoxal 5′-phosphate. There is no direct access to the site from the bulk solvent. Access to this site can be only achieved through a narrow channel which is some 5 Å wide and 12 to 15 Å long, but this channel is blocked mainly by the loop 280 to 292. Although the catalytic site itself is not at the subunit interface there are indirect contacts from the catalytic site to the interface.
2. The allosteric (or nucleotide) site is placed at the subunit interface and is some 32 Å from the catalytic site. Ligands which bind at this site either activate (AMP, IMP, P_i, SO_4^{2-}) or inhibit (ATP, glucose-6-P, UDP-Glc, β-glycerophosphate) the

---

* Color plates follow page 118.

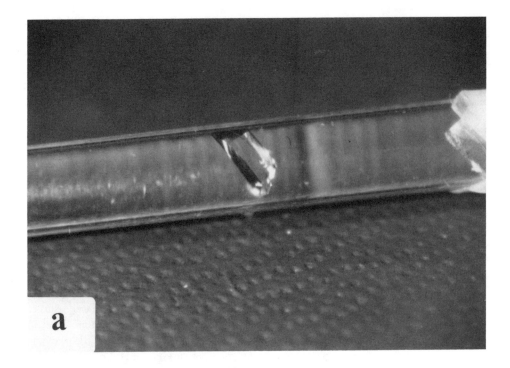

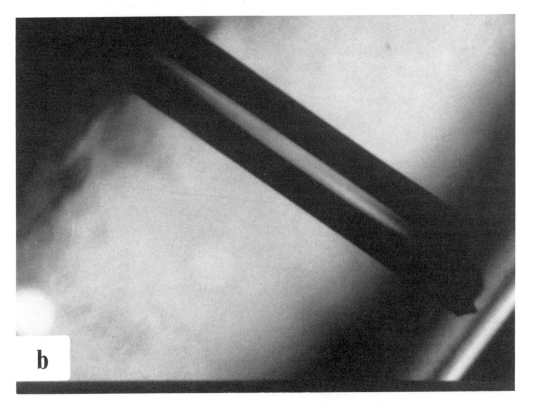

FIGURE 3.   Crystals of rabbit muscle glycogen phosphorylase *b*. (a) Native tetragonal crystal, (b) Tetragonal crystal of phosphorylase *b* reconstituted with 5′-deoxypyridoxal 5′-methylphosphonate, (c) Monoclinic crystals produced in the presence of 1.4 *M* ammonium sulphate and 6 m*M* IMP at pH 7.5, (d) Monoclinic crystals grown from 1.0 *M* phosphate buffer (pH 6.8).

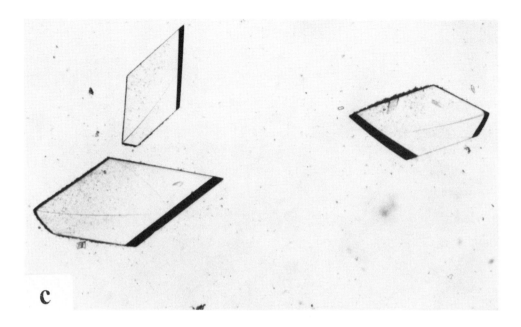

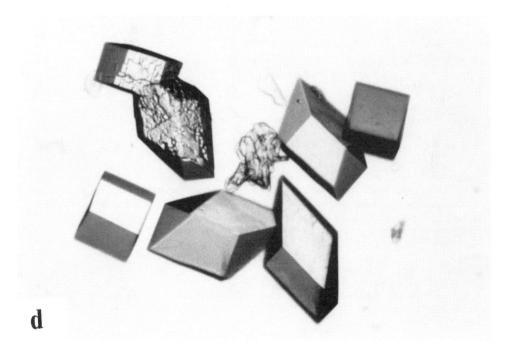

FIGURE 3 C and D.

enzyme.[9,116,132,158,200,202,243,250] The allosteric activator, AMP, when bound is located between the two α helices, α2 and α8 (residues 47 to 78 and 289 to 314, respectively) and is close to the subunit interface.

3. The phosphorylation, Ser (P), site is developed during activation of GPb by phosphorylation.[244] The phosphate group is bound by two arginine side chains, one, Arg 69 from its own subunit, and the other Arg 43', from the other subunit. It also interacts with the main chain nitrogen of Val 15 and Gln 72 through a water molecule.

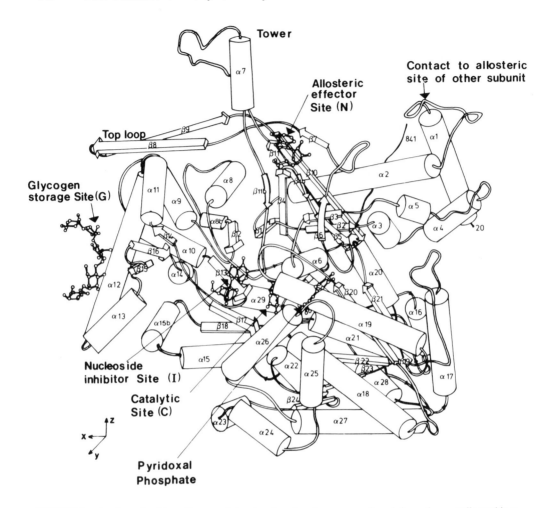

FIGURE 4.   A schematic ribbon diagram of the phosphorylase *b* monomer viewed down the crystallographic y axis. α Helices and β strands are represented by cylinders and arrows respectively. The regions of polypeptide chain corresponding to α1 helix, cap, start of α2 helix and the tower helix, 280s loop and start of α8 helix together with the loop connecting β7 and β8 are important in subunit-subunit contacts. The essential cofactor, pyridoxal 5′-phosphate, is buried at the centre of the subunit, linked via a Schiff base to Lys 680 in α21 helix. The catalytic site, shown here with glucose-1-P, is close to the cofactor and accessible to the bulk solvent through a channel some 15 Å long. The allosteric site with AMP bound is located at the subunit-subunit interface between the α2 helix and the α8 helix. The glycogen storage site with 5 sugar molecules bound and associated with the α12 helix is on the surface of the enzyme and removed from the allosteric and catalytic sites. The nucleoside inhibitor site is situated at the entrance to the catalytic site channel. Purines or nucleosides or nucleotides (at high concentrations) bound at this site make contacts with Phe 285 from the 280s loop and with Tyr 613 at the start of α19 helix. Occupancy of this site stabilizes the T state and inhibits the enzyme.

4.   The inhibitor (or nucleoside) site is located on the surface of the enzyme some 10 Å from the catalytic site and, in the T state, obstructs the entrance to the catalytic site. Ligands which bind to this site include purines, nucleosides or nucleotides at high concentrations and are stabilized by intercalating between the aromatic side chains of Phe 285 (domain 1) and Tyr 613 (domain 2).[242,250] Binding at this site stabilizes the T state and is synergistic with glucose binding.[132,133]

5.   The oligosaccharide binding site is also located on the exterior of the protein some 30 Å from the catalytic site and 40 Å from the allosteric site.[134,269] The site has been termed the glycogen storage site and appears to be the region through which GP is attached to glycogen particles *in vivo*.[122]

C. CRYSTAL STRUCTURE OF THE R-STATE CONFORMATION

The growth of monoclinic crystals of GP*b* and GP*a* in the presence of 1.0 to 1.2 *M* of ammonium sulphate has been reported a long time ago.[54,165,177] It is well known that high concentrations of substrate anions and several anions high on the Hofmeister series such as sulphate are able to active GP*b* to a considerable extent[20,50] and that this activation could be further stimulated by AMP.[237] The three-dimensional crystal structures of both R-state phosphorylases *b* and *a* were recently described.[8,9] The X-ray analysis at a resolution of 2.9 Å revealed tertiary structural changes in response to sulphate binding. These changes, concomitant with conversion of the enzyme to the active R-state conformation, are coupled to large changes in quaternary structure.[8] The quaternary structural changes can be described in terms of a 10° rotation of one subunit relative to the other about an axis approximately normal to the molecular 2-fold axis and bisecting the axis in the region of the cap'/α2 interface. The cap'/α2 interface is drawn closer together by 1 Å and the interface in the vicinity of the tower helices, (α7), is drawn apart by 6 Å. This change in turn imposes tertiary structural changes on both subunits such that a high affinity AMP allosteric site is developed adjacent to the cap'/α2 region. In the R-state GP*b*, sulphate is bound close (2 Å) to the Ser-P 14 of GP*a*, making a hydrogen bond to the side-chain of Ser 14 and salt bridges with the guanidinium groups of Arg 43' and Arg 69 following rotations of their side-chains. Residues 10 to 18, which are disordered in the T-state, become ordered. A change in the position of the tower helices induces also structural changes at the catalytic site. At the catalytic site, the side chain of Arg 569 has rotated so that the guanidinium group has moved 7 Å towards a strong isolated density feature representing the sulphate of crystallization. Movement of Arg 569 and concomitant disorder of residues 282 to 286 which removes Asp 283 from the vicinity of the catalytic site develops a high affinity phosphate binding site which is close (hydrogen bonding distance) to the 5'-phosphate of the essential cofactor. The importance of this residue for GP catalysis and control has been previously recognized from chemical modification experiments.[48,258] The movement of Arg 569 and disorder of the 280s loop is linked to the motion of the tower helix through displacement of residues Asn 133 and Pro 281 which pack against Ile 265. Ile 265 forms part of the β-sheet with the 276 to 279 at the C terminus of the tower helix. Movement of the tower helix perturbs this sheet which is propagated to Asn 133 and Pro 281. Hydrogen bonds made to Arg 569 by Asn 133 and Pro 281 in the T state are ruptured.

The quaternary and tertiary structures of R-state GP*a* are essentially identical to those of R-state GP*b*, except for small differences in the vicinity of the Ser (P) site. The overall structures of R-state GP*b* and GP*a* are similar (rms differences in Cα positions = 0.67 Å). At the phosphorylation site, Ser 14 is displaced by 0.75 Å as compared to that of GP*b* and the phosphate of the Ser 14 is positioned 1.0 Å from the sulphate in GP*b*. In this position the phosphate makes interactions with the guanidinium side-chains of the two arginines Arg 43' and Arg 69 which are also displaced by 1.0 Å relative to GP*b* in order to optimize interactions with the phosphate. A more homogenous constellation of atoms at the catalytic site was also observed at the catalytic site. There is strong density to suggest the new position of Arg 569 and the occurrence of sulphate in all 4 subunits.[9] Thus phosphorylation of GP*b* enhances affinity by means of long-range conformational changes, in antithesis to isocitrate dehydrogenase — another phospho-regulated protein for which structural data are available[110,111] — in which phosphorylation can act directly at the catalytic site without inducing large conformational changes.

D. ALLOSTERIC TRANSITION

GP has four distinct phosphate recognition sites (the PLP, substrate, AMP, and Ser 14 phosphates, respectively) and the crystallographic studies with both GP*b* and GP*a* show tertiary structural changes in response to ligand binding. A simplistic but unifying principle was suggested by Johnson et al.,[120] namely, that GP action and control can be viewed in terms of conformational changes which either create or change electrostatically these phosphate recognition sites and by so doing produce additional conformational changes that affect the subunit-subunit contacts. At the catalytic site, the movement of Arg 569 contributes to the stabilization of the product phosphate-coenzyme phosphate contact, which is an essential part of the catalytic mechanism. At the Ser (P) site, Arg 43' and Arg 69, one from each of the subunits, move to create the Ser (P) site and to order the basic N terminal tail. The phosphate of AMP, which is essential for activation and binds to a high affinity allosteric site developed by movement of the cap'/α2 interface, also induces movement of Arg 309 of α8 helix. A superposition of the R-state subunit onto a T-state subunit reveals the extent of the structural differences between the 2 states, the rms deviation in Cα positions being 1.3 Å. Those regions which deviate significantly are either involved in the allosteric response (e.g., the tower helices) or form part of the new tetramer interface (helices α12, α13 and α23 to α27) but relative movements between the two domains do not occur. A superposition of the R-state dimer onto a T-state dimer (Plate 2*) reveals an extensive quaternary change, the rms deviation in Cα positions being 6 Å.[8] The change in quaternary structure directly affects the allosteric effector site and the Ser (P) site, which are located at the cap'/α2 interface, and transmits a signal to the catalytic site through conformational changes that affect the relative disposition of the two tower helices. Because the geometry of one helix is constrained by that of the symmetry related helix, molecular symmetry is preserved and a concerned conformational change of both subunits occurs. The close coupling between tertiary and quaternary conformational changes and the conservation of symmetry suggests therefore a mechanism for the homotropic and heterotropic allosteric properties of the enzyme.[125] This mechanism based on the tighter subunit interactions at the cap'/α2 interface (Figure 5) and the tighter subunit association at the tower helices triggered by phosphorylation (or AMP) is consistent with solution studies of limited proteolysis of GP by subtilisin. It is well known that subtilisin hydrolyzes GPa at two sites:[218,253] either (1) between Arg 16-Gly 17, a reaction also catalyzed by trypsin,[82] giving an enzyme form (GPb') that is still active in the presence of AMP but in which homotropic and heterotropic interactions are highly suppressed relative to GPb, or (2) between Gln 264-Ala 265 resulting in a total loss of activity and producing two proteolytic fragments that are attached through noncovalent bonds to form a species called GPa*.[72] In this species the communication between the allosteric and catalytic site is broken.[72] Studies of the effect of various ligands on the rate of proteolysis have also shown that AMP decreases the rate of proteolysis at the Arg 16-Gly 17 site but increases the rate of proteolysis of the site Gln 264-Ala 265 by subtilisin; the opposite effects are observed with glucose-6-P (Dombradi et al.[45,46] and references therein).

E. ENZYME CONFORMATION IN THE CRYSTAL

Demonstration of enzymic activity in the crystalline state, comparable to that observed in solution, represents the most rigorous criterion for the definition of enzyme conformation in the crystal. The tetragonal crystals of GPb are obtained in the presence of IMP, a weak activator which at high concentrations (e.g., 2 mM) increases catalytic activity (V_{max}) without affecting enzyme affinity (K_m = 37 mM) for glucose-1-P.[14] In contrast, AMP activates with a K_α = 50 μM and concomitant decrease of K_m value for the substrate to 3 mM. The crystals are stable in the presence of either glucose-1-P or maltoheptaose, but when soaked with both substrates undergo severe disruption. Kinetic studies on the activity of glutaraldehyde cross-linked tetragonal crystals (performed under the conditions of crystallization) have shown that the catalytic rate decreases markedly (from 10- to 50-fold).[131] On going from solution to crystal the K_ms for the substrates glucose-1-P and maltoheptaose do not change

---

* Color plates follow page 118.

as compared to those of the soluble enzyme but V_{max}s decrease considerably.[131] Apparently, crystal lattice forces lock the enzyme in a conformation stabilized against conformational changes required for the transition state.[131] Loss of catalytic efficiency can therefore be attributed to changes in the catalytic site caused by crystallization. No binding of oligo-saccharides at this site has been observed. Oligosaccharides, (even at a concentration of 0.7 M), bind only to the glycogen storage site and not to the catalytic site.[222] Low affinity of the crystal for oligosaccharide is associated with the lack of access to the catalytic site from the solvent.[7,89] However, formation of products was observed in crystallographic ex-periments in which mixtures of substrates were diffused rapidly into crystals and data collected within 1 h using the bright Synchrotron Radiation Source in Daresbury, U.K.[89] In view of the kinetic and crystallographic studies on activity in crystal, conformational changes can take place in the crystalline GPb that do allow both substrates of the reaction to visit the catalytic site transiently.[124,222] Another test for characterizing the enzyme conformation in crystal is to soak the crystal in saturating concentrations of an activator.[221] If the crystal cracks (as a result of conformational changes occurring during activation which are stronger than the lattice forces) it probably contains the enzyme molecules in the inactive form. Conversely, if soaking the crystal in an inhibitor results in cracking, it probably contains the enzyme in the active form. Crystals of GPa are resilient to cracking when AMP is introduced,[67] while AMP has a tendency to shatter GPb crystals.[250] Soaking of GPa crystals with glucose-1-P (or the R-state inhibitor glucose-1,2-P, a possible transition state analogue) results in substantial conformational changes sufficient to disorder the crystals.[166,275] In contrast to the situation in crystals of GPa, both glucose-1-P and glu-cose-1,2-P bind tightly at the catalytic site of GPb with no disruption of the crystals. On addition of glucose-6-P (a T-state inhibitor), native GPb crystals (but not GPa crystals) initially crack and then reanneal slowly,[284] but different conformational changes from those with AMP are observed.[158] It is apparent that the transition from the T to R state cannot be accommodated in the tetragonal crystal lattice and on the basis of these observations, it has been suggested that GP in the crystal adopts an intermediate conformation in the T $\rightarrow$ R transition.[158,244]

Preliminary kinetic experiments with P2$_1$ monoclinic microcrystals of GPa and GPb produced by bulk crystallization in the presence of ammonium sulphate[285] indicate that both crystalline enzymes, when assayed in the direction of glycogen synthesis and under the conditions of crystallization (1.0 M ammonium sulphate) possess intrinsic activity, not confined to the surface molecules.

The enzyme:ligand interactions at the corresponding binding sites are described below in some detail. The crystallographic binding studies relate to the T-state conformation of GP with the exception of AMP, sulphate and substrate glucose-1-P binding to the tetrameric R-state GPb.

IV. THE SER (P) SITE

The Ser (P) site is created from the guanidinium side-chains of Arg 43′ from the cap′ and Arg 69 of the α2 helix. The X-ray analysis of R-state GPb at a resolution of 2.9 Å revealed that sulphate mimics the substrate phosphate by binding to the catalytic site, the AMP allosteric site, and the Ser (P) site of GPb, resulting in significant localized changes in tertiary structure.[8] These tertiary changes are coupled to large changes in quaternary structure which directly affect the AMP and the Ser (P) site and indirectly the catalytic site. It appears that ammonium sulphate works as an activator in place of phosphate at the phosphorylation site at Ser 14. A detailed analysis of the effect of ammonium sulphate on the solution properties of GPb[155] prompted by the crystallographic results, showed that ammonium sulphate activated GPb displays characteristics of GPa such as an enhanced

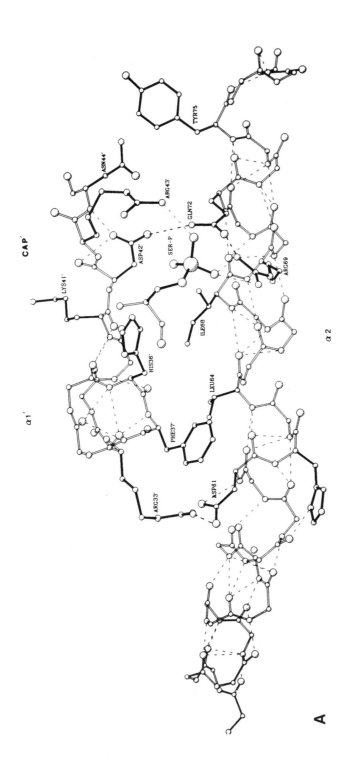

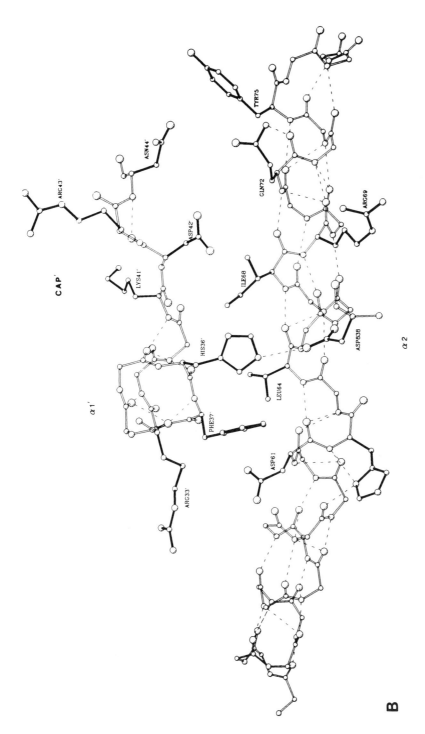

FIGURE 5. Conformational changes at the cap'/α2 interface. (A) R-state phosphorylase a; (B) T-state phosphorylase b. The view is down the twofold axis of the dimer. In Figure 5a the Ser-14 (P) contacts Arg 69 (from α2 helix) and Arg 43' (from cap' of the other subunit) and the contacts are made possible by tertiary and quaternary structural changes on the T to R transition. Val 15 docks against Ile 68 (not shown) and there is a hydrogen bond from Asp 42' to Gln 72. In Figure 5b His 36' from cap' is in ionic contact with Asp 838 from the C terminal region and adopts a different conformation. (From Barford, D., Hu, S.-H., and Johnson, L. N., *J. Mol. Biol.*, 218, 233, 1991. With permission.)

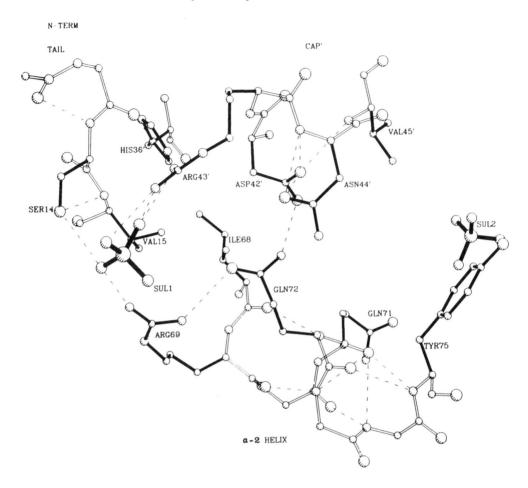

FIGURE 6.   Interactions between sulphate and R-state phosphorylase *b* at the Ser 14 (P) site. The closer association of the cap′ to the α2 helix is stabilized by salt bridges to sulphate (SUL1) from Arg 43′ and Arg 69 and interactions between Asp 42′ and Gln 72. The sulphate (SUL2) shown on the right is bound in a position close to that observed for the phosphate of AMP. (Reprinted by permission from Barford, D. and Johnson, L. N., *Nature,* 340, 609, 1989. Copyright © 1989 Macmillan Magazines Ltd.)

affinity for AMP, a reversal of the enzyme inhibition by glucose-6-P and enzyme tetra-merization. Under the conditions of crystallization, the tetrameric enzyme also exhibits a high affinity for AMP (apparent $K_d = 4 \mu M$).[155] In addition to the binding of sulphate at the Ser (P) site (Figure 6), the N terminal residues 10 to 18, which are poorly ordered and adopt an irregular extended conformation in the T-state GP*b*, become ordered. The crys-tallographic studies with T-state GP*a* and R-state GP*a*[9,244] also show that phosphorylation of Ser 14 causes a radical restructuring of the N terminal tail: residues 10 to 22, forming an extended chain, are orientated in the opposite direction to that of T-state GP*b*. Binding of sulphate (or phosphorylation) causes a shift of Arg 43′ that creates a binding site for Ile 13 and Arg 10 and a shift of Arg 69 that creates a binding site for Val 15. These shifts change the disposition of the cap′ relative to α2 helix favouring T to R transition. The crystallographic results suggest therefore that the sulphate or phosphate, in the absence of the N terminal tail, will not stabilize the R-state. Kinetic experiments performed on GP*b*′, a proteolytic species which lacks the first 16 residues[62] and is inactive in the absence of nucleotides, showed that this form of the enzyme is no longer activated by ammonium sulphate.[156] Lack of activation of GP*b*′ by ammonium sulphate indicates that sulphate ac-tivation of GP*b* occurs through binding to the Ser (P) site with concomitant and essential ordering of the N terminus and not as a result of sulphate binding to the allosteric and catalytic sites. Thus, the N terminus plays an active as opposed to a passive role, of simply supporting Ser-P, in the process of activation of GP*b* by phosphorylation.

Regarding the role played by the N terminus in the process of activation of GP*b* by AMP, it has been shown by cross-linking experiments[86] that AMP activation of the enzyme, unlike phosphorylation, does not cause ordering of the highly basic N terminus. In accordance, location of the N terminus in the position observed by GP*a* or ammonium sulphate activated GP*b*, but without the phosphate group, would result in the unfavourable opposition of positive charges.[8,125,24] A major difference between GP*b* and GP*b*′ is that the strong homotropic cooperativity displayed by GP*b* is suppressed in GP*b*′.[82] Since both enzymic forms can be activated by AMP, the interactions that the N terminus makes with GP*b* should play an important role in allosteric control.

V. THE ALLOSTERIC SITE

The amount of biochemical papers dealing with the properties of the allosteric effector site is vast. Some of the allosteric properties of the enzyme were described in the introduction. A number of excellent reviews may be consulted for details.[22,43,66,83] GP*b* is activated by AMP (K_a approximately 50 μM) and some analogues and weakly by IMP (K_a approximately 2 mM).[14,187,191,192,204] It is inhibited by ATP, ADP, glucose-6-P and UDP-Glc.[161,189,211] Phosphorylation increases the affinity of the enzyme for AMP by nearly two orders of magnitude[100] and causes a further activation (10 to 20%). However, GP*a* is less inhibited by ATP, ADP and glucose-6-P and it is generally considered that it has escaped some of the allosteric properties that GP*b* exhibits.[66,169] According to Fletterick and Madsen[66] and Madsen[169] it is the extra AMP-induced activity that ATP inhibits only.

The allosteric effector site is well characterized in the three-dimensional structure of both GP*b* and GP*a*. Structural studies[66,116,117,132,158,200,201,243] showed that the allosteric site is close to the subunit interface and consists of three subsites: the phosphoryl, the sugar, and the base. It can accommodate AMP or IMP, ADP, ATP and larger molecules such as UDP-Glc and NADH. The phosphoryl and sugar subsites can bind P_i, SO_4^{2-}, glycerol-2-P, glucose-1-P and glucose-6-P.

A. BINDING STUDIES WITH T-STATE GP*b*
1. Binding of AMP and IMP
The activator AMP binds strongly at the allosteric site to produce a large increase in V_{max} and substrate binding affinity[99,162] although this site is some 32 Å from the catalytic site. The dissociation contant of AMP from T-state GP*b* has been measured by using a number of physicochemical techniques such as equilibrium dialysis (0.37 mM),[129] fluorescence spectroscopy (0.3-0.7 mM),[191] equilibrium dialysis and calorimetry (microscopic dissociation constants K_{d1} = 0.04 mM and K_{d2} = 0.25 mM for the two allosteric binding sites per dimer, and K_d = 3.1 mM for the inhibitor site.[176] Although IMP has a very similar structure to AMP, IMP activation results only in an increase in V_{max} leaving substrate binding affinity unaffected.[14] In addition to the kinetic studies, calorimetric, fluorescence, ESR, modification and cross-linking experiments strongly suggest that IMP as compared with AMP induces different conformational changes in the enzyme.[43,84,88,105,187,286] For example, IMP as compared to AMP induces a quite different ESR ratio in an active spin-labeled GP*b*[84] by increasing rather than decreasing the apparent mobility of the spin label. IMP binds at 25° with K_{d1} = 0.5 mM and K_{d2} = 1.4 mM for the allosteric site and with K_d = 1.7 mM for the inhibitor site.[176] However it can be made to behave like AMP if the binding at the inhibitor site is abolished by addition of hydrophobic solvents.[256] An explanation for these differences in behaviour was sought from binding studies of these ligands to the crystalline

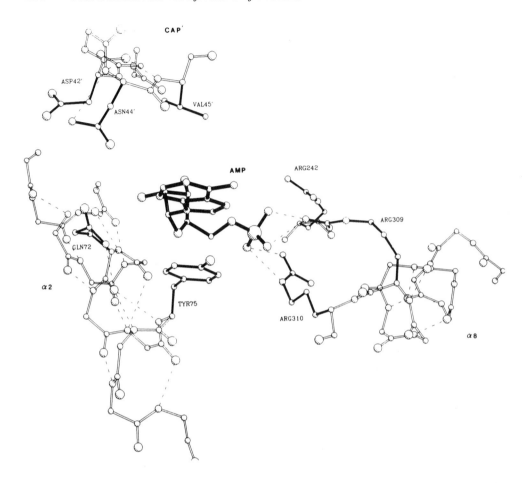

FIGURE 7.   AMP interactions to T-state phosphorylase *b* at the allosteric site. (From Barford, D., Hu, S.-H., and Johnson, L. N., *J. Mol. Biol.*, 218, 233, 1991. With permission.)

enzyme. AMP bound to the allosteric site is located at the interface of the cap′ and C terminus of α2 and α8 helices of the symmetry related subunit. When AMP (Figure 7) is bound at the allosteric site adopts a 3′-*endo* conformation at the sugar. Arg 309 is positioned after conformational change to form a salt bridge to the phosphate of AMP. A few specific contacts are formed between AMP and the enzyme, consistent with the observation that AMP is bound with a low affinity by the T-state GP*b*. The interactions made consist of salt bridges between the side-chains of Arg 309 and Arg 310 of α8 helix to the phosphate of AMP. No specific interactions are made to the base (except from an indirect polar contact of N3 to Gln 72 through solvent) or sugar. Tyr 75 of α2 helix is also stacking coparallel with the adenine base at a distance of 3.5 Å. The importance of Tyr 75 for the AMP binding had been anticipated by chemical modification experiments using potassium ferrate.[154]

2. Binding of Glucose-6-P

Glucose-6-P is a potent inhibitor of GP*b*[189] but only inhibits GP*a* under extreme conditions.[181] It is bound cooperatively by both forms of the enzyme. Kinetic studies have demonstrated that glucose-6-P acts as a nonlinear competitive inhibitor with respect to glucose-1-P[163,181] and stabilizes the catalytically inactive T-state conformation. An inhibition constant of $K_i = 0.52$ mM (with respect to glucose-1-P) has been recently determined in the presence of 1 mM AMP.[155] In the absence of other ligands, the K_d for α-D-glucose-6-P is 19 μM.[10] Glucose-6-P is partially competitive with AMP with a $K_i = 0.3$ mM[189] or with a

$K_i = 0.9$ mM.[266] Strong antagonistic (or negative heterotropic) effects between glucose-6-P and AMP binding to GPb have been also observed[11,21,187] and the physiological significance of these effects discussed.[124,158,169] A previous crystallographic study[158] had shown that glucose-6-P bound at the allosteric site. Both glucose-6-P and AMP phosphates contact the two arginines, Arg 309 and Arg 310, but the altered position of the glucose-6-P phosphate (P-P separation = 2.7 Å) allows it to contact a third arginine Arg 242. The glucopyranose ring occupied a site which is distinct from AMP[158] and this explained the partial competition between the activator AMP and the inhibitor glucose-6-P observed in solution studies.[266] In this early study the crystals had been cross-linked by dimethylmalondi-imidate and the data were recorded to only 3 Å resolution. In a recent crystallographic experiment the binding of glucose-6-P (100 mM) to T state GPb has been studied at 2.2 Å resolution and the structure of the complex refined using molecular dynamics programme XPLOR.[287] The polar contacts between glucose-6-P and the enzyme are shown in Figure 8. The glucose-6-P molecule is situated at the subunit-subunit interface and participates in hydrogen bonds to both subunits. The largest conformational change involves Arg 309 which moves 6 Å to contact the phosphate group. Arg 242 and Arg 310 also shift slightly to optimize the phosphate contacts. The main chain of residues 192-198 move by 0.5 to 1.5 Å with significant shifts in Phe 196 and Arg 193. The cap region of the symmetry related molecule also shifts closer to glucose-6-P. The hydrogen bond between the main chain carbonyl oxygen of Val 40′ to the O2 hydroxyl of the sugar results in loss of a protein subunit-subunit interaction between Arg 193 to Val 40′ and instead a hydrogen bond is made between Lys 191 and Thr 38′. In addition these changes are correlated with movements of the α2 helix and there is an overall global rotation of one subunit with respect of the other. In GPa the sugar-binding subsite is not well formed.[244] In response to conformational changes induced by phosphorylation the side-chain of Gln 71 rotates into the sugar-binding subsite and weakens glucose-6-P binding. It has been reported that in the GPa-glucose-6-P complex, the phosphate group binds as in GPb but the sugar ring is completely disordered.[244]

3. Binding of UDP-Glc

In a crystallographic study with GPb at 100 mM UDP-Glc[200] the occupancy at the allosteric site was about the same at that as the catalytic site (see Section VIII). In solution, the binding of UDP-Glc at the allosteric site of GPb has been implied by its ability to compete with the binding of glucose-6-P and glycerol-2-P.[84,85] Their results indicated that UDP-Glc binds to GPb with an apparent $K_d = 4.0$ mM. Spectroscopic measurements with a sensitive fluorescein derivative[197] have also indicated an apparent K_d of 9 mM for binding of UDP-Glc to the allosteric site. The overall conformation of UDP-Glc at the allosteric site (Figure 9) is in a well defined folded conformation with the UMP component of UDP-Glc occupying a similar position to that of AMP (Figure 10) bound to GPb but with the ribose in a C2′-endo conformation, in antithesis with the ribose of AMP which adopts a C3′-endo conformation. The uracil and the glucose moieties stack against the aromatic side chains of Tyr 75 and Phe 196, respectively. Binding of UDP-Glc at the allosteric site promotes structural changes across the subunit-subunit interface which are different from those observed with AMP or glucose-6-P. Thus, in UDP-Glc binding the subunit contact of Glu-195 and Lys-41′ is perturbed by the movement of Phe 196 in response to the interaction with the glucose component but no movement of Asp 42′ in the cap region is observed. The pyrophosphate component overlaps the phosphate-binding site for glucose-6-P but the binding sites for the rest of these molecules do not overlap. As previously pointed out the mode of glucose-6-P inhibition is different from UDP-Glc in that glucose-6-P is able to inhibit AMP binding directly and glucose-1-P binding indirectly by stabilizing the T state.[11,163,266] The interactions at the allosteric site with UDP-Glc are more extensive for the glucose and pyrophosphate components than for the uracil. The contacts include good ionic interactions between basic

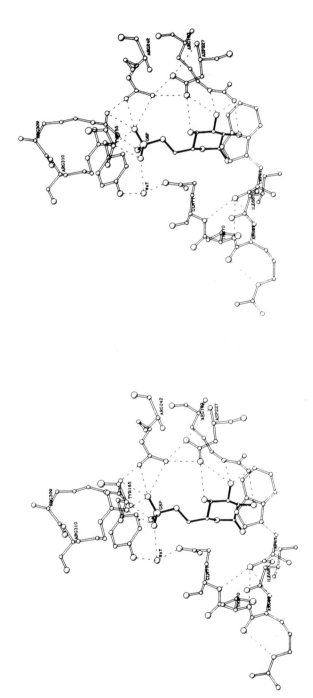

FIGURE 8.   A stereo diagram of glucose-6-P interactions to T-state phosphorylase *b* at the allosteric site. (Snape, P., Martin, J. L., and Johnson, L. N., unpublished results. With permission.)

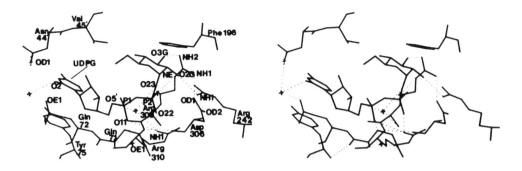

FIGURE 9. A stereo diagram of the contacts between UDP-Glc and T-state phosphorylase *b* at the allosteric site.

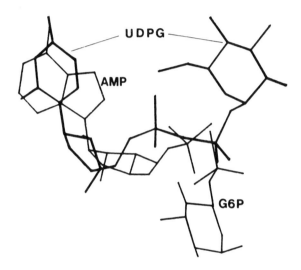

FIGURE 10. Comparison of the positions of AMP, glucose-6-P and UDP-Glc at the allosteric site of T-state phosphorylase *b*. (From Oikonomakos, N. G., Acharya, K. R., Stuart, D. I., Melpidou, A. E., McLaughlin, P. J., and Johnson, L. N., *Eur. J. Biochem.*, 173, 569, 1988. With permission.)

groups on the enzyme (Arg 309, Arg 310 and Arg 242) and the phosphates, a total of three hydrogen bonds to the glucose and a striking packing of the nonpolar part of the glucopyranose ring against the aromatic ring of Phe-196. In general, the results indicate that the conformational changes observed with UDP-Glc bound to the allosteric site of T-state GP*b* do not appear to be on the path towards the R-state.

4. Binding of P_i

In the GP*b*-P_i complex, phosphate binds at a site some 2 Å away from that observed for AMP and in this position it makes contacts to Arg 310 and to some extent to Arg 242.[258] There was no indication for movement of the side chain of Val 40′ or any of the other associated cap residues.

5. Binding of Glycerol-2-P

Glycerol-2-P a commonly used buffer in GP work and an inhibitor of the enzyme at high concentrations was shown to bind to the allosteric site of GP*b* with its phosphate position close to that observed for P_i and glucose-6-P.[202] In this position, glycerol-2-P is stabilized through interactions of its phosphate moiety with the guanidinium groups of Arg 309 and Arg 310 and the hydroxyl group of Tyr 75. The glycerol moiety was well located making a few van der Waals interactions with the side-chain carbon atoms of Tyr 75 and solvent.

B. BINDING STUDIES WITH T-STATE GP*a*

The structures of the GP*a*-AMP-glucose and GP*a*-ATP-glucose complexes at 2.5 Å resolution have been described by Sprang et al.[243]

1. Binding of AMP

Comparison with AMP binding to GP*b* shows some similarities and some differences.[244] The contacts to the phosphate are similar but the phosphate itself is in a different position in the two complexes. In GP*a* the ribose makes much more extensive interactions with the cap region and with Gln-71. Relative to its orientation in GP*b*, AMP bound to GP*a* is rotated by 45° about an axis perpendicular to the ribose plane and translated by 1.0 Å toward the interior of the protein. Tyr 75 undergoes a large conformational change to accommodate the base in a different orientation from that observed in GP*b*. There are also extensive changes in the positions and contacts of Arg 69, Gln 72, Asn 44', Asp 42'. Thus Asn 44', Asp 42' and Gln 71 all form hydrogen bonds with AMP in GP*a*, whereas in GP*b* only Asp 42' makes a contact to the protein. These are all consistent with the notion that T-state GP*a* is more characteristic of the activated (R) state of the enzyme since it makes more extensive contacts and binds nucleotide tightly while T-state GP*b* makes few contacts to AMP and binds the nucleotide weakly.[243] The key conformational changes involve several residues whose different positions are triggered by phosphorylation of Ser 14.[243,244]

2. Binding of ATP

ATP occupies the same site as does AMP but only the triphosphate group and one hydroxyl group (O3') make contacts with the enzyme (Arg 309, Arg 310 and Gln 71). ATP competes with AMP for the nucleotide site but does not stabilize the active form of the enzyme. In GP*a* ATP appears to adopt a binding mode that is similar to the interaction between GP*b* and AMP or ATP. The glucosyl conformation is *anti* in both AMP and ATP but the molecules fitted with a C2'-*endo* in AMP and a C3'-*endo* ribose pucker in ATP.

C. BINDING STUDIES WITH R-STATE GP*b*
1. Binding of AMP

As a consequence of the structural changes induced by sulphate binding to the Ser (P) site a high affinity AMP binding site is created in the R-state GP*b*.[8] The binding of AMP to GP*b* in the absence or presence of ammonium sulphate has been recently studied using ESR spectroscopic experiments.[155] In the absence of ammonium sulphate an apparent $K_d = 200$ μM was calculated with a Hill coefficient $n = 1.5$. However, in the presence of 0.9 M ammonium sulphate, AMP binds with a 50-fold higher affinity (apparent $K_d = 4.0$ μM). Also, in the presence of ammonium sulphate, the addition of 1 mM AMP did not change the aggregation state of the tetrameric enzyme.[155] The global tertiary and quaternary structures of the R-state GP*b*-SO$_4$-AMP complex (produced after soaking native R-state crystals in a solution containing 100 mM AMP and 1.0 M ammonium sulphate) are almost identical to those of unliganded R-state GP*b* and R-state GP*a*.[9] The interior of the allosteric site cleft becomes more packed following a small motion of the cap' and the binding of AMP. The R-state structure exhibits complementarity with the AMP in the side-chain positions of residues from the cap' and α2 and α8 helices.[9] The contacts are shown in Figure 11. On transition from the T state to R state, AMP is displaced towards the interior of the cleft by 2 Å and the ligand is rotated. The phosphate group is more firmly bound. It makes two

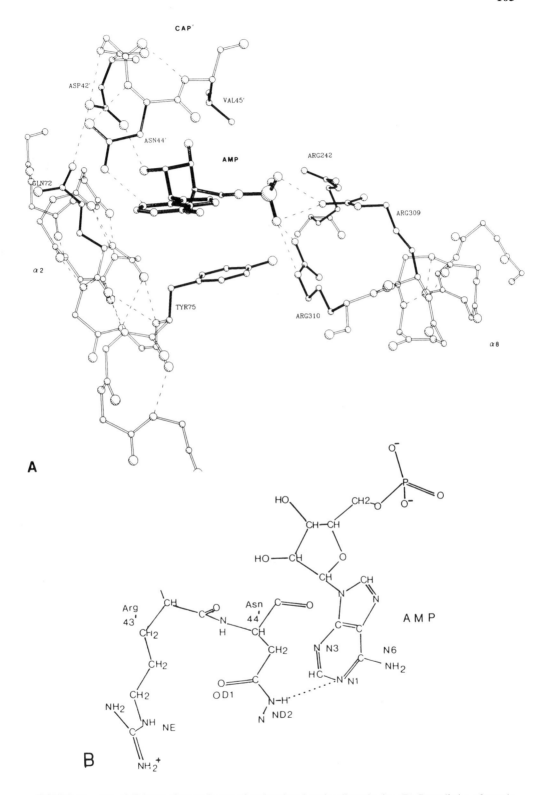

FIGURE 11. (A) AMP interactions to R-state phosphorylase *b* at the allosteric site. (B) Constellation of protein atoms around the N1 position of AMP in the R-state phosphorylase *b*-AMP complex. The observed hydrogen bond between N1 and ND2 of Asn 44', assumed to differentiate between the affinity of AMP and IMP (see text) at the allosteric site, is shown. (From Barford, D., Hu, S.-H., and Johnson, L. N., *J. Mol. Biol.*, 218, 233, 1991. With permission.)

contacts to Arg 309 and Arg 310 and one contact to the OH group of Tyr 75 (which undergoes a rotation of 70° about Cα-Cβ, turning in towards the cleft), while the distance between Arg 242 and the phosphate group is shortened from 5 Å in the T-state complex to 4 Å in the R state. Withers and Madsen[302] found that the monoanionic phosphate analogues of AMP, adenosine 5'-monophosphofluoridate and adenosine 5'-monophosphoamidate, could not activate GP*b*. Also the arsonomethyl analogue of AMP does not activate catalysis, whilst it competitively inhibits AMP activation.[288] In contrast, AMPS, in which the 5'-phosphorothioate group is more acidic than the 5'-phosphate group of AMP and thus exists solely as a dianion, is a more powerful activator than AMP.[192] The structure shows a hydrogen bond between O2' and Asp 42' but no contacts between O3' and the protein. This may explain why 3'-deoxy-AMP is as effective as AMP in binding and activation of GP*b*, whereas the 2'-deoxy-AMP binds with K_d value 43-fold higher than AMP and achieves a maximal velocity 59% that of AMP.[204] There remains the question of why AMP is a strong activator and IMP a weak activator.[142,187] Morange et al.[188] proposed that weak activators like IMP bind to GP*b* in the *syn* conformation, while strong activators like AMP in the *anti* conformation. Furthermore the *syn* conformation of weak activators changes into *anti* in the presence of substrates.[24] For T-state GP*b* crystals, the binding of these two nucleotides to site N appears to be identical.[289] In the R-state GP*b* bound AMP, N1 can form a hydrogen bond to the ND2 group of Asn 44' which is a hydrogen bond donor. IMP, in which the amino group at N6 is replaced with a carbonyl group and the N1 is protonated, could not make this hydrogen bond.[9] As discussed by Barford et al.[9] this appears to be a rather weak determinant to account for the different structural changes induced in the enzyme by the two nucleotides. In this respect, crystals of GP*b* in the presence of ammonium sulphate and IMP have been obtained[290] and will be analyzed to investigate in detail the site of interaction of R-state GP*b* with IMP.

2. Binding of SO_4^{2-}

The sulphate bound in the allosteric site of R-state GP*b* is in a position close to the phosphoryl subsite of AMP and stabilized by the three arginines Arg 242, Arg 309 and Arg 310.[8] In the R-state GP*b*-SO_4^{2-}-AMP complex, this position is occupied by the phosphate of AMP and binding of sulphate is not observed.

VI. THE GLYCOGEN STORAGE SITE

Early kinetic studies were consistent with the notion that there are two glycogen binding sites per GP subunit, the catalytic and a high affinity separate binding site. Preincubation of GP with glycogen enhances activity[263] and promotes dissociation of inactive tetramers to active dimers,[182] indicating the existence of a separate glycogen binding site distinct from the catalytic site. This hypothesis was supported by the discovery that a substantial amount of GP is tightly bound to glycogen particles, isolated from muscle, together with substantial amounts of most of the enzymes involved in glycogen metabolism;[22,23,29,92,93,96,184] GP intimately associated with the glycogen particles was able to be activated and controlled. Crystallographic binding studies with GP*b* and GP*a* have confirmed the existence of a separate glycogen binding site located some 30 Å from the catalytic site.[134,269] In a detailed kinetic study, Kasvinsky et al.[134] showed that the dissociation constants for maltoheptaose, a seven α-1,4 linked oligomer of α-D-glucose and a substrate of the enzyme, at the glycogen storage site and the catalytic site were 1 m*M* and 22 m*M,* respectively. Since the former is similar to the dissociation constant for glycogen,[66] the kinetically calculated K_m values for glycogen should largely reflect binding at the glycogen storage site. It appears that the high affinity binding site is filled first in the catalytic cycle and that prior occupation of this site is an obligatory part of the activation process and the kinetic mechanism. Its proposed role in muscle GP is to serve as an attachment point to the glycogen particles *in vivo* to produce an effective high concentration of end groups and to act as an additional control site whereby occupation by glycogen oligosaccharide results in an increase in the rate of catalysis (positive heterotropic interactions between the glycogen and catalytic sites).[122,134,216,236]

A. BINDING OF OLIGOSACCHARIDES

The site is located on the surface of the molecule, well removed from other sites and subunit-subunit interface. In GP*b*, maltoheptaose can bind to this site by filling 5 major and 2 minor subsites.[122] Binding energy for the stabilization of the enzyme-oligosaccharide complex derives from the formation of extensive hydrogen bonding networks between the oligosaccharide and protein side-chains and main chain atoms and nonpolar interactions between the oligosaccharide and solvent-accessible hydrophobic residues. The major site consists of five glucosyl residues labeled S3-S4-S5-S6-S7 and is made up principally of α12 and α13 helices (residues 396 to 418 and 420 to 429, respectively) and the prominent loop connecting the two antiparallel strands β15 (residues 430 to 432) and β16 (residues 437 to 411). These α-1,4-linked glucosyl residues of maltoheptaose ligand adopt a left-handed helical secondary structure with the two ends of the helix (S3 and S7) curling away from the protein surface. This conformation is also stabilized by a series of intramolecular O2-O3′ hydrogen bonding between adjacent sugars. The minor site consists of only two glycosyl residues S8-S9 and it lies above the nonreducing end of the major site making polar interactions, through water molecules, to the top of the α12 helix, the loop of antiparallel β sheet from β8 to β9 and one nonpolar contact to Val 354 from α9. The major conformational changes in the protein on binding oligosaccharide involve residues Glu 433 and Lys 437, which move to allow optimal contact with the oligosaccharide in subsites S4 and S5. Small side-chain movements are also observed for the residues Gln 401, Gln 408 and Arg 409.[122] In GP*a*, the structures of the inhibited GP*a*-maltoheptaose-caffeine-glucose and activated GP*a*-maltopentaose-P$_i$ complexes revealed seven glucosyl residues (S1 to S7) bound at the major site and four glucosyl residues ordered (S8-S9-S10-S11) at the minor site.[75,76,78] In contrast, maltopentaose in the complex GP*a*-maltopentaose-caffeine-glucose is well-ordered at only S4-S5-S6, while only one single glucosyl residue is observed in the minor site (S10),[78] suggesting that occupancy of the minor site is induced either by formation of the R state, or by long chain oligosaccharides. In general, the contacts observed for oligosaccharide binding to the major site of GP*a* are similar but not identical to those observed in GP*b*-maltoheptaose complex. However, the conformational changes in GP*a* are more extensive than in GP*b*. For example, Arg 426 is displaced by S3 and a salt pair between Arg 426 and Asp 423 present in the parent structure is lost. The structural changes induced by P$_i$ and maltopentaose in the GP*a*-maltopentaose-P$_i$ complex involve local conformational changes to the major and minor sites, domain separation and significant changes in the subunit interface.[78,79] Model-building studies[179] have shown that the minor site might be consistent with position of oligosaccharide at a α-1,6 branch position. This result has implications for the increased affinity of GP for branched polysaccharides like glycogen and amylopectin over linear glucans like amylose and maltodextrins.[69,107] The contacts between the oligosaccharide and protein as deduced from the refined structure of the GP*b*-AMP-maltoheptaose-heptulose-2-P complex[127] are shown in Figure 12. The specificity of the glycogen storage site has been discussed.[122,127] Glucose and glucose-1-P do not bind to the site in the crystals of GP*b*. Maltose, maltotriose, maltotetraose and maltopentaose bind respectively to subsites S4-S5, S4-S5-S6, S3-S4-S5-S6, S3-S4-S5-S6-S7, indicating that at least a disaccharide is required for recognition at the storage site and that the central S4-S5-S6 subsites are stronger than those at the termini. There was no binding of oligosaccharides to the catalytic site. In the absence of binding of oligosaccharides to the catalytic site it is difficult to describe the mechanism by which glycogen promotes the binding of substrates

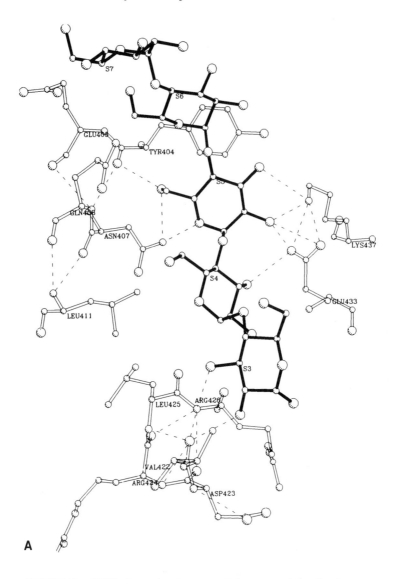

A

FIGURE 12. (A) The interactions between maltoheptaose and phosphorylase *b* at the major glycogen storage site in the phosphorylase *b*-AMP-maltoheptaose-heptulose-2-P complex. Only 5 sugar subsites labeled S3 to S7 are occupied. (B) The nonpolar residues that contribute to the oligosaccharide recognition site. (From Johnson, L. N., Acharya, K. R., Jordan, M. D., and McLaughlin, P. J., *J. Mol. Biol.,* 211, 645, 1990. With permission.)

through positive heterotropic interactions between the catalytic site and the storage site.[134,216] However, a possible pathway for the transmission of conformational changes, between the two sites, seen in the structures of the activated GP*a*-maltopentaose-P$_i$ and GP*b*-AMP-maltoheptaose-heptulose-2-P complexes has been suggested.[78,127] The movement of Glu 433 and Lys 437 promoted by oligosaccharide binding at the storage site could be correlated with opening of the catalytic site channel through the hydrogen-bonding network Glu 432 . . . Arg 386 . . . O of Leu 380. The adjacent Glu 382 forms a salt bridge with Arg 770 in the domain interface which stabilizes the closed conformation in the T state. The van der Waals interaction between Glu 382 and Phe 286 is destroyed because of the significant shift in the side-chain position of Phe 286 or loss of the inhibitor site (Phe 285-Tyr 613) induced by heptulose-2-P and P$_i$, respectively.

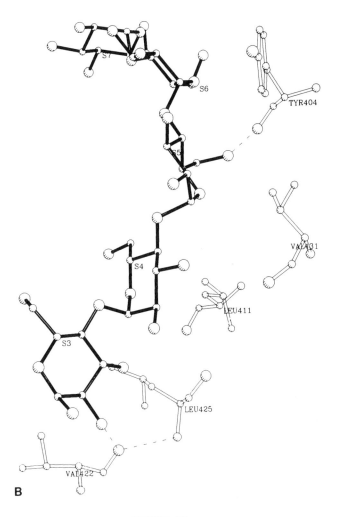

FIGURE 12B.

B. BINDING OF ACARBOSE

Acarbose, a pseudo-maltotetraose with the cyclitol unit α-1,4-linked to a 4-amino-4,6-dideoxyglucose residue, is a potent inhibitor of many α-glucosidases, α-amylases and sucrase-isomaltase[254] and considered to be an analogue of a glucosyl cation like transition state. For example, it is a powerful inhibitor of the rabbit small intestinal sucrase (K_i = 0.47 μM at pH 6.8).[91] Binding studies in the crystal of GPa[77] and GPb[6] showed that it bound at the glycogen storage site in an analogous way to maltoheptaose (but with a slight difference in conformation for the valienamine ring at the nonreducing end) by occupying the S4-S5-S6-S7 subsites and not the S3-S4-S5-S6 subsites as would expected if it were to bind like maltotetraose because ring B, which has no hydroxyl at C6, is unable to make a hydrogen bond with the main-chain carbonyl O of Tyr 404 at the subsite S5. Kinetic studies with GPa have shown that acarbose is a reasonable inhibitor, K_{iapp} = 26 mM, when assayed in the direction of glycogen degradation at saturating concentrations of P_i and 30 mM maltopentaose (K_m = 40 mM) but no inhibition could be detected in the reverse direction (with 6 mM glucose-1-P and 10 mM maltopentaose).[77]

In the tetrameric R-state GPb the interface that is formed because of the association of the dimers to tetramers involves contacts between residues from the glycogen storage site (prominent loop representing the β15-β16 turn) and from the bundle of helices α23, α24, α26 residues 724-767.[8] The involvement of a key residue from the glycogen storage site (Glu 433) in a hydrogen bond at this interface explains why glycogen promotes dissociation of tetramers to dimers.[109,182,263] Binding studies of oligosaccharide to R-state GP crystals are in progress and the results of these studies should provide further information on the role of the glycogen storage site.

VII. THE INHIBITOR (NUCLEOSIDE) SITE

In addition to the allosteric (nucleotide) site, T-state GP has an inhibitor (or nucleoside) site near the catalytic site and domain interface. The existence of two binding sites per GP subunit for AMP (or IMP) was originally proposed based on the results from calorimetric,[105,267] equilibrium dialysis,[187] and spectroscopic studies.[197] The inhibitor site is a hydrophobic binding pocket of relatively low specificity. It is located at the entrance to the catalytic site and comprises residues from both domains 1 and 2. In the T state, Phe 285 is stacked close to Tyr 613 and together these two hydrophobic aromatic residues form the inhibitor site. In addition to AMP, purines such as adenine and caffeine, nucleosides such as adenosine and inosine, nucleotides such as IMP and ATP, NADH and certain related heterocyclic compounds such as FMN (flavin mono-nucleotide) have been shown to bind at this site.[132,133,242,252] Ligands occupying this site are all inhibitors competitive with glucose-1-P, they act in synergism with glucose and dissociate the tetrameric forms of GP[132,133,247,271] by stabilizing the T-state conformation. Thus, caffeine competitively inhibits both GP*b* and GP*a* activity with respect to glucose-1-P (K_i = 0.1-0.2 mM) and functions with glucose in a strongly synergistic mode, with an interaction constant of α = 0.2.[133,168] The notion that purines and their analogues inhibit GP*a* synergistically with glucose by stabilizing the T-state structure was also supported by cross-linking experiments.[44,88] Crystallographic analysis by the difference Fourier method showed that purines and purine nucleotides stabilize the T-state conformation by intercalating between the aromatic side chains of Phe 285 and Tyr 613 at the domain interface.[221,242,250] The ligands exhibit wide variation in the orientation of the base but little variation in the interplanar ring separation from the stacking plane. There are no strong hydrogen bonds to the protein. No contacts are made to the ribose or phosphate(s) moieties with the density being quite poor, and hence these groups do not seem to be important for binding. Sprang et al.[242] from binding and thermodynamic analysis have concluded that the major source of binding energy derives from the three-ring stacking interaction itself. In support of this, Soman and Philip[235] have shown that the ability of aromatic compounds to inhibit GP is related to their hydrophobicity. Binding at the inhibitor site (which is some 10 Å from the catalytic site) is essentially inhibitory since it prevents binding of the substrate by anchoring the 280s loop. Because this loop, containing Phe 285, must move to accommodate substrates or R-state substrate analogues, purine inhibitors and glucose-1-P or R-state substrate analogues show mutually exclusive binding (competition). This has been demonstrated kinetically for the pairs IMP and glucose-1-P,[14] caffeine, and glucose-1-P, caffeine and UDP-Glc and IMP and UDP-Glc.[132,133,168,200]

A physiological role for this site in the muscle has not yet been demonstrated. In contrast, liver GP*a* shows a more stringent selectivity for inhibitors.[133] It has been suggested[132] that the inhibitory effect of insulin on liver GP*a* activity in cells (Witters and Avruch)[280] may be mediated, in response to changing glucose concentrations, by an as yet unidentified compound which may interact with the inhibitor site (also discussed by Johnson et al.,[124] Newgard et al.[196]).

The rearrangement of the active site 280s loop reveals the structural basis for the negative cooperativity observed between purines and substrates.[132,133,242] Withers et al.[274] have shown that binding of glucose-1,2-P to GP*a* disordered the 284-286 loop. The binding of UDP-Glc to the catalytic site promotes extensive and concerted conformational changes[200] and a striking feature is the occurrence of continuous negative density over the main and side

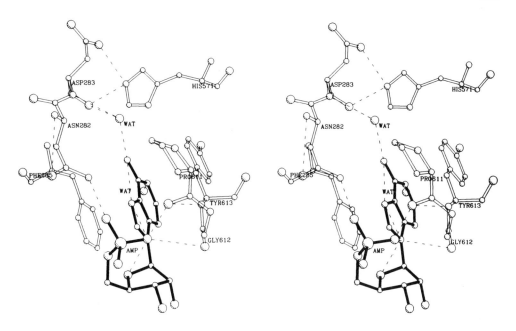

FIGURE 13. A stereo diagram of the inhibitor site of T-state phosphorylase *b* showing AMP bound between Phe 285 and Tyr 613.

chains of the loop 279 to 289 (Figure 27), which links the catalytic site to the inhibitor site, indicating displacement of this loop and abolition of Tyr 613. In the GP*a*-maltopentaose-P$_i$ complex,[79] displacement of the active site gate by P$_i$ disorders Phe 285 and the purine effector site is destroyed. In the native T-state GP*b* and glucose bound structure, Phe 285 stacks in an approximately planar orientation, some 7 Å away from Tyr 613.[174] However, when glucose-1-P and R-state phosphorylated ligands such as glucose-1-methylene phosphonate and 2-deoxy-2-fluoro-glucose-1-P bind at the catalytic site the induced 280s loop motion causes Phe 285 to slide past Tyr 613 and the final interaction is perpendicular.[174] This position would prevent inhibitors from binding at the inhibitor site.

A. BINDING OF AMP AND IMP

In a recent crystallographic experiment with 100 m*M* AMP[9] the structure of AMP bound to the T-state GP*b* at 2.2 Å resolution showed that at the inhibitor site, the adenine moiety is observed sandwiched between the planar side-chains of Phe 285 and Tyr 613 (Figure 13). A water mediated hydrogen bond occurs between the N6 amino group and main-chain N and O of Asp 283. The ribose and phosphate groups project into solvent and density extends only partially over the ribose, with weak density corresponding to the phosphate group, implying that these groups are mobile. In the conformation resulting from the complex refinement the phosphate group forms a hydrogen bond to the side-chain of Asn 282 and to main chain O of Pro 611. No conformational changes at the inhibitor or neighbouring catalytic site were observed. Structural results[259] suggest that the binding of IMP to the inhibitor site appears to be largely similar to that of AMP. However, for IMP there is a clear indication that His 571, situated approximately equidistant from the catalytic and inhibitor sites, moves away from the IMP base. This does not appear to happen for AMP. No contacts are made to the ribose or phosphate with the density being quite poor, and hence these groups do not seem to be important for binding.[259] IMP at 25° C binds to GP*b* with a slightly greater affinity to this site (K_d = 1.7 m*M*) than AMP (K_d = 3.1 m*M*).[176] Kinetic experiments for the inhibition of GP*a* (with respect to glucose-1-P) showed a K_i = 1.5 m*M* for IMP and K_i = 6.0 m*M* for AMP.[132] Uhing et al.[256] have shown that in the presence of organic solvents, which can activate GP through stabilization of its most active conformation (Uhing et al. [257]) the binding of IMP at the inhibitor site is weakened.

In the tetrameric R-state GP*b* there is no visible electron density to account for residues 282 to 286 and it is assumed that these residues become mobile.[8] Disorder of the 280s loop suggests that the gate to the catalytic site is removed, allowing oligosaccharides to bind and perturbing the inhibitor site, thus abolishing inhibition by nucleosides. His 571 rotates about its Cα-Cβ bond, following rupture of a hydrogen bond to Asp 283 made in the T state, and forms a new hydrogen bond to Tyr 613. Therefore as the evidence suggests that the inhibitor site does not exist in the R state, the task of distinguishing between AMP and IMP falls upon the allosteric (activator) site in the R state.

VIII. THE CATALYTIC SITE

A. ROLE OF PYRIDOXAL 5'-PHOSPHATE

The mechanism by which GP uses its essential cofactor, PLP, has long been a mystery. GP contains 1 mole per mole of subunit PLP[5,128] linked through its 4-aldehyde group to the enzyme (Lys 680) by a Schiff base.[61] The coenzyme is indispensable for activity; its removal from the enzyme molecule following a prior distortion with deforming agents such as imidazole citrate leads to total inactivation[228] but the apoenzyme can be fully reactivated by reconstitution with PLP.[94] Reduction of the Schiff base with $NaBH_4$ results in an enzyme derivative which still retains activity.[61] All other classical B_6 enzymes involved in the metabolism of amino acids, notably the transaminases, where PLP is directly involved in catalysis are totally inactivated by this reduction.[64] Reconstitution experiments with a great number of PLP analogues[26,28,129,142,212,214,215,229,231,251,262,273,282] have demonstrated that the functional groups at positions 1, 2, 3, 4 and 6 of PLP are not essential for catalysis but only the 5'-phosphate group of the coenzyme can play an important role in GP mechanism of action. Moreover, only PLP analogues with a protonatable dianion such as $-OPO_3^{2-}$ or $-CH_2PO_3^{2-}$ in position 5 were able to reconstitute an active enzyme. On the contrary, in the classical B_6 enzyme aspartate transaminase whose crystal structure has been determined[140] the 5'-phosphate group is acting mainly as an anchor point.

In GP, ^{31}P-NMR studies have shown that the 5'-phosphate group of PLP is a monoanion in the inactive T conformation (with a chemical shift of about 3 ppm upfield) and a dianion in its activated R conformation (with a chemical shift to about 0 ppm downfield)[57,101,143,205] and that upon binding of substrates the coenzyme phosphate dianionic form is most likely either in a partially protonated monoanionic form or as a tightly coordinated and distorted dianion.[170,272,291] These studies combined with biochemical and crystallographic evidence led to two main proposals for the role of the coenzyme 5'-phosphate in the catalytic mechanism, i.e., that it functions (1) as a proton donor-acceptor shuttle in general acid-base catalysis or, (2) as an electrophilic "constrained" dianion. In the acid-base mechanism, it is proposed that upon formation of the ternary complex the 5'-phosphate becomes protonated and functions as an acid serving to promote through a direct hydrogen bond interaction, general acid attack of the substrate phosphate on the α-Cl-Ol glycosidic bond leading to a carbonium ion intermediate. The carbonium ion is then stabilized by the substrate phosphate dianion itself, and the reaction is completed by nucleophilic attack of the 4-OH of the terminal glucose of the acceptor polysaccharide on the carbonium ion.[89,114,117,124,144,178,208] In the electrophilic mechanism, when the substrate is bound, the coenzyme 5'-phosphate is tightly coordinated by positive charges and constrained into a trigonal bipyramidal configuration, where the empty apical position is carrying a positive charge; this position is attacked by a nucleophilic O of the substrate phosphate and the resulting pseudo-pyrophosphate bond is expected to withdraw electrons from the substrate phosphate and thus labilize the glycosidic bond.[70,170,251,273] The proposals require the coenzyme 5'-phosphate and the substrate phosphate to be directly interacting. The interaction of the two phosphates was first noted by Parrish et al.[212] from kinetic studies with pyridoxal (PL) reconstituted GP. This enzyme derivative can be significantly activated by noncovalent anions such as phosphite, fluorophosphate, phosphate and some analogues,[25,212] indicating that these anions may bind at the subsite occupied by the coenzyme phosphate in the native GP. The observation that phosphite activation of PL-GP*b* can be reversed by pyrophosphate, a potent inhibitor of the enzyme strictly competitive with both anion and glucose-1-P,[212] implied the adjacent positions of activator anion and substrate phosphate in the catalytic site. Another convincing evidence in favour of this hypothesis came from studies with pyridoxal pyrophosphate glucose reconstituted GP.[251,273] This enzyme derivative was shown to catalyse glucosyl transfer from the bound coenzyme analogue to oligosaccharide, suggesting a similar binding mode of the substrate glucose-1-P in the native GP. The interacting "phosphates hypothesis" was confirmed crystallographically by McLaughlin et al.[178] in the analysis of the transition-state intermediate-like heptulose-2-P-GP*b* complex which showed a close P-P distance (of phosphorus atoms of phosphate to cofactor phosphate) of 4.8 Å. The acid-base mechanism is supported by the direct experimental proof of protonation of glycosylic substrates, compounds of nonglycosidic structure with the anomeric C1 linked via an electron-rich bond.[142-144,206,208,291] Thus it has been shown that an enzyme-derived proton attacks *endo*- or *exo*-cyclic double bond of D-glucal or heptenitol and the activated sugar moiety is subsequently transferred to an acceptor.[291] A main criticism against proton donor-acceptor function for the 5'-phosphate of the cofactor is based upon the finding that fluorophosphate ($pK_a = 4.8$) which induces significant activity in PL-GP cannot become protonated at pH 6.8 where the enzyme is active.[212] The pH dependencies of the PL-GP*b*s using phosphite and fluorophosphate are also unaltered, despite the difference in the pK_a values of the activator anions.[276] ^{19}F-NMR spectra of the bound fluorophosphate activator also showed that no change in its ionization state occurred on binding or during catalysis.[25] However the possibility exists that PL-GP (+ activator anion) may follow a different mechanism by using an amino acid side chain as an acid-base group.[143] On the other hand, recent kinetic and ^{19}F-NMR experiments of GP*b* reconstituted with 6-fluoro-pyridoxal[28] suggest that the cofactor phosphate may also have a structural role in holding other groups in the correct orientation for catalysis.

A detailed stereochemical understanding of how the coenzyme interacts with the enzyme active site is therefore of key importance in considering possible catalytic mechanisms. A different approach to unravel the role of coenzyme in GP was by using GPs reconstituted with synthetic analogues of PLP in place of natural coenzyme for crystal structure analysis.[199,203] The procedure consisted of the following stages: (1) removal of PLP, (2) reconstitution of apoenzyme with the analogue, (3) crystallization of the modified enzyme derivative, and (4) crystal structure analysis at 2.4 Å resolution. The analogues that gave appropriate crystals were 5'-deoxypyridoxal 5'-methylenephosphonate (PDMP), 6-fluoro-pyridoxal 5'-phosphate (FPLP) and PL. Crystals (Figure 3) isomorphous to the native tetragonal ones were readily obtained under conditions that favoured T-state GP*b*. Their three-dimensional structure has been determined.[199,203] The interactions of PLP with the enzyme will be described in some detail before the interactions of PLP derivatives are examined.

1. The Interactions of Coenzyme with GP*b*

The PLP is buried in the center of the molecule at a site some 15 Å from the surface where domains 1 and 2 come together and is close to the substrate binding site. The finding is in good agreement with deductions made previously by spectroscopic studies[39,40,55,71,126,260] which had indicated the environment of the cofactor to be highly hydrophobic and distant from the subunit-subunit interface. The electron density for the PLP molecule as bound to T-state GP*b* is shown in Figure 14. The main chain atoms of Lys 680, the plain of the pyridoxal and the positions of the phophate oxygens are very well defined. The cofactor is

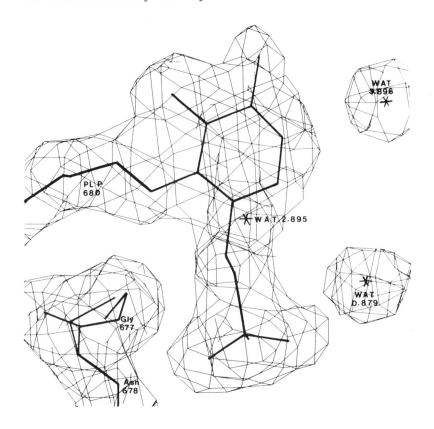

FIGURE 14.   Electron density for pyridoxal 5′-phosphate (PLP) bound to phosphorylase *b*. The map was calculated by using as coefficients $2F_o$-F_c. (From Acharya, K. R., Stuart, D. I., Varvill, K. M., and Johnson, L. N., World Scientific Publishing, Monogr., 1991. With permission.)

approximately 90% buried with the 5′-phosphate partly surrounded by water molecules and only one of the 5′-phosphate oxygen atoms (O22) accessible to the bulk solvent.[121,221-223] There are 3 aromatic rings (Figure 15) in the vicinity, Tyr 648, Tyr 90 and Trp 491, of which only the OH of Tyr 648 can interact indirectly with O5′ of PLP through water OH2.895. Lys 680 is located on the short α21 helix (residues 676 to 686) which is equivalent to the αE helix of the nucleotide binding domain.[220] The N1 atom interacts via a water molecule to the hydrogen bond network OH3.896...OD1 Asn 133... NH1 Arg 569 and makes a few contacts to Gly 134 and Tyr 648. The role of the pyridine N1 in the GP reaction has been rather difficult to establish.[18] N-methyl-PLP did not bind to apoenzyme[229] indicating a significant steric effect due to the presence of a -CH₃ group in the coenzyme. Recent results with GP*b* reconstituted with FPLP have shown that the enzyme is active with the N1 in the neutral form and that the state of ionization of the N1 is unlikely to be affected by the binding of substrates.[26,27] The C2′ methyl group makes a great number of close van der Waals contacts with protein. This feature could explain the inability of ω-methyl-pyridoxal 5′-phosphate (where a larger group, a -CH₂CH₃, has been introduced in position 2) to activate apoenzyme, while 2-nor-pyridoxal 5′-phosphate (in which the -CH₃ is replaced by an H-atom) can activate apoenzyme by 65%.[229] The Schiff base is planar and *trans* with torsion angles CE-NZ-C4′-C4 and NZ-C4-C4′-C5 179° and -178°, respectively. There is a hydrogen bond between the 3′-OH and the Schiff base nitrogen. Methylation of 3′-OH results in partial activity even though the 3′-OH...NZ of Lys 680 hydrogen bond must be disrupted. However, since the aliphatic side chain of Lys 680 is relatively flexible, any conformational changes could be "buffered".[221] The C6 position is relatively open with van der Waals

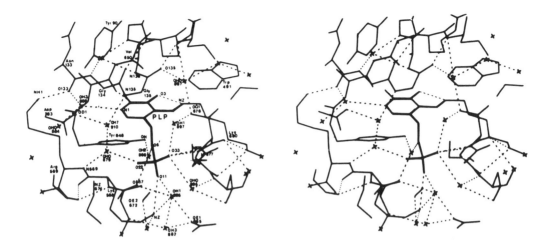

FIGURE 15. A stereo diagram showing the interactions between pyridoxal 5'-phosphate (PLP) and T-state phosphorylase *b* residues of the catalytic site. Hydrogen bonds less than 3.3 Å are indicated. Water molecules are shown as crosses. (Reprinted with permission from Oikonomakos, N. G., Johnson, L. N., Acharya, K. R., Stuart, D. I., Barford, D., Hajdu, J., Varvill, K. M., Melpidou, A. E., Papageorgiou, A. C., Graves, D. J., and Palm, D., *Biochemistry*, 26, 8381, 1987. Copyright 1987 American Chemical Society.)

contacts only to Gly 134 and three water molecules. The torsion angles for the phosphate ester, C4-C5-C5'-O5' = 94° and C5-C5'-O5'-P = 167°, are respectively *trans, trans* with the C5'-O5' bond approximately perpendicular to the pyridoxal ring. This contrasts with the PLP of aspartate aminotransferase in which the 5'-phosphate group is in a "strained" *cis, trans* conformation.[68] Phosphate oxygen O11 interacts with the basic group NZ of Lys 568. Phosphate oxygen O22, the most exposed, interacts via a water molecule to the main chain NH of Arg 569 and via a second water to Asp 283 and main chain NH of Gly 135. This oxygen makes a hydrogen bond to the product phosphate in the enzyme-heptulose-2-P.[89,127,178] The third oxygen O33 makes two direct contacts to the main chain NH groups of Thr 677 and Gly 677 at the start of the α21 helix and two more contacts with two water molecules which are involved in additional hydrogen bond networks. The contacts to the phosphate of the PLP are esssentially the same in the R-state GP*b*.[8] The details of the coenzyme contacts to the enzyme are shown in Figure 15. The structural results suggest that the 5'-phosphate of the bound cofactor, due to its interactions with a basic group (NZ of Lys 568), the helix dipole of the α21 helix and Asp 283 indirectly through water molecules[199] remains in a monoanionic state. Other ionic groups in the vicinity are Lys 574 (4.5 Å) and Glu 672 (4.8 Å), where distance refers to the closest approach of charged group to the cofactor 5'-phosphate oxygens. Asp 283 (6.6 Å), however, interacts with the cofactor 5'-phosphate oxygens by a route that involves two water molecules. The presence of Asp 283, which is also hydrogen bonded through its other oxygen to His 571 and Asn 284, would therefore tend to diminish the net positive charge. On activation by heptulose-2-P[89] the acidic group of Asp 283 is replaced by Arg 569 which would favour the dianionic state. Similarly, in the ammonium sulphate activated GP*b* (R-state),[8] Asp 283 is displaced by Arg 569, and the additional positive charge favours the dianionic phosphate of the coenzyme. On ligation of sulphate (or phosphate) the 5'-phosphate reverts to a monoanion. The structural changes provide therefore an explanation for the change in state of ionization of the PLP phosphate as a result of activation.[101] All the protein residues mentioned above are also conserved in *E. coli* maltodextrin phosphorylase[207] and potato phosphorylase.[193] This similarity, in addition to the ^{31}P-NMR evidence[208] and recent kinetic evidence,[279] leads one to assume that all the three α-1,4-glucan phosphorylases share the same catalytic mechanism.

2. The Interactions of Coenzyme Analogues with GP*b*

PDMP-GP*b* is the only enzyme derivative that possesses significant activity (25%) despite the fact that the cofactor phosphate has been replaced by phosphonate.[262] ^{31}P-NMR experiments[143] indicate that, in contrast to the 5′-phosphate in native enzyme, the 5′-phosphonate group is dianionic in the T state and monoanionic in the R state. This reversal cannot be attributed to a change in pK_a values of the two cofactor phosphates because pK_a for PLP is 6.2 and pK_a for PDMP is 7.35. In the structure of PDMP-GP*b*, as a result of the different geometry of the phosphonate linkage (C-P and O-P bond lengths are 1.8 and 1.5 Å, respectively) and movement of Lys 574, the phosphonate oxygens are placed over 1 Å closer to the NZ of Lys 574. Thus, the NZ of Lys 574 comes close to the 5′-phosphonate oxygen O22 (3.4 Å), a feature different from that in the native enzyme, where the separation is 4.5 Å. This closer proximity could explain why the ionization in this derivative is reversed. The substitution of the polar O5′ by the nonpolar -CH$_2$-group caused movements of the water OH2.895 and its hydrogen-bonded OH of Tyr 648 by 0.9 and 0.6 Å respectively.[199] The pyridoxal ring was also translated about 0.4 Å to follow the movement of the tyrosine. In general, the protein appeared to be able to accommodate these shifts, without any further adjustments.

FPLP has been used as a sensitive reporter molecule in ^{19}F NMR studies.[26,27] FPLP-GP*b* exhibits approximately 28% activity compared with the native enzyme but essentially no change in other properties such as pH optimum, reconstitution energy, K_m values for glucose-1-P and AMP and cooperativity. The electron density for the fluorine atom was easily identified in the difference Fourier synthesis for FPLP-GP*b*[199] demonstrating the power of X-ray analysis to distinguish a single F atom among over 7000 protein and water molecules. In this enzyme derivative no significant conformational changes were found. The extra F atom can be accommodated without any steric hindrance. It makes van der Waals contacts to Gly 134, Gly 135 and 3 water molecules. The pyridoxal ring has altered its position slightly but the shifts were too small to change the van der Waals contacts or hydrogen bonds. The presence of the electronegative F atom in position 6 changes the pK_a of the ring N1 from 8.7 in free PLP to less than 1 in FPLP and Chang and Graves[26] suggested that the lower activity of FPLP-GP*b* could be attributed to a change in the interactions of N1. The N1 atom makes no direct polar contacts with protein atoms. It participates however in a hydrogen bond network through OH3.896 to OD1 of Asp 133 and NH1 of Arg 569 (Figure 17). Movement of the Arg 569 side chain is a key feature in the R-state enzyme-heptulose-2-P complex.[89,127] The lowered pK_a of N1 in the FPLP-GP*b* could lead therefore to a strengthening of this network, making it less easy for the arginine to move.

PL-GP*b* is an allosteric enzyme whose R state is stabilized by pyrophosphate (a powerful inhibitor) and T state is stabilized by phosphite and glucose.[275] In PL-GP*b*-cocrystallized with phosphite anion and glucose, the anion binds to a site that is close to but separate from the 5′-phosphate site of the coenzyme in the native GP*b* (P-P separation = 1.6 Å) (Figure 16). It is remarkable that the enzyme can tolerate a shift of 1.6 Å in an essential catalytic group and exhibit a reduction in rate by a factor of about 5.[25,212] Although some of the contacts to the enzyme are similar to the 5′-phosphate of PLP, the anion is significantly closer to Lys 574 and Glu 672 and also interacts with glucose via solvent. Further work on the structure of PL-GP*b* cocrystallized with phosphate or fluorophosphate and glucose to 2.4 Å resolution has been recently presented.[203] The positions of the peaks were similar but not identical to those observed in the difference Fourier synthesis for the PL-GP*b*-phosphite-glucose complex. Furthermore, the double difference maps (against the latter complex) showed strong positive peaks suggesting where one of the phosphate oxygen or fluorine might be located. The crystallographic results show that there exists a degree of flexibility that is not in accord with the electrophilic hypothesis for catalysis which requires a more intimate association of the 5′-phosphate (or anion) with the enzyme.

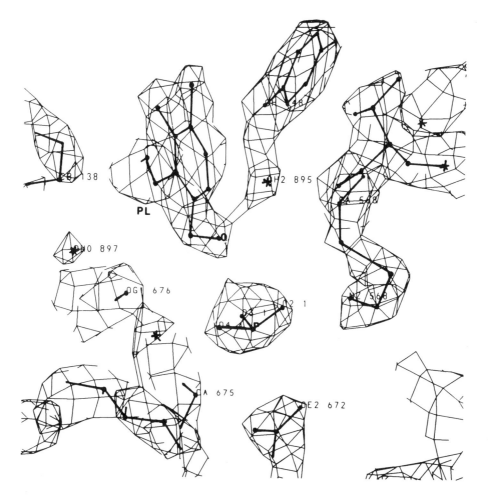

FIGURE 16. Electron density in the region of the catalytic site of pyridoxal (PL)-phosphorylase *b*-phosphite-glucose complex, where phosphite is bound. The map was calculated by using as coefficients $2F_o$-F_c. The X-PLOR refined coordinates for the PL-phosphorylase *b*-phosphite-glucose complex are shown. (N. G. Oikonomakos et al., unpublished.)

No information is available concerning the environment of the modified cofactors in the R-state enzyme. Crystals of GP*b* reconstituted with pyridoxal 5′-diphosphate (an R-state stabilizing cofactor analogue) have been obtained by Withers et al.[275] in the presence of AMP and maltopentaose (space group $P2_12_12$) and a structural analysis is in progress. R-state crystals of GP*b* reconstituted with pyridoxal, 5′-deoxypyridoxal and pyridoxal 5′-diphosphate in place of the natural cofactor have been recently produced.[202] The crystals are isomorphous to those of the native tetrameric R-state GP*b*.[8] They diffract X-rays to a resolution of at least 2.8 Å and are suitable for X-ray structure analysis. The results of the X-ray analyses of the enzyme derivatives would help to understand the molecular events of these modifications and allow a comparison with those of the T-state GP*b* derivatives.

B. CHEMICAL MECHANISM

An understanding of the catalytic mechanism of an enzyme requires the description of the various enzyme-intermediate complexes and transition states, that are formed during the course of the reaction.[90,157] The nature of possible intermediates in the GP reaction is still under investigation, in spite of extensive studies on the mode of catalysis of this enzyme. In the absence of the second substrate, glycogen, there is no isotope exchange between glucose-1-P and labeled P_i[33] or positional isotope exchange between the peripheral and ester oxygens of glucose-1-P[74] in the presence of the enzyme. Thus no observable breakage of the C1-O bond of glucose-1-P occurs in the binary complex with GP. Supportive evidence for a glucosyl-enzyme intermediate exists and comes from studies on potato phosphorylase. The finding by Kokesh and Kakuda[145] of an exchange of the sugar ester oxygens and the phosphoryl oxygens of glucose-1-P in the ternary complex with cyclodextrin could be interpreted in favour of a covalent glucosyl-enzyme intermediate. Klein et al.[141] have provided evidence for some kind of glucosyl-enzyme intermediate to be formed in the reaction between potato phosphorylase and D-glucal. No evidence for such an enzyme-bound intermediate has been obtained for rabbit muscle GP. A secondary kinetic isotope effect ($k_H/k_D = 1.1$) has been observed in the GP reaction[255] consistent with a carbonium ion intermediate, but later studies by Firsov et al.[58] gave a ratio of k_H/k_D or k_H/k_T of 1.0 with glucose-1-P, C-1[H], C-1[^{2}H] and C-1[^{3}H] a result consistent with a covalent intermediate. The absence of a significant α-deuterium kinetic isotope effect on GP reaction does not however rule out a mechanism involving oxonium-ion-like transition states since a different step may be rate-determining.[274] The possible value of glucose analogues with a half-chair conformation in elucidating the catalytic mechanism was first pointed out by Tu et al.[255] and Gold et al.[73] who reported that 1,5-gluconolactone is a potent inhibitor of the enzyme and binds to the ternary enzyme-glycogen-phosphate complex. Since this compound can form a structure similar to the half-chair conformation of an oxonium ion,[87] it was suggested that the GP catalyzed conversion of glycogen to α-D-glucose-1-P probably proceeds through a carbonium ion intermediate. However direct evidence for such an intermediate is less clear than for the β-glucosidase.[234] Although D-1,5-gluconolactone is an inhibitor of GP*b* ($K_i = 1$ m*M*), it is considerably less potent than anticipated for a transition state analogue. Ligands which resemble the half-chair geometry of such an intermediate should be expected to bind several orders of magnitude higher than those which exhibit geometry of the ground state substrate.[281] The enzymic reactions of a series of deoxy and fluorodeoxy glucosyl phosphates were recently examined.[249,279] All were found to be utilized by GP and potato phosphorylase, but at rates 30 to 300000 times slower than glucose-1-P itself. Since their V_{max} values correlated reasonably ($p = 0.9$) with their nonenzymic rates of acid-catalyzed hydrolysis,[277] a reaction known to proceed via oxocarbonium-ion-like transition states, it was suggested that the GP reaction must proceed through a mechanism involving oxocarbonium-ion-like transition states.[249]

Stereoelectronic theory[80,139,217] predicts that there is a stereoelectronic barrier to the cleavage of an equatorial glycoside, which is absent for the axial isomer. Thus for the aglycon-glycon (α-γλυκον the opposite of γλυκον = sweet) bond to break, the sugar ring must assume a conformation such that this bond is antiperiplanar to an sp^3 lone pair on the ring oxygen atom. In the ground state conformation of a β-glycoside the aglycon is equatorial, antiperiplanar only to ring bonds so that C-OR cleavage is predicted to be unfavourable, a reaction that can be made more favourable by distortion to a half-chair conformation. For α-glycosides, on the other hand, such as the substrates of GP, a lone pair of electrons in the ring oxygen is available antiperiplanar to the C-OR bond so C-OR cleavage is stereoelectronically favourable in the ground state conformation. Hence there is not the obligation for α-glycoside recognition enzymes to favour the distorted conformation of the glucopyranose ring in contrast to the situation for β-glycoside recognition enzymes where, from the example of lysozyme, distortion plays an important role.[293] Therefore it is uncertain whether compounds with trigonal geometry at C1 should act as transition state analogues and be powerful inhibitors of GP catalysis.

Various biochemical studies in GP have revealed that a glucose-like T-state inhibitor causes (1) homotropic cooperativity between substrate sites (giving sigmoid curves such as that shown in Figure 2), (2) exhibits a weakening of binding to AMP activated GP,

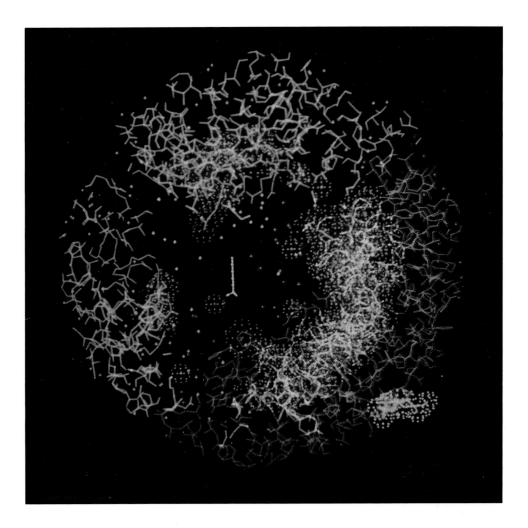

PLATE 1. A view of the channel which runs the whole length of the crystallographic c axis in tetragonal phosphorylase b crystals. The slab shown is about 8 Å thick. All atoms within 27 Å of the center are shown, and those within 18 Å have their van der Waals surface dotted. Water molecules are shown as crosses. Glucose-1-P with van der Waals surface shown in yellow is at the bottom right of the figure. The ruler at the center of the pore is 5 Å. One subunit is shown orange and the symmetry related subunits in blue. (From Johnson, L. N. and Hajdu, J., *Biophysics and Synchrotron Radiation,* Hasnain, S., Ed., Ellis Horwood, Chichester, 1989, 142. With permission.)

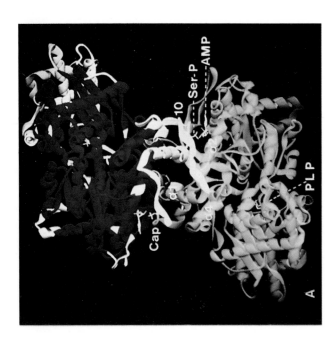

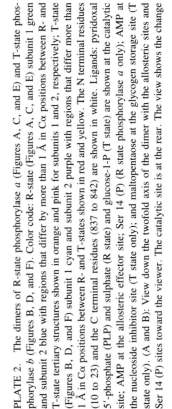

PLATE 2.   The dimers of R-state phosphorylase *a* (Figures A, C, and E) and T-state phosphorylase *b* (Figures B, D, and F). Color code: R-state (Figures A, C, and E) subunit 1 green and subunit 2 blue with regions that differ by more than 1 Å in Cα positions between R- and T-state tertiary structures shown in orange and pink for subunits 1 and 2, respectively; T-state (Figures B, D, and F) subunit 1 cyan and subunit 2 purple with regions that differ more than 1 Å in Cα positions between R- and T-states shown in red and yellow. The N terminal residues (10 to 23) and the C terminal residues (837 to 842) are shown in white. Ligands: pyridoxal 5'-phosphate (PLP) and sulphate (R state) and glucose-1-P (T state) are shown at the catalytic site; AMP at the allosteric effector site; Ser 14 (P) (R state phosphorylase *a* only); AMP at the nucleoside inhibitor site (T state only); and maltopentaose at the glycogen storage site (T state only). (A and B): View down the twofold axis of the dimer with the allosteric sites and Ser 14 (P) sites toward the viewer. The catalytic site is at the rear. The view shows the change in conformation of the N terminal tail from intra- to inter-subunit contacts and the shifts of the α1 helix-cap-α2 helix and α4-α5 inter-helix loop. (C and D): View down the twofold axis of symmetry with the catalytic site and tower helices towards the viewer. The allosteric site is at the rear. The view illustrates the changes in the tower helices and the 280s loop. The movements in the helix bundle (lower right, orange and red in R and T states) are part of the dimer-dimer contact in the tetramer in which these helices pack against part of the glycogen site (left). (C and F): View normal to the twofold axis of the dimer. The view shows the change in quaternary structure in which one subunit (subunit 2, top) rotates 10° with respect to the other subunit (subunit 1, bottom) about an axis normal to the twofold axis that intercepts the axis at a point near the cap/α2 interface. The view shows changes at the subunit-subunit interface and the shifts of the N terminal and C terminal residues. (From Barford, D., Hu, S.-H., and Johnson, L. N., *J. Mol. Biol.*, 218, 233, 1991. With permission.)

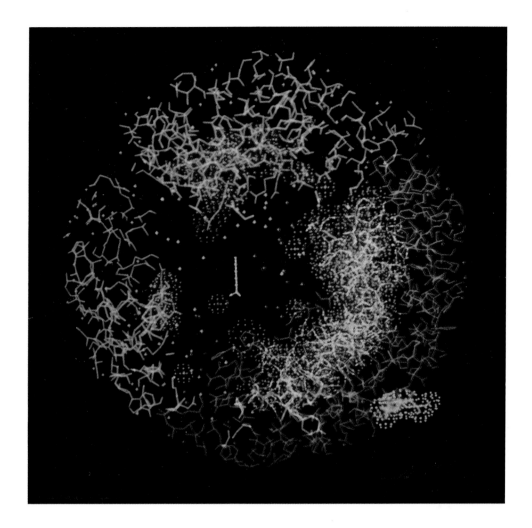

PLATE 1. A view of the channel which runs the whole length of the crystallographic *c* axis in tetragonal phosphorylase *b* crystals. The slab shown is about 8 Å thick. All atoms within 27 Å of the center are shown, and those within 18 Å have their van der Waals surface dotted. Water molecules are shown as crosses. Glucose-1-P with van der Waals surface shown in yellow is at the bottom right of the figure. The ruler at the center of the pore is 5 Å. One subunit is shown orange and the symmetry related subunits in blue. (From Johnson, L. N. and Hajdu, J., *Biophysics and Synchrotron Radiation,* Hasnain, S., Ed., Ellis Horwood, Chichester, 1989, 142. With permission.)

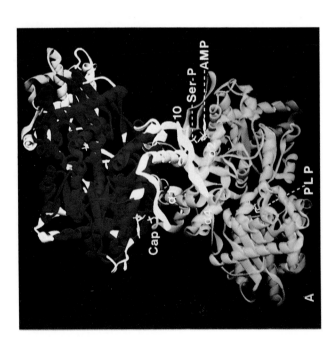

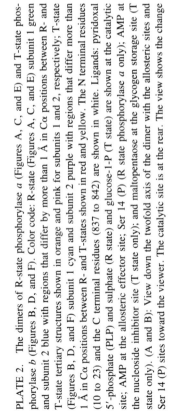

PLATE 2.   The dimers of R-state phosphorylase *a* (Figures A, C, and E) and T-state phosphorylase *b* (Figures B, D, and F). Color code: R-state (Figures A, C, and E) subunit 1 green and subunit 2 blue with regions that differ by more than 1 Å in Cα positions between R- and T-state tertiary structures shown in orange and pink for subunits 1 and 2, respectively; T-state (Figures B, D, and F) subunit 1 cyan and subunit 2 purple with regions that differ more than 1 Å in Cα positions between R- and T-states shown in red and yellow. The N terminal residues (10 to 23) and the C terminal residues (837 to 842) are shown in white. Ligands: pyridoxal 5′-phosphate (PLP) and sulphate (R state) and glucose-1-P (T state) are shown at the catalytic site; AMP at the allosteric effector site; Ser 14 (P) (R state phosphorylase *a* only): AMP at the nucleoside inhibitor site (T state only); and maltopentaose at the glycogen storage site (T state only). (A and B): View down the twofold axis of the dimer with the allosteric sites and Ser 14 (P) sites toward the viewer. The catalytic site is at the rear. The view shows the change in conformation of the N terminal tail from intra- to inter-subunit contacts and the shifts of the α1 helix-cap-α2 helix and α4-α5 inter-helix loop. (C and D): View down the twofold axis of symmetry with the catalytic site and tower helices towards the viewer. The allosteric site is at the rear. The view illustrates the changes in the tower helices and the 280s loop. The movements in the helix bundle (lower right, orange and red in R and T states) are part of the dimer-dimer contact in the tetramer in which these helices pack against part of the glycogen site (left). (C and F): View normal to the twofold axis of the dimer. The view shows the change in quaternary structure in which one subunit (subunit 2, top) rotates 10° with respect to the other subunit (subunit 1, bottom) about an axis normal to the twofold axis that intercepts the axis at a point near the cap′/α2 interface. The view shows changes at the subunit-subunit interface and the shifts of the N terminal and C terminal residues. (From Barford, D., Hu, S.-H., and Johnson, L. N., *J. Mol. Biol.*, 218, 233, 1991. With permission.)

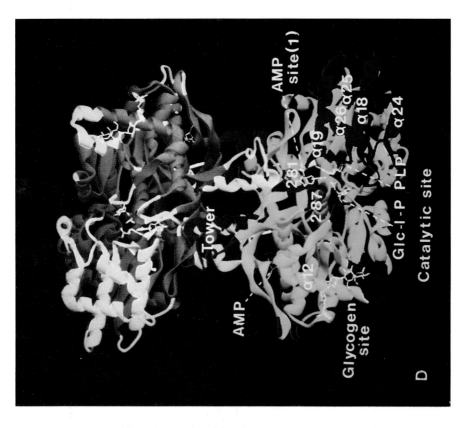

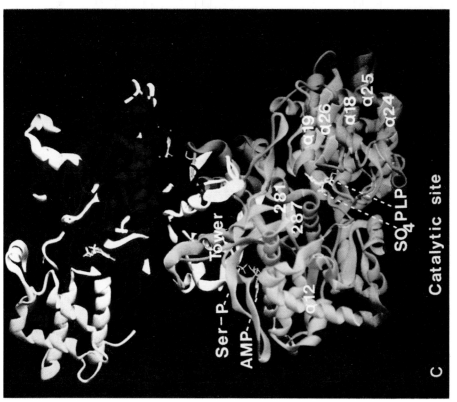

PLATE 2 (continued).

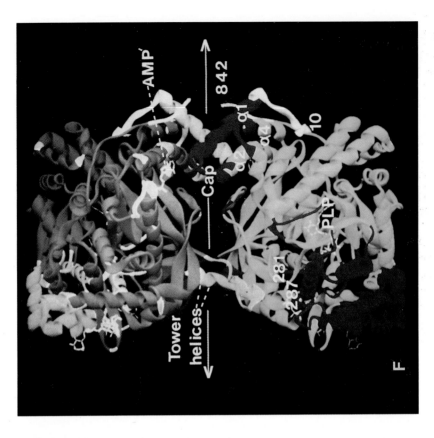

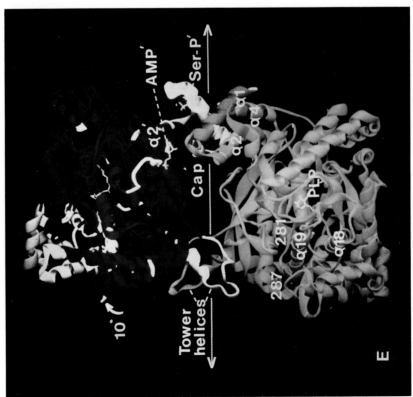

PLATE 2 (continued).

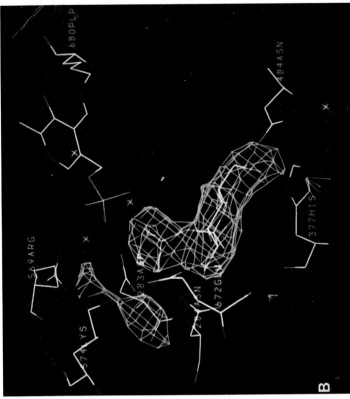

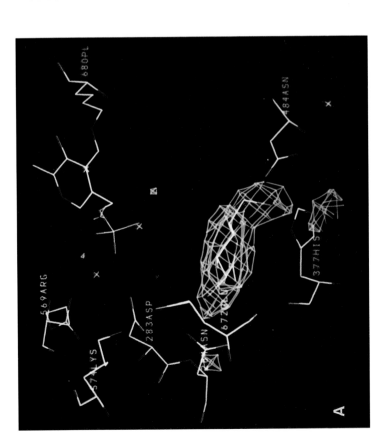

PLATE 3. Difference Fourier syntheses in the vicinity of the catalytic site for phosphorylase *b* catalysing the conversion of heptenitol to heptulose-2-P, observed in the crystal directly. A single positive contour level is shown (corresponding to 300 arbitrary units). Selected amino acids and the pyridoxal 5'-phosphate in the native enzyme structure are shown. Water molecules are shown as crosses. (A) 100 m*M* heptenitol, 50 m*M* arsenate, 2.5 m*M* AMP. Only heptenitol is observed bound. His 377 is displaced slightly as indicated by additional positive contours. (B) 100 m*M* heptenitol, 50 m*M* P$_i$, 50 m*M* maltoheptaose, 2.5 m*M* AMP, and soaking time 50 h. The product heptulose-2-P is apparent from the electron density and there are additional positive contours indicating movement of Arg 569 at upper left. (From Hajdu, J., Acharya, K. R., Stuart, D. I., McLaughlin, P. J., Barford, D., Oikonomakos, N. G., Klein, H., and Johnson, L. N., *EMBO J.*, 6, 539, 1987. With permission of Oxford University Press.)

(3) shows synergistic inhibition with a purine inhibitor such as caffeine, which is obviously a strong ($K_i = 0.1$ to 0.2 mM) T-state inhibitor and (4) promotes dissociation of the tetrameric enzyme. On the contrary, a substrate-like R-state inhibitor showing a preference for the R state such as glucose-1,2-P and UDP-Glc gives a rectangular hyperbola for the substrate saturation curve as would be expected for an enzyme obeying Michaelis-Menten kinetics, exhibits an increased affinity for the AMP activated enzyme, weakens binding of purine inhibitors and stabilizes the tetrameric species.[2,3,100,108,133,136,168,181,209,210,264,271,272]

C. BINDING STUDIES WITH T-STATE GP

A series of crystallographic studies of complexes of GPb and GPa with various substrates and substrate analogues have been undertaken to elucidate the mode of binding in the crystal. To date, in chronological order, the binding of glucose,[241,242] glucose-1,2-P,[274] and the substrate P_i[79] at the catalytic site has been reported for GPa, while the following analogues have been reported at the catalytic site of GPb: glucose-1-P,[117,174] glucose-1,2-P,[114] heptenitol and heptulose-2-P,[89,127,178] glucose,[174,199] UDP-Glc,[200] D-gluconohydroximo-1,5-lactone-N-phenylurethane,[7] 2-deoxy-2-fluoro-glucose-1-P and glucose-1-methylene-phosphonate,[174] and D-gluconohydroximo-1,5-lactone.[210] In addition a great number of GPb-inhibitor (α- and β-anomeric derivatives of D-glucose) complexes is under study.[175,294]

1. Binding of Glucose and Related Inhibitors

α-D-*Glucose* — The crystallographic study of the binding of α-D-glucose to the catalytic site of GPb has been recently described at 2.3 Å resolution by Martin et al.[174] The GPb-glucose structure is very similar to the native enzyme structure and the interactions formed by the functional groups of the glucosyl pyranoside moiety at the catalytic site were essentially the same as those described for the interactions of glucose with GPa by Sprang et al.[242] and subsequently by Street et al.[248] using refined coordinates. α-D-Glucose on binding to GPb adopts a well-defined $^4C_1$ chair conformation (Figure 17). It is interesting that glucose and all glucosyl compounds studied so far bind to GPb with their sugar rings in a gg conformation about the C5-C6 bond.[210] In contrast, glucose (or glucose-1,2-P) on binding to GPa in the crystal adopt a gt conformation[242,274] about C5-C6. In these complexes His 377 moves toward the sugar while in the GPb complexes it moves away from the sugar position. The crystallographic data showed that every potential hydrogen bonding group of glucose makes at least one hydrogen bond to GPb.[174] O1 hydroxyl and O5 ether oxygen appear to be not so important for binding, although a hydrogen bond to O1 of glucose from a water molecule, OH8 872, is present. O2 hydroxyl makes two hydrogen bonds to OE1 of Glu 672 (2.9 Å) and ND2 of Asn 284 (3.2 Å). O3 hydroxyl makes three hydrogen bonds to OE1 of Glu 672, N of Ser 674 and N of Gly 675. O4 hydroxyl makes two hydrogen bonds to N of Gly 675 and OD1 of Asn 284. O6 hydroxyl makes two hydrogen bonds to ND1 of His 377 and OD1 of Asn 284. Extensive hydrogen bonding between enzyme and ligand polar groups is therefore important for the binding of the pyranose ring at the catalytic site. Sprang et al.,[242] Kasvinsky,[136] and Street et al.[248] have tested a number of deoxy and fluoro-substituted glucose analogues as inhibitors of both GPb and GPa. They showed that α-D-glucose was a more powerful inhibitor of GPb than the β-anomer and more recent data supported enzyme α-specificity.[17] Loss of either the O1 or O2 hydroxyl results in a loss of 1.5 kcal/mol in binding energy, but replacement of these hydroxyls by fluorine causes no change, indicating that the hydrogen bonds at O1 and O2 both involve the hydroxyl as a proton acceptor. However, inversion of configuration at C2 results (e.g., D-mannose) in a considerable loss in affinity. Deoxygenation at C3, C4, or C6 leads to large loss of affinity, which is not surprising in view of the interactions shown in Figure 17.

α-D- *and* β-D-*Glucose Related Compounds* — The structure of T-state GPb has been exploited in the design of inhibitors that may be more potent than the physiological inhibitor

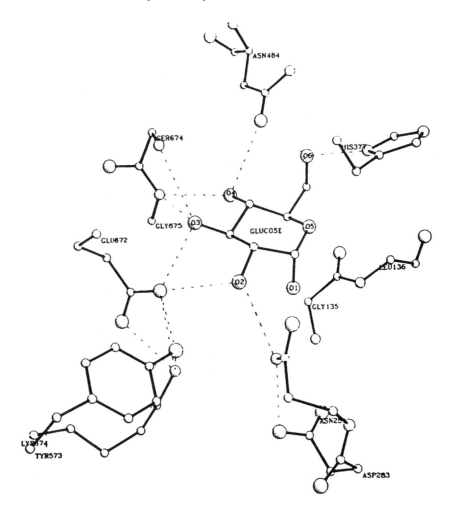

FIGURE 17.   The interactions between α-D-glucose and phosphorylase *b* at the catalytic site. (From Martin, J. L., Ph.D. thesis, University of Oxford, 1990. With permission.)

α-D-glucose. Some of them were synthesized and tested for binding to, and inhibition of GP*b*.[175] The ligands were structurally similar in that they all possessed glucosyl moieties but differed from each other in their substituent at C1. Thus, the binding of the glucose derivatives αCH$_2$OH-C1-βH (1), αH-C1-βCH$_2$OH (2), αH-C1-βCH$_2$N$_3$ (3), αOCH$_2$CH$_2$OH-C1-βH (4), and αH-C1-βOCH$_2$CH$_2$OH (5), at the catalytic site has been investigated at 2.3 Å resolution. All the glucose derivatives bind to the catalytic site and appear to make similar interactions to those formed by glucose. For compounds 1, 2, 3 and 5 there was strong difference Fourier density for both the glucosyl portion and for the C1 substituent. However for compound 4 density for the C1 substituent was very poor though density for the glucosyl portion was strong. Structure refinement of the complexes GP*b*-ligand is underway. Kinetic experiments indicate that compounds 1, 2, 3 and 5 inhibit GP*b* (with respect to glucose-1-P) with K$_i$ values of 2.0 m*M*, 22 m*M*, 15 m*M* and 25 m*M*, respectively. In contrast, compound 4 (which differs from 5 only in the orientation of the C1 substitutent) was found to be a weak activator of the enzyme, suggesting that the flexible anomeric substitutent destabilizes the 282 to 285 loop. Binding of the β-substituted compounds has not been demonstrated previously by crystallographic methods.

2. Binding of Substrates and R-State Inhibitors

Glucose-1-P — The crystallographic analysis of the enzyme-substrate complex[117] led to the first stereochemical proposal for the catalytic mechanism. It was based on the importance of the dianionic state of the coenzyme phosphate, the need for a nucleophile to stabilize the transition state carbonium ion and the need to accommodate an oligosaccharide molecule at the catalytic site. It was then thought that the observed glucose-1-P binding mode in the crystal was "nonproductive". Thus, on the basis of model building studies to permit fitting of oligosaccharide at the catalytic site, an alternative "productive" mode of substrate binding and a simple catalytic mechanism was suggested.[114,117] According to this mechanism, the coenzyme phosphate could act as a nucleophile to stabilize the carbonium ion, with the adjacent His 377 residue acting as general acid catalyst. However, the proposal for the "productive" mode suffers from the absence of any direct evidence in support of it. All substrates and substrate analogues studied since have their glucosyl moiety in the same orientation. In addition, Takagi et al.[295] showed that GP reconstituted with PLP-β-D-glucose was unable to transfer glucose to glycogen as would have been expected if a PLP-β-glucosyl was a possible intermediate of proposed reaction mechanism. Furthermore, Withers et al.[274] showed that β-glucose-1-P when tested up to a concentration of 10 mM acted neither as substrate nor as an inhibitor of PL-GPb.

Following the greater precision of the phases of the native GPb structure, binding of glucose-1-P and the R-state stabilizing phosphorylated ligands 2-deoxy-2-fluoro-glucose-1-P (2-F-glucose-1-P) and glucose-1-methylene phosphonate (glucose-1-me-P) was studied recently at 2.3 Å resolution.[174] 2-Deoxy-2-fluoro-glucose-1-P can be utilized as a substrate by the enzyme,[249] but glucose-1-methylene phosphonate is a competitive inhibitor (K$_i$ = 0.7 mM) and cannot be turned over by the enzyme. The interactions formed by the glucosyl portions of the three ligands to the catalytic site were essentially the same as those to glucose with GPb. Binding of the phosphorylated ligands (but not glucose) is accompanied by movement of catalytic site residues, mainly a shift of the 280s loop, a loop which in the T state is thought to block access of oligosaccharide, out of the catalytic site and towards the exterior of the protein. The loop becomes much more mobile and shifts to allow room for, and to optimize contacts with the phosphate group of these ligands. The phosphates of the phosphorylated ligands point away from that of PLP, with a P-P separation of 5.5 Å for glucose-1-me-P, 6.1 Å for glucose-1-P and 6.9 Å for 2-F-glucose-1-P. The contacts between the protein and the phosphate group of glucose-1-P are shown in Figure 18. A striking feature of the bound ligand structures is that the conformations of the phosphate groups at the catalytic site are different from each other and from the conformation found for heptulose-2-P (see below). In the structures reported Arg 569, which moves from its buried position to interact with the heptulose-2-P phosphate, is buried and thus inaccessible for interaction with the ligand phosphates. In this respect, it is of interest to note that Dreyfus et al.[48] have shown that Arg 569 is inaccessible to modification by arginine specific reagents in the T state of the enzyme, but becomes exposed upon activation. A full development of the phosphate site requires conformational changes resulting in exposure of Arg 569. Such a proposal is consistent with the known T-state GPb structure in which Arg 569 is close to the catalytic site but with the guanidinium group directed away from the ligands and coenzyme.

Binding of P$_i$ — Binding of P$_i$ at the catalytic site of GPb has not been observed in the crystal. P$_i$ binds to the allosteric effector site (Section V) and not at the catalytic site. Even at very high concentrations (0.5 M) the allosteric site is the only site where P$_i$ binds. This inability of P$_i$ to bind to the T-state GPb at the catalytic site apparently suggests that, in the tetragonal crystals of GPb the substrate P$_i$ binding site is not fully developed. In contrast, Goldsmith et al.[79] have recently solved the structure of a crystal of GPa soaked in a solution of P$_i$ and maltopentaose and found that P$_i$ was bound at the catalytic and allosteric sites,

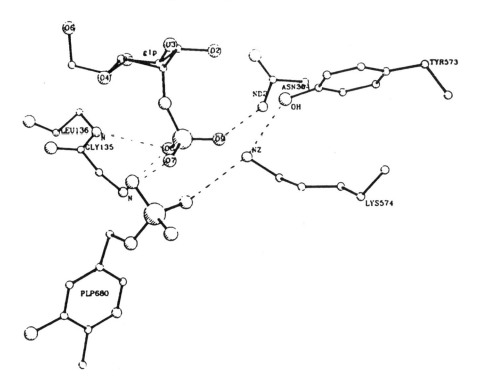

FIGURE 18.   The contacts between substrate glucose-1-P phosphate and T-state phosphorylase *b* at the catalytic site. (From Martin, J. L., Ph.D. thesis, University of Oxford, 1990. With permission.)

while maltopentaose was bound at the glycogen storage rather than the catalytic site. The binding of these molecules induce local conformational changes in enzyme structure. P_i is bound between the 280s loop (282 to 285) and the amino terminus of the α6 helix (133 to 150) forming hydrogen bonds with the imino groups of Gly 135 and Leu 136. At the catalytic site P_i displaced Asp 283 and formed hydrogen bonds with imino groups at the end of thehelix containing residues 133 to 149. The binding of P_i and oligosaccharide caused the C terminal catalytic domain to turn 1° and shift by 0.95 Å away from the N terminal domain; the tower helices moved toward each other. Their movement was much smaller than, and different from, the one seen by Barford and Johnson[8] in the R structure of GP*b*. Goldsmith et al.[79] suggested that their structure may be an intermediate between the inactive T and the fully active R structure. In contrast to the glucose-1-P, 2-F-glucose-1-P, glucose-1-me-P and heptulose-2-P complexes with GP*b* the P_i is well outside (7.7 Å P-P distance) of van der Waals contact distance of the coenzyme 5′-phosphate.

Glucal — The binding of glucal has been recently investigated[173] to gain more knowledge of how glucosyl structures bind to the catalytic site, how a phosphate ion interacts with this substrate, and how the catalytic reaction proceeds, in this case to give the postulated product 2-deoxy-glucose-1-P.[142] Glucal structure has a double bond between C1 and C2 of the glucosyl ring, and both of these carbon atoms in the glucal structure are unsubstituted. Thus, unlike glucose it cannot stabilize the inactive T state of the enzyme by forming a hydrogen bond to Asn 284 since glucal lacks a O2 hydroxyl group. This was thought to be an advantage in studying the reaction in the crystal since the R state of the enzyme is now less unfavourable. In a binding study with 100 mM glucal the contour map showed the highest peak at the catalytic site of GP*b* and this density fitted well to the molecular structure of glucal (calculated using the semi-empirical molecular orbital program AM1, in the AMPAC package). The density indicated that glucal was bound in the same orientation as other glucosyl structures studied and makes similar bonding interactions.

Glucose-1,2-P — This compound is a potent inhibitor of GP*a* competitive with glucose-1-P (K_i = 0.3 mM and 0.5 mM in the absence and presence of AMP, respectively, Hu and Gold[108]) and of GP*b* (K_i = 0.9 mM in the presence of AMP, Withers et al.[274]). Glucose-1,2-P inhibition pattern (linearity in the double reciprocal plots), its tightening of nucleotide binding and structural observations (stabilization of tetrameric GP*a*) (Withers et al.[272]) argue for glucose-1,2-P acting as an R state inhibitor, therefore mimicking glucose-1-P binding.[170] Analysis of the difference Fourier synthesis of the GP*b*-AMP-glucose-1,2-P at 3 Å resolution for cross-linked crystals of GP*b* soaked in 10 mM glucose-1,2-P showed binding at the catalytic site with the glucose moiety in approximately the same position as that observed for glucose-1-P but with the phosphate in a different position.[114] The phosphate of glucose-1,2-P made better contacts with enzyme and pointed towards the phosphate of the essential cofactor pyridoxal 5'-phosphate (P-P separation = 5.3 Å).[222] Withers et al.[272] in their ^{31}P NMR experiments have shown that on binding glucose-1,2-P to the GP*b*-AMPS-maltopentaose complex, the cofactor 5'-phosphate resonates at 0 ppm in the position of the monoanion but it is considerably broadened, which may be interpreted either by a change to a partially protonated state or by a more tightly bound constrained dianion. No large conformational changes were observed in the electron density for the cross-linked complex. In contrast, the binding of glucose-1,2-P to GP*a* resulted in numerous and quite extensive conformational changes including shifting of the loop 282 to 286, His 376, Arg 569, Lys 574 and translation of solvent molecules.[274]

Glucose-1-Phosphonate — This is a further novel compound, synthesized by Meuwly and Vasella[183] to explore possible phosphate recognition sites in the crystal. Strong binding was observed at the catalytic site and weak binding at the allosteric site.[296] At the catalytic site the glucose moiety was close to that observed for glucose-1-P showing that the major specificity at this site is determined by the glycopyranosyl ring. The distance of the phosphorus atom from the phosphorus atoms of glucose-1-P and heptulose-2-P was 1.4 Å and 1.6 Å respectively. Substantial changes in the vicinity of residues 282 to 285 were observed. Kinetic experiments indicated an apparent K_i value of 12 mM (with respect to glucose-1-P) for GP*b*.[297]

Deoxynojirimycin (1-Deoxy-5-Amino-D-Glucose) — This compound is a potent inhibitor of many glucosidases (e.g., K_i = 0.032 µM for rabbit small intestinal sucrase)[91] but only weakly inhibits GP*b* (K_i = 55 mM).[3] The high affinity for the glucosidases is thought to arise from the ring nitrogen, features which allow deoxynojirimycin to mimic a carbonium ion transition state. In the crystals of GP*b*, deoxynojirimycin binds weakly at the catalytic site.[288] The weak binding can be rationalized by the proximity (hydrogen bonding distance) of the main chain NH of Leu 136 and the ND1 atom of His 377 to the position of the ring oxygen of glucosyl sugars which is now replaced by the ring nitrogen in deoxynojirimycin. These features emphasize the important differences at the catalytic site between GP and glucosidases (including lysozyme). In GP there is no negatively charged group (corresponding to Asp-52 in lysozyme)[15,16] that can act as an electrostatic stabilizing group for the transition state of the catalysed reaction.

Heptenitol and Heptulose-2-P — The most informative study from a mechanistic point of view was the kinetics of the reaction,

$$\text{Heptenitol} + P_i \rightarrow \text{Heptulose-2-P}$$

in the crystal.[89,178] GP, despite being highly specific in recognizing glucose-1-P, can also utilize compounds such as D-glucal,[142] glucose-1-F,[206] and heptenitol.[143,144] The latter compound[95] which carries an exocyclic bond (Figure 19), can serve as a substrate for GP exclusively in the degradative direction (the reaction itself does not require a primer). Heptulose-2-P, the dead end product, is the most potent inhibitor of the enzyme (K_i = 14 µM)

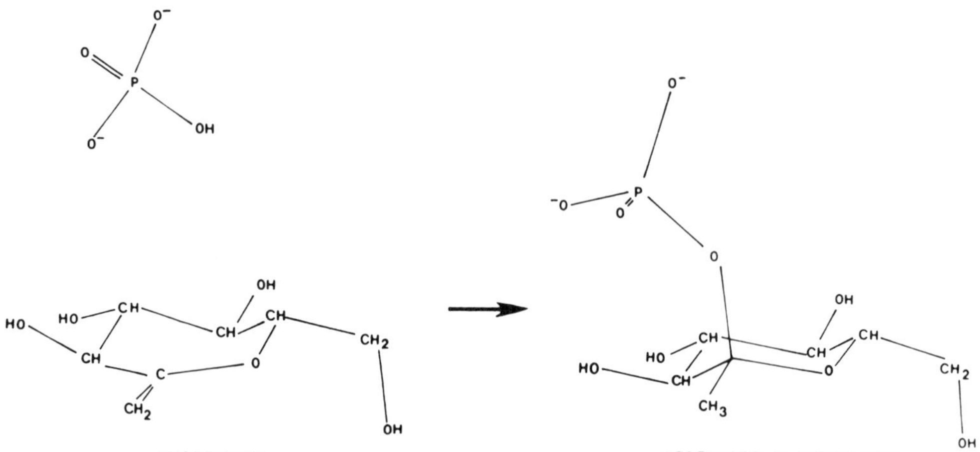

FIGURE 19. Conversion of heptenitol to heptulose-2-P by phosphorylase in the presence of inorganic phosphate.

yet discovered.[89,144] The reaction is an attractive system for study since it does not require a primer and because it leads to a high affinity product which may resemble the transition state of the natural reaction. Due to the development of high intensity synchrotron radiation sources of X-rays[97,98] which greatly decrease exposure times, direct observation of the progress of the conversion of heptenitol to heptulose-2-P has been made possible in the crystal.[89] A series of time resolved experiments were carried out but only the control and the end result is shown here (Plate 3*). The product heptulose-2-P has been formed in the crystal and a direct interaction of the product with the cofactor 5'-phosphate was observed. The enzyme-heptulose-2-P complex has recently been refined[127] and the polar contacts between GP*b* and heptulose-2-P are shown in Figure 20. All the polar groups of the sugar are observed to form hydrogen bonds with the enzyme. The exception is the ring oxygen, O5, whose separation from the main chain N of Leu 136 (3.5 Å) is long for a hydrogen bond. There were pronounced conformational changes associated with the formation of the product, showing that movement of the side chain of Arg 569 ($\approx$ 7 Å), from a buried position to another that brings the NH1 atom close to O9 of heptulose-2-P, is critical for the formation of the substrate phosphate recognition site. In the native structure Arg 569 is turned away from the catalytic site and buried with its charged group in contact with main chain oxygens of Pro 281 and Asn 133. In the product complex formed by catalysis with GP*b* crystals the basic group of Arg 569 swings into the catalytic site to make contact with the phosphate of heptulose-2-P. This movement causes a simultaneous displacement of an acidic residue, Asp 283, from the catalytic site and results in shifts of the whole loop from 281 to 285, the loop that blocks access to the site. The changes observed for heptulose-2-P are much the same as those observed on binding of glucose-1-P, 2-F-glucose-1-P and glucose-1-me-P except that the Arg 569, which remains in the native conformation in all glucose-1-P, 2-F-glucose-1-P and glucose-1-me-P bound enzymes, rotates to interact with the phosphate of heptulose-2-P. Heptulose-2-P binds so that the product phosphate oxygen (O7) is within hydrogen bonding distance of the phosphate oxygen (O22) of coenzyme. Such a close approach must perturb the electron distribution of the 5'-phosphate of the coenzyme, as reflected in the concomitant change in its ^{31}P NMR spectrum[143] and rule out earlier proposal for the cofactor 5'-phosphate as a nucleophile.[117] The proximity of the product phosphate to the cofactor phosphate provides a satisfactory mechanism for catalysis that involves the cofactor phosphate to first act as a general acid in promoting attack by the substrate phosphate

---

*   Color plates follow page 118.

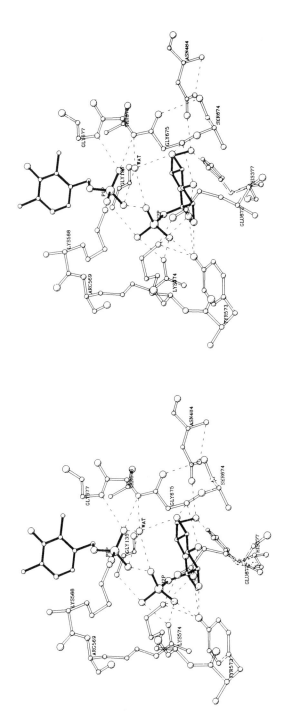

FIGURE 20. A stereo diagram showing the intermolecular interactions between phosphorylase *b* and the product heptulose-2-P at the catalytic site. (From Johnson L. N., Acharya, K. R., Jordan, M. D., and McLaughlin, P. J., *J. Mol. Biol.*, 211, 645, 1990. With permission.)

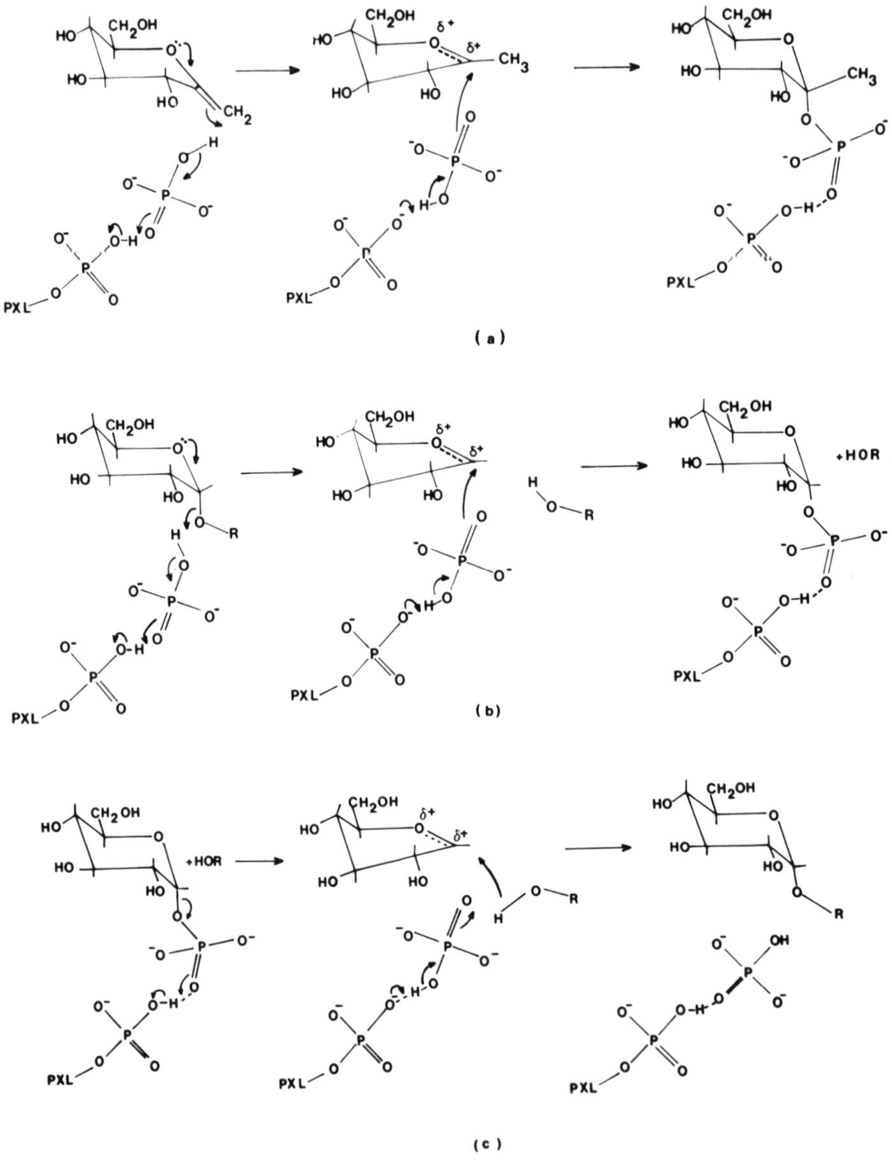

FIGURE 21.   The proposed mechanism of phosphorylase catalysis for (a) phosphorylysis of heptenitol, (b) phosphorylysis of oligosaccharide and (c) oligosaccharide synthesis. (From Johnson L. N., Acharya, K. R., Jordan, M. D., and McLaughlin, P. J., *J. Mol. Biol.*, 211, 645, 1990. With permission.)

on the heptenitol.[89,124,127,144,178] After protonation of the methylene group, the glucosyl carbonium ion intermediate is stabilized by the negatively charged substrate phosphate group. The reaction is then completed by nucleophilic attack of the phosphate on the carbonium ion to give the product heptulose-2-P. The results provide therefore a stereochemical basis for a plausible mechanism (Figure 21) of catalysis which is consistent with proposals made on the basis of ^{31}P-NMR experiments. The refined structure of the heptulose-2-P complex[127] indicates that there are no ionizable groups in the vicinity of the C1 atom of the glucopyranose ring that could act to stabilize the carbonium ion transition state, nor are there any groups in the vicinity of the glycosidic oxygen that could protonate the substrate directly. The proposed mechanism depends on the correct orientation of the substrate phosphate and sugar

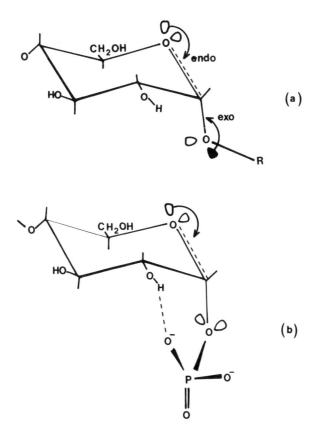

FIGURE 22. Stereo-electronic effects for α-glycosides. (a) In the preferred conformation for α-glycosides both C1-O5 (*endo* effect) and C1-O1 (*exo* effect) bonds are stabilized. (b) In the conformation observed for heptulose-2-P, the *exo* anomeric effect is weakened. (From Johnson L. N., Acharya, K. R., Jordan, M. D., and McLaughlin, P. J., *J. Mol. Biol.*, 211, 645, 1990. With permission.)

groups with respect to the 5'-phosphate group of the cofactor. The conformation observed for heptulose-2-P (Figure 22) is similar to that predicted from stereoelectronic theory to weaken the *exo*-anomeric effect and promote cleavage of the α-glycosidic bond.[127,217] Thus, heptulose-2-P formed on the reaction pathway, exhibits a conformation that is anticipated to promote cleavage of the α-glycosidic bond. In this respect heptulose-2-P resembles a reactive intermediate on the reaction pathway. However, in heptulose-2-P the absence of reaction in the direction of glycogen synthesis is probably explained by the presence of the -CH_3 group on C1 which prevents access of the oligosaccharide acceptor. In the GP*b*-heptulose-2-P complex an internal hydrogen bond between the O2 hydroxyl and O9 oxygen of the phosphate is observed. This conformation is favoured by the methyl group attached in the β configuration to C1 and brings the phosphate to different position from that observed with glucose-1-P. The favoured position for the "bent-down-towards O2" position (with respect to *z*-axis) of the phosphate also explains why glucose-1,2-P is also a good inhibitor, but not as good as heptulose-2-P. Heptulose-2-P can be therefore considered as one of the best inhibitors known and represents the closest structure to the transition state that has been seen in the crystal studies.

UDP-Glc — A substrate analogue which belongs to a category of inhibitors that bind preferentially to the R conformation,[168] i.e., it binds more strongly to the AMP-activated

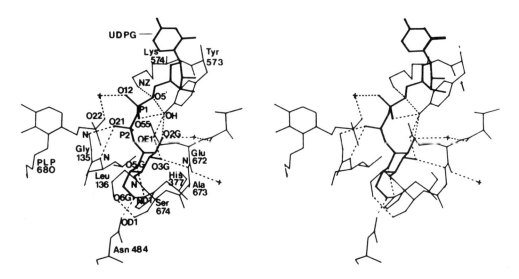

FIGURE 23. A stereo diagram of the contacts between the glucose and pyrophosphate components of UDP-Glc at the catalytic site. Residues which are displaced are not included. (From Oikonomakos, N. G., Acharya, K. R., Stuart, D. I., Melpidou, A. E., McLaughlin, P. J., and Johnson, L. N., *Eur. J. Biochem.*, 173, 569, 1988. With permission.)

form of the enzyme than the free enzyme,[164] or improves the binding of AMP by promoting the R conformation.[22] In all GPs so far examined UDP-Glc is competitive with respect to glucose-1-P (or P$_i$) with a K$_i$ of approximately 1 mM.[168] Kinetic studies have shown that, with rabbit GP*b*, UDP-Glc is a noncompetitive inhibitor for AMP but these two compounds have an interaction constant of 0.25 because both stabilize the R conformation.[168] Thus, in GP*b*, AMP improves the binding of UDP-Glc by fourfold. UDP-Glc (the largest ligand tested so far) at a concentration of 100 mM can bind at both the allosteric and the catalytic site.[200] An electron difference density map for UDP-Glc showed clear density for glucose at the glucose-1-P site. Density contiguous with this, corresponding to diphosphate, was situated adjacent to Asn 284 directed towards channel 1 (see VIII.D). Both phosphates of the pyrophosphate are distinguishable and these were approximately on either site of the phosphate position observed for glucose-1-P. No definite density corresponding to the uridine was observed. The position of the diphosphate dictates that ribose is placed over the side chain of His 571. A lack of density for uridine is explained since it replaces protein residues. A plausible position for the ligand molecule bound at the catalytic site is presented in Figure 23. The overall conformation of the UDP-Glc molecule at the catalytic site is extended. The binding of its glucose-1-P moiety is almost identical to that observed for glucose-1-P itself,[117] a result in good agreement with the strictly competitive nature of these two compounds, demonstrated in kinetic studies.[161] Kinetic experiments also indicate that UDP-Glc is able to relieve the inhibitory effect of caffeine[168] and high levels of IMP.[200] Binding of UDP-Glc to the catalytic site promotes extensive conformational changes. The loop 279 to 288 which links the catalytic site to the inhibitor site is displaced and becomes mobile (Figure 24). Associative positive peaks were not interpretable and a definitive relocation of the loop could not be made. Positive and negative peaks in the vicinity of Arg 569 and His 571 side chains suggests new positions. Arg 569 is relocated breaking interactions to the main chain carbonyl of Asn 133 and the Asp 283 side chain to form a new one to the phosphate of UDP-Glc. The torsion angle Cα-Cβ in His 571 rotates shifting the pyrrole ring towards the inhibitor site which disrupts interactions to the main chain carbonyl of Asn 283. Thus the conformational changes are concerted. Negative density overlying side chains of residues Asn 133 and Leu 136 implicate these in conformational change. The consequences of these

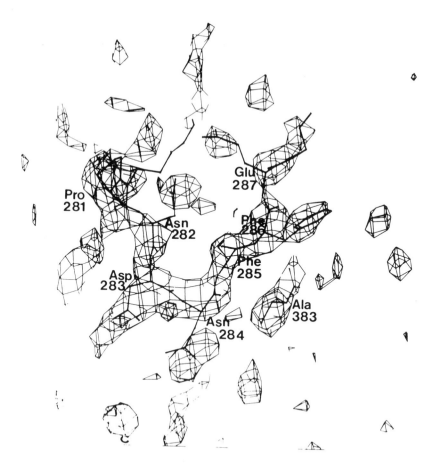

FIGURE 24. Negative contours (at -300 arbitrary units) in the UDP-Glc difference map indicating displacement of residues 281 to 287 from the catalytic site. Positive contours have been omitted for clarity. These are not extensive and give no clear indication of the new position of the loop. (From Oikonomakos, N. G., Acharya, K. R., Stuart, D. I., Melpidou, A. E., McLaughlin, P. J., and Johnson, L. N., *Eur. J. Biochem.*, 173, 569, 1988. With permission.)

changes is to open up a channel (channel 1) for UDP-Glc. As discussed by Barford[6] "displacement of loop 279 to 289 occurs for two reasons: one is that interactions to Arg 569 and His 571 which "pin" the loop to the C terminal domain are broken and secondly, the residues of the two domains are prised apart on forming a binding site for uridine". Thus, the loop 279 to 288 appears to play a major role in maintaining the GP molecule in a low-affinity state. Since Phe 285 consists of a basic structural component of the inhibitor site, this site is therefore destroyed. Opening and closing of the channel, when the corresponding complexes of enzyme-UDP-Glc and enzyme-caffeine are formed, can provide therefore an explanation for the biochemical studies. In general, the electron density map for the enzyme-UDP-Glc complex shows almost identical conformational changes to those noted to occur in the enzyme-heptulose-2-P map.[89,127,178] Arg 569 moves to interact with the sugar phosphate, triggering displacement of the loop 279 to 289. Similar conformational changes occur at the side chains of Asp 133, Leu 136, and His 571. These changes represent salient features in the production of phosphate and oligosaccharide binding sites and transition to the R state.

3. Binding of Hydroximolactones

A number of D-gluconohydroximo-1,5-lactones have been designed and synthesized to produce stabilization of the half-chair conformation at the C1.[12] Unlike the 1,5-gluconolactone which decomposes rapidly in aqueous solutions, these compounds resist hydrolysis under physiological conditions. Kinetic studies with the β-glucosidase emulsin showed that D-gluconohydroximo-1,5-lactones were also potent inhibitors of the enzyme,[13] suggesting that they might act as transition state analogues for the enzymic cleavage of β-D-glucosides by mimicking the trigonal geometry required at C1 for the formation of the carbonium ion intermediate. D-gluconohydroximo-1,5-lactone-*N*-phenylurethane with a K_i = 2.3 μ*M* proved a more potent inhibitor than D-1,5-gluconolactone (K_i = 37 μ*M*).[13] Combined kinetic, ultracentrifugation and X-ray crystallographic studies have characterized the effect of these inhibitors on the catalytic and structural properties of GP to probe more deeply the stereo-chemical consequences of a half-chair glucosyl residue at the active site. In addition, a compound such as D-gluconohydroximo-1,5-lactone-*N*-phenylurethane with a bulk substi-tuent at the trigonal C1 position should mimic an oligosaccharide transition state analogue and lead to the identification of the active site channel. Kinetic studies with this analogue and the unsubstituted D-gluconohydroximo-1,5-lactone showed that these compounds were good inhibitors of GP*b* (with K_i values of 0.4 m*M* and 0.9 m*M*, respectively).[7,209,210] However, in spite of their structural similarities and the trigonal geometry at C1, the two inhibitors acted differently. Thus, D-gluconohydroximo-1,5-lactone-*N*-phenylurethane was a linear competitive inhibitor for glucose-1-P and exhibited a weak synergistic effect with caffeine, while D-gluconohydroximo-1,5-lactone was a nonlinear competitive inhibitor with respect to glucose-1-P and exhibited a strong synergistic effect with caffeine. Moreover, D-gluconohydroximo-1,5-lactone-*N*-phenylurethane did not promote the dissociation of GP*a* to dimers, while D-gluconohydroximo-1,5-lactone did. Also, the latter compound can induce in GP*a* the formation of tetragonal crystals, a result consistent with the effect of the T-state inhibitor glucose.[65] A similarity between the two compounds existed only in their inhibitory pattern in the reverse reaction. Both compounds were found to be linear noncompetitive inhibitors with respect to phosphate.[209,210] Binding studies in the crystals of the T-state GP*b* at 2.4 Å resolution revealed that the sugar moiety of the molecules is located within the glucose-1-P binding site but D-gluconohydroximo-1,5-lactone adopts a chair conformation while its phenylurethane derivative adopts a half-chair conformation[7,210] (Figure 25). Their glucosyl rings therefore occupy a similar but not identical position (C1-C1 separation is 1 Å) with the hydroximo group occupying a different position (N1-N1 separation is 2 Å, O7-O7 separation is 2.5 Å) and the phenylurethane moiety extended to a cavity next to the catalytic site with little conformational change of the enzyme. This cavity is fortuitously available in the T-state GP*b*. In addition to the usual contacts made to the glucose-like component the phenylurethane makes two good hydrogen bonds to Asn 284 and one to Asp 339 and there are good van der Waals interactions between the phenyl group and His 341 and Leu 136, showing the ability of proteins to provide binding sites for nonphysiological ligands. Both ligands stabilize the geometry of the 280s loop occurring in the T state since they interact with residues of this loop. This feature could explain the behaviour of D-gluconohydroximo-1,5-lactone in the kinetics (in the direction of glycogen synthesis), in the ultracentrifugation and crystallization experiments but is not consistent with some of the solution properties observed with D-gluconohydroximo-1,5-lactone-*N*-phenylurethane, i.e., the linearity in the double-reciprocal plots with respect to substrate and the stabilization of the tetrameric GP*a*. The major difference between the two analogues is the bulky phenyl-urethane group. In order to accommodate this in the channel, D-gluconohydroximo-1,5-lactone-*N*-phenylurethane has to adopt a half-chair conformation. Since the energy difference between chair and half-chair conformation for these compounds is not very great,[298] it may be suggested that the enzyme is able to influence ligand conformation. Thus, the ability of D-gluconohydroximo-1,5-lactone-*N*-phenylurethane to favour T- or R-state enzyme has to do with ability to interact with the 280s loop. If the chair conformation of this analogue is the most stable one in solution, it would displace the 280s loop giving rise to R-state-like structure. It cannot do this in the crystal structure where the enzyme is locked in the T state and the 280s loop blocks the binding of the chair conformation of this analogue. In order for the chair conformation of D-gluconohydroximo-1,5-lactone-N-phenylurethane to bind, steric constraints have to be overcome. This steric hindrance is relieved during the structural rearrangement of the 280s loop, when UDP-Glc (with a chair geometry for the glucosyl residue and a bulky substituent at C1) binds to the enzyme.[200] Also, the crystallographic refinement of the heptulose-2-P complex[127] does show that the 280s loop could move into the channel which becomes occupied by phenylurethane. These studies lead to the conclusion that the geometrical change chair → half-chair by itself does not lead to a good transition-state model and indeed would direct an oligosaccharide into the wrong channel. The channel into which phenylurethane is directed does not lead to the solvent but is blocked by the protein in the region of Pro 280, Glu 287 and Arg 292. If a polymer were to reach the surface by this route it would need to pass through the loop of chain 279 to 293. Thus the binding of D-gluconohydroximo-1,5-lactone-N-phenylurethane cannot be regarded as being analogous to oligosaccharide. Kinetic studies[209] further suggest that this compound does not bind to the active site analogously to oligosaccharide. Comparison of the binding in the crystal of compounds with half-chair geometry (e.g., heptenitol, glucal and D-gluconohydroximo-1,5-lactone-N-phenylurethane) with those that exhibit chair geometry (e.g., glucose, glucose-1-P, heptulose-2-P, UDP-Glc, D-gluconohydroximo-1,5-lactone) show no interactions that either favour or destabilize one conformation with respect to the other. The results therefore support the notion, which is consistent with stereoelectronic theory that distortion of the glucopyranose ring towards half-chair geometry and promotion of trigonal geometry at C1 is not an obligatory first step in the catalytic reaction. Rather it can be considered in agreement with Street et al.[249] that it is the electronic character of the oxo-carbonium ion that is important and its stabilization by phosphate.

D. CHANNELS AT THE CATALYTIC SITE

Several binding studies including 1 M maltose, 0.6 M maltoheptaose and 100 mM α-phenyl-glycoside have shown that even at these very high concentrations none of these sugars bound at the catalytic site of GP, strongly suggesting that conformational changes are required to bind more than one sugar. Since catalysis does occur in the crystal[89,131,178] it is assumed that the oligosaccharide channel to the catalytic site can open, at least transiently, but not for sufficiently long periods to allow binding to be observed in crystallographic experiments. In an attempt to define routes to the catalytic site, a computer program that allows channels to be contoured within a protein has been written.[299] Model building to channels is therefore facilitated since a molecule can be fitted to the contours in a similar way as to electron density map.[7] These studies showed that there are two clear channels that could allow access for substrates. Each of these channels would require however conformational changes in the enzyme in order to reach the surface. The channels are running on opposite sides of the loop 280 to 289 (Figure 26) and have a length to the surface that could accommodate 4 to 5 glucosyl residues. Channel 1 passes between the two domains of the enzyme in a fairly direct path to the inhibitor site. It leads from the phosphate of glucose-1-P through the side chains of Asp 283, Asn 284 and Arg 569 to reach the solvent interface. This channel is blocked by the side chains of Asp 283, Asn 284 and Arg 569. Channel 2 passes through the 280 to 289 loop and is thus partially within a domain. It extends between Asp 283, Phe 285 and Phe 286 on one side and Asp 339, His 341 and Arg 292 on the other side, but access to the surface is blocked by the loop of chain 287 to 292. A substrate at the active site with tetrahedral geometry around C1 (e.g., UDP-Glc) is directed into channel 1, but a substrate analogue such as D-gluconohydroximo-1,5-lactone-N-phenylurethane with a trigonal geometry around C1 is directed into a channel 2. The low affinity for oligosaccharide exhibited by the T state GPb in the tetragonal crystal is undoubtedly associated with the lack of access to the catalytic site from the solvent. Both channels 1 and

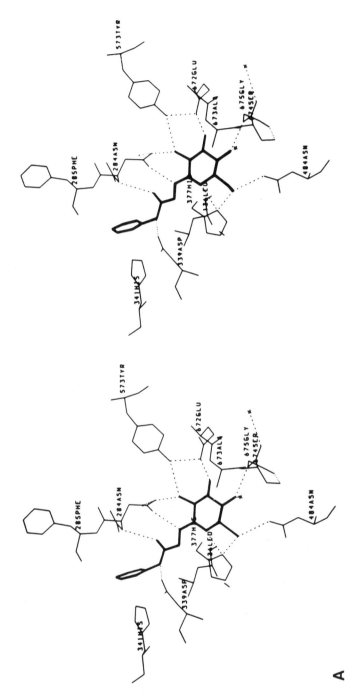

FIGURE 25. A stereo diagram showing the contacts between phosphorylase *b* and D-gluconohydroximo-1,5-lactone-*N*-phenylurethane (A) and D-gluconohydroximo-1,5-lactone (B). (A, reprinted by permission from Barford, D., Schwabe, J. W. R., Oikonomakos, N. G., Acharya, K. R., Hajdu, J., Papageorgiou, A. C., Martin, J. L., Knott, J. C. A., Vasella, A., and Johnson, L. N. *Biochemistry*, 27, 6733, 1988. Copyright 1988 American Chemical Society; B, from Papageorgiou, A. C., Oikonomakos, N. G., Leonidas, D. D., Bernet, B., Beer, D., and Vasella, A., *Biochem. J.*, 274, 329, 1991. With permission.)

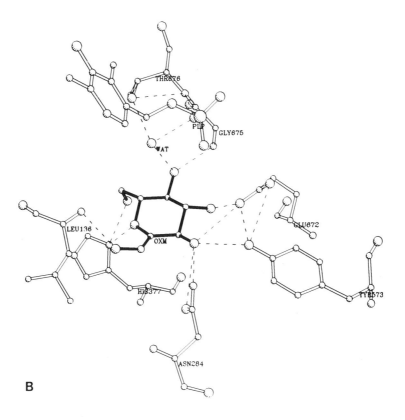

B

FIGURE 25B.

2 are blocked by the loop of chain 278 to 292. This loop is connected, on one side, via a short β sheet (β11b-residues 276 to 279) to the tower helix (α7) and on the other side, to another helix (α8), one of the helices making up the allosteric site. The crystallographic studies with heptulose-2-P and UDP-Glc provide strong evidence that the major feature of the T to R transition in GP involves displacement of the loop and, through movement of Phe 285, subsequent destruction of the inhibitor site.

Neither of the channels located at the active site adequately accommodate an α-1,4-linked oligosaccharide and it is apparent that to do so requires substantial protein conformational changes. Thus, in a model building study with GPa[274] it was possible to model maltopentaose into the catalytic site of GPa structure with the terminal glucose in the glucose binding site. The model for maltopentaose however required an alternative conformation about the linkage between the first two residues in which the O2 to O3′ hydrogen bond is broken and also revealed a few steric conflicts between protein and oligosaccharide that could not be relieved by minor conformation changes in the protein. Thus oligosaccharide binding would require reordering of the loop 282 to 285, a shift of the strand 133 to 135 toward the protein interior and rotations of Tyr 613 and His 571 about the Cα to Cβ bonds.[274] Modeling studies in GPb structure also suggest that maltopentaose can only be accommodated within the D-gluconohydroximo-1,5-lactone-N-phenylurethane channel (channel 2) if it adopts unrealistic conformations, with either a trigonal geometry at C1 or unfavorable torsion angles.[6,226]

Observation of oligosaccharide bound in a channel to the catalytic site is necessary before we can be certain of the catalytic channel. Since the structure of the R state is now available, there is the chance that a ligand-binding study might reveal oligosaccharide bound. It is now clear that the 280s loop in the R state of the enzyme differs considerably from that in the

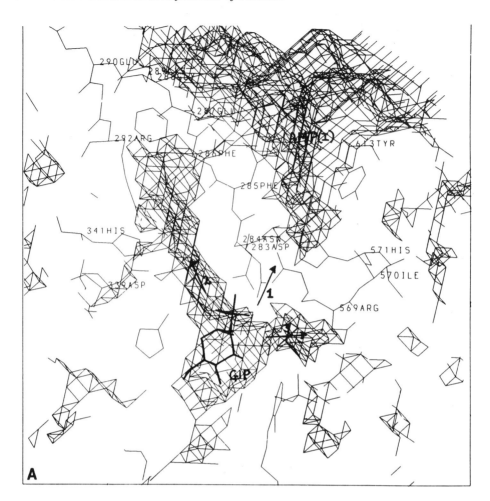

FIGURE 26.   Channels at the catalytic site of phosphorylase *b* generated at distances greater than 2.5 Å from any atom. The substrate glucose-1-P (G1P) is shown. The channel to its left accommodates D-gluconohydroximo-1,5-lactone-*N*-phenylurethane. The inhibitor site at the protein surface is at the top, with access to the active site blocked by His-571. The program to contour protein channels and cavities was written by D. Barford and L. N. Johnson. (From Barford, D., Schwabe, J. W. R., Oikonomakos, N. G., Acharya, K. R., Hajdu, J., Papageorgiou, A. C., Martin, J. L., Knott, J. C. A., Vasella, A., and Johnson, L. N., *Biochemistry, 27,* 6733, 1988. With permission.)

T state;[8] certainly one end of the loop is connected to one of the helices ($\alpha8$) that forms the allosteric site. This could facilitate glycogen binding and thus partially explain the greater activity of the R state.

E. BINDING STUDIES WITH R-STATE GP

SO_4^{2-} — The sulphate binding site observed in the R-state GP[8] is similar to the phosphate-recognition site developed in the GP*b*-heptulose-2-P complex.[89,178,292] Similar movement of Arg 569, triggered indirectly by movement of the tower helix, was also observed. In R-state GP*b* Arg 569 contacts the sulphate in only 2 of the subunits[8] but in the R-state GP*a* Arg 569 interacts with sulphate in all 4 subunits.[9] The binding site for the sulphate is formed from the side chains of basic residues Arg 569 and Lys 574 and the main-chain nitrogen of Gly 135 (Figure 27) and is situated adjacent to the 5′-phosphate of PLP (S-P separation = 5 Å).

Glucose-1-P — A crystallographic study with R-state GP*b* crystals soaked in 200 m*M* glucose-1-P, 50 m*M* ammonium sulphate and 1.0 *M* tartrate (replacing sulphate) showed

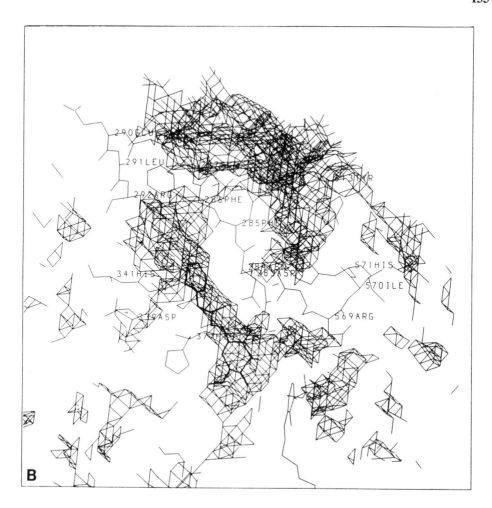

FIGURE 26B.

clear density corresponding to glucose-1-P at the catalytic site with a negative peak super-imposing over the phosphate recognition site, indicating displacement of sulphate.[300] The phosphate group of glucose-1-P occupies the catalytically inactive "down" position (Figure 28).

F. CATALYTIC MECHANISM

The strongest evidence for the direct phosphate-phosphate interaction between the co-factor and product phosphate comes from the crystallographic analysis of the GP*b*-heptulose-2-P complex. The results from this study showed a constellation of hydrogen bonds that could be consistent with a role of the cofactor either acting as a proton donor itself or as a stabilising group in a mechanism in which the proton is donated by another group. A striking feature of this arrangement is that there is no group in the vicinity of C1 and O5 capable to directly protonate the glycosidic oxygen, nor is there any group in the correct orientation to stabilize the carbonium ion intermediate. Glu 672 is a favoured candidate but it is on the wrong side and too far away. This residue interacts with the O2 and O3 hydroxyl groups of the sugar and is undoubtedly important in locating the sugar in the right orientation, perhaps also serving to create an approximate electrostatic environment and to repel the enzyme-product complex in the natural reaction. Therefore the proposal that it is the phos-phate group itself which stabilizes the transition-state is favoured. Nevertheless the cluster

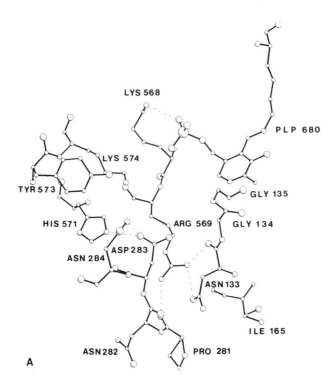

FIGURE 27.   Constellation of residues in the vicinity of the phosphate/
sulphate recognition site in the T-state (A) and R-state of phosphorylase
b (B). (C) Superimposition of catalytic site residues from T-state (thin
lines) and R-state phosphorylase *b* (solid lines). The recognition site is
created by adjustments in the torsion angles in the Arg 569 side chain,
shifting the guanidinium group by 7 Å so that it swings into the catalytic
site and displaces Asp 283. Motion of Arg 569 is triggered by the displacement of Asn 133 and Pro 281, induced by motion of Ile 165, weakening interactions to Arg 569, and by disorder of the 280s loop. Residues
Gly 134 to 135, Lys 574 and PLP 680 superimpose approximately. (Reprinted by permission from Barford, D. and Johnson, L. N., *Nature,* 340,
609, 1989. Copyright © 1989 Macmillan Magazines Ltd.)

of groups around O2 and O9 in the enzyme-heptulose-2-P complex is intriguing. They
undoubtedly serve to promote the binding and at the same time allow easy access for
oligosaccharide but is not clear if they could play a role in catalysis. In a recent paper[106] it
has been shown that GP*b* reconstituted with PLDP-mannose cannot break the phosphate
link, in contrast to PLDP-glucose, also supporting the importance of O2. Glucosamine-1-P
which can substitute for glucose-1-P in the GP reaction[219] binds at the catalytic site of GP*b*
about twice as strongly as glucose-1-P with its phosphate and glucose positions identical to
those of glucose-1-P.[195] This indicates that replacement of the O2 hydroxyl of glucose-1-P
by an amino group leads to tighter binding with Glu 672 but not to "down" position of the
phosphate in glucosamine-1-P. Kinetic experiments have shown that glucosamine-1-P will
act as a competitive inhibitor of GP*b* (with respect to glucose-1-P) with a $K_i = 0.3$ m*M*.
This may be compared with the K_m for glucose-1-P of 3 m*M* (in the presence of AMP) and
37 m*M* (in the presence of IMP). 2-Deoxy-2-fluoro-α-D-glucose-1-P which can be utilized
as a substrate by GP ($K_m = 2.6$ m*M*) exhibits a catalytic rate 10^5 times slower than glucose-
1-P.[249] Proposals for the involvement of certain amino acid residues in the binding of substrate
or in catalysis can now be tested in site-directed mutagenesis experiments. Cloning and
expression of *E. coli* maltodextrin phosphorylase has been achieved and these studies are

FIGURES 27B and 27C.

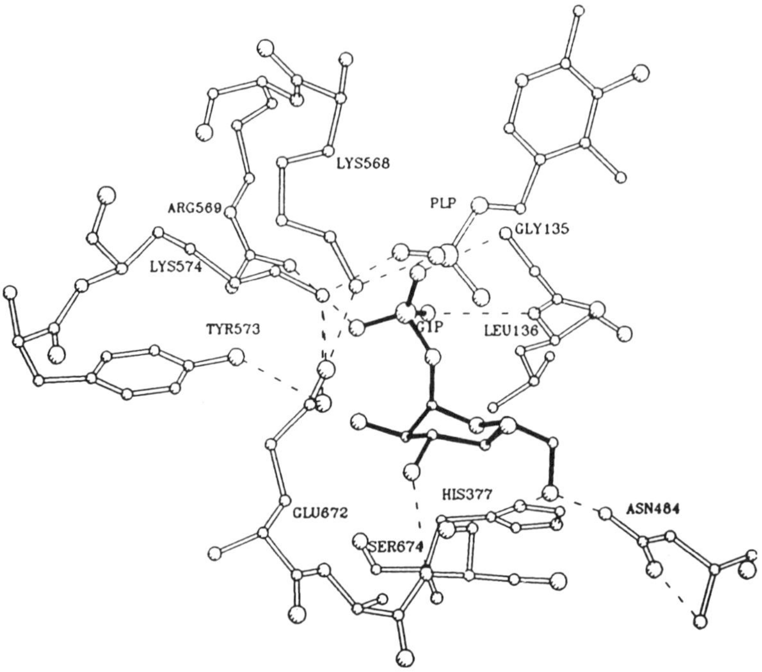

FIGURE 28.   The interactions of glucose-1-P (G1P) to the R-state phosphorylase *b*. (Hu, S.-H. and Johnson, L. N., unpublished results.)

at an advanced stage.[208,224] *E. coli* maltodextrin phosphorylase is always active; it requires neither allosteric effector nor covalent modification and has allowed the study of catalysis without the complication of control. Sequence studies[207,207a] have shown that the bacterial enzyme is nearly 100% homologous to the rabbit muscle enzyme in the vicinity of the catalytic site but differs elsewhere especially in the N terminal portion of the polypeptide chain, observations that explain the lack of control properties. Site-directed mutagenesis studies on several key residues identified in the catalytic site have shown the contributions of these to the kinetic properties of the enzyme.[208] Substitution of Glu 637 by an Asp or Gln residue reduced k_{cat} to about 0.2% of native enzyme activity, while the K_m values for P_i, glucose-1-P and maltoheptaose were affected to a minor extent. ^{31}P-NMR experiments also indicated that the coenzyme 5′-phosphate in the Glu637Asp mutant enzyme was in a more protonated state than in wild-type enzyme. However, substitution of Tyr 538 by a Phe resulted in a mutant enzyme that retained more than 10% activity and a slightly decreased affinity for the substrates.[224] In the GP*b*-heptulose-2-P complex, Tyr 573, like Glu 672, is in hydrogen-bond contact to O2 hydroxyl of the glucosyl residue and in addition makes direct contacts to the substrate phosphate. From the consequences of these mutations it was concluded[224] that the side chain of Glu 637 predominantly stabilizes the transition-state intermediate probably through the O3 hydroxyl of the glucosyl pyranose and also promotes the active dianionic state of cofactor phosphate, while Tyr 538 by excluding the insertion of a water molecule from the active site, thus decreasing the error rate of the enzyme (hydrolysis/phosphorylysis), is mainly involved in the binding of both the glucopyranose ring and the substrate phosphate. The conformation observed for the heptulose-2-P molecule when bound to GP*b* is different from that observed for glucose-1-P in the single-crystal structure.[301] In the crystal structure of glucose-1-P the torsion angle (φ) O5-C1-O1-P = 87°. In the heptulose-2-P complex the corresponding angle is 224° and there is an increase in bond angle at O1 from 120° to 135°. The change in torsion angle appears to be directed by the additional methyl group in the β-configuration and there is an internal hydrogen bond between a phosphate oxygen and the O2 hydroxyl. The conformation of the glucosyl portion of heptulose-2-P is similar to that found for glucose-1-P, and in GP*b* both heptulose-2-P and glucose-1-P bind with their glucosyl groups in similar positions. However, the positions of the phosphate groups are quite different as a result of the change in torsion angle φ between heptulose-2-P and glucose-1-P. In the glucose-1-P complex with T-state GP*b* φ = 140°, the phosphate group makes no interactions with the cofactor phosphate and the movement of Arg 569 is not observed.[174] In the heptulose-2-P complex the phosphate group is turned towards the cofactor phosphate and the cofactor-P to sugar-P separation decreases from 6.1 Å in the glucose-1-P complex to 4.8 Å in the heptulose-2-P complex. The difference in phosphate position is due almost entirely to the difference in torsion angle O5-C1-O1-P between heptulose-2-P and glucose-1-P. The change directs the heptulose-2-P phosphate towards the cofactor phosphate and achieves a conformation that is relevant for catalysis. How a "down" position for glucose-1-P phosphate (i.e., phosphate in contact with the cofactor phosphate) is accomplished is still unresolved. It might be anticipated that the reactive conformation is favoured by the binding of oligosaccharide to the catalytic site channel, i.e., the active conformation of the enzyme is stabilized by the presence of both substrates. Thus, when the reaction proceeds in the direction of glycogen degradation the glycosidic bond is weakened by direct protonation, whereas when the reaction proceeds in the direction of glycogen synthesis the glycosidic bond is weakened by steric factors which facilitate the development of the carbonium ion and ensure that the phosphate is able to act as a good base to abstract a proton from the oligosaccharide. In both mechanisms (Figure 24) the substrate phosphate plays a crucial role in stabilization of the transition state.[127]

IX. EPILOGUE

The structural results to date have provided the events leading from the low affinity dimeric T state to the activated tetrameric R-state conformation and some understanding of the catalytic mechanism. In the absence of structural data concerning oligosaccharide binding to the catalytic site channel, all proposals for the catalytic mechanism must remain vague concerning the events involving oligosaccharide substrate. The solution of the structure of R-state phosphorylase opens up the possibility for direct observation of oligosaccharide bound at the catalytic site and these studies, which are in progress, should allow further tests of the proposals for the catalytic mechanism and the nature of the recognition processes. The continued characterization of new phosphorylase-inhibitor complexes by crystallographic techniques aided by molecular calculations and modeling work might allow the design and synthesis of more potent inhibitors that may make it possible to define the structure of the transition state. A detailed structural description of the catalytic properties of the fully active dimer must also await structure determination of the enzyme in the dimeric R-state conformation. Rabbit muscle GP*b* has been recently successfully expressed in *E. coli*.[19] This achievement undoubtedly sets the stage for new developments in the field. Crystals of both wild-type and Pro48Thr GP*b* have been obtained which are isomorphous with those of the rabbit muscle GP*b* ($P4_32_12$, $a = b = 128.4$ Å, $c = 116.6$ Å) and an investigation of the structural mechanism of allosteric regulation is therefore possible. The crystal structure of the nonallosteric phosphorylase from potato tubers which has been crystallized in the trigonal space group $P3_121$ ($a = 128.30$ Å, $c = 117.45$ Å with a monomer in the asymmetric unit) is now well advanced in Prof. M. Buehner's laboratory in Wurzburg[270] using molecular replacements methods with R-state rabbit muscle phosphorylase as a starting structure, while attempts are being made to determine conditions favourable to the crystallization of *E. coli* maltodextrin phosphorylase and some of its mutants recently developed by site-directed mutagenesis.[208,225] Both potato and *E. coli* phosphorylases are always active and require neither allosteric effectors nor covalent modification. The enzymes also prefer linear oligosaccharides as substrates and have no a glycogen storage site. For these properties potato and *E. coli* phosphorylases provide attractive systems to study catalysis without the complications of allosteric transitions.

ACKNOWLEDGMENTS

We wish to thank our collaborators D. Barford, D. Beer, B. Bernet, L. Duke, J. Hajdu, S.-H. Hu, D. D. Leonidas, J. L. Martin, A. E. Melpidou, P. J. McLaughlin, A. C. Papageorgiou, M. S. P. Samsom, D. I. Stuart, A. Vasella, and K. Veluraja for their outstanding contributions to the research work on phosphorylase. We especially thank D. D. Leonidas and H. S. Tsitoura for drawing Figure 2. We are grateful to G. Bot, M. F. Browner, R. J. Fletterick, C. B. Newgard, S. Shaltiel, S. R. Sprang, W. Stalmans, S. G. Withers, and W. J. Whelan who sent us reprints and preprints and for allowing us to describe their results before publication. In addition we are grateful to the staff at the SERC's Synchrotron Radiation Source, Daresbury for the synchrotron facilities.

REFERENCES

1. **Acharya, K. R., Stuart, D. I., Varvill, K. M., and Johnson, L. N.,** Glycogen phosphorylase *b:* description of the protein structure. Monogr., World Scientific Publishing, Singapore, 1991.
2. **Ariki, M. and Fukui, T.,** Inhibition of α-glucan phosphorylase by α-D-glucopyranosyl fluoride, *J. Biochem.,* 78, 1191, 1975.
3. **Ariki, M. and Fukui, T.,** Affinity of glucose analogues for α-glucan phosphorylases from rabbit muscle and potato tubers, *J. Biochem.,* 81, 1017, 1977.
4. **Bailey, J. M. and Whelan, W. J.,** The roles of glucose and AMP in regulating the conversion of phosphorylase *a* into phosphorylase *b, Biochem. Biophys. Res. Commun.,* 46, 191, 1972.
5. **Baranowski, T., Illingworth, B., Brown, O. H., and Cori, C. F.,** The isolation of pyridoxal 5'-phosphate from crystalline muscle phosphorylase, *Biochim. Biophys. Acta,* 25, 16, 1957.
6. **Barford, D.,** Crystallographic studies on glycogen phosphorylase *b,* Ph.D. thesis, Oxford University, 1988.
7. **Barford, D., Schwabe, J. W. R., Oikonomakos, N. G., Acharya, K. R., Hajdu, J., Papageorgiou, A. C., Martin, J. L., Knott, J. C. A., Vasella, A., and Johnson, L. N.,** Channels at the catalytic site of glycogen phosphorylase: binding and kinetic studies with the β-glycosidase inhibitor D-Gluconohydroximo-1,5-lactone *N*-phenylurethane, *Biochemistry,* 27, 6733, 1988.
8. **Barford, D. and Johnson, L. N.,** The allosteric transition of glycogen phosphorylase, *Nature,* 340, 609, 1989.
9. **Barford, D., Hu, S.-H., and Johnson, L. N.,** The structural mechanism for glycogen phosphorylase control by phosphorylation and AMP, *J. Mol. Biol.,* 218, 233, 1991.
10. **Battersby, M. K. and Radda, G. K.,** The stereospecificity of glucose-6-phosphate binding site of glycogen phosphorylase *b, FEBS Lett.,* 72, 319, 1976.
11. **Battersby, M. K. and Radda, G. K.,** Intersubunit transmission of ligand effects in the glycogen phosphorylase *b* dimer, *Biochemistry,* 18, 3774, 1979.
12. **Beer, D. and Vasella, A.,** Aldonhydroximo-lactones. Preparation and determination of configuration, *Helv. Chim. Acta,* 68, 2254, 1985.
13. **Beer, D. and Vasella, A.,** Inhibition of emulsin by D-gluconohydroximo-1,5-lactone and related compounds, *Helv. Chim. Acta,* 69, 267, 1986.
14. **Black, W. J. and Wang, J. H.,** Studies on the allosteric activation of glycogen phosphorylase *b* by nucleotides. I. Activation of phosphorylase *b* by inosine monophosphate, *J. Biol. Chem.,* 243, 5892, 1968.
15. **Blake, C. C. F., Mair, G. A., North, A. C. T., Phillips, D. C., and Sarma, V. R.,** On the conformation of the hen egg-white lysozyme molecule, *Proc. R. Soc. London, (Biol.) Ser. B,* 167, 365, 1967a.
16. **Blake, C. C. F., Johnson, L. N., Mair, G. A., North, A. C. T., Phillips, D. C., and Sarma, V. R.,** Crystallographic studies of the activity of hen egg-white lysozyme, *Proc. R. Soc. London, (Biol.) Ser. B,* 167, 378, 1967b.
17. **Bollen, M., Malaisse-Lagae, F., Malaisse, W., and Stalmans, W.,** The interaction of phosphorylase *a* with D-glucose displays α-stereospecificity, *Biochim. Biophys. Acta,* 1038, 141, 1990.

18. **Bresler, S. and Firsov, L.,** Spectroscopic study of the enzyme-substrate complex of phosphorylase *b, J. Mol. Biol.,* 35, 131, 1968.

19. **Browner, M. F., Rasor, P., Tugendreich, S., and Fletterick, R. J.,** Temperature-sensitive production of rabbit muscle glycogen phosphorylase in *Escherichia Coli, Prot. Eng.,* 4, 351, 1991.

20. **Buc, H.,** On the allosteric interaction between 5′-AMP and orthophosphate on phosphorylase *b, Biochem. Biophys. Res. Commun.,* 28, 59, 1967.

21. **Buc, H., Buc, M. H., Oudin, L. C., and Winkler, H.,** Conformational changes in glycogen phosphorylase *b* from rabbit skeletal muscle. Equilibrium and kinetic studies, in *Dynamic Aspects on Conformational Changes in Macromolecules,* Sadron, C., Ed., Reidel, Dordrecht, 1973, 225.

22. **Busby, S. J. W. and Radda, G. K.,** Regulation of the glycogen phosphorylase system—From physical measurements to biological speculations, *Curr. Top. Cell. Regul.,* 10, 89, 1976.

23. **Caudwell, B., Antoniw, J. F., and Cohen, P.,** Calsequestrin, myosin and the components of the protein-glycogen complex in rabbit skeletal muscle, *Eur. J. Biochem.,* 86, 511, 1978.

24. **Chachaty, C., Forchioni, A., Morange, M., and Buc, H.,** Conformation of purine ribosyl 5′-nucleotides bound to glycogen phosphorylase. Nuclear-magnetic-resonance and electron-spin-resonance investigations of the effect of substrates, *Eur. J. Biochem.,* 82, 363, 1978.

25. **Chang, Y. C., McCalmont, T., and Graves, D. J.,** Functions of the 5′-phosphoryl group of pyridoxal 5′-phosphate in phosphorylase: a study using pyridoxal-reconstituted enzyme as a model system, *Biochemistry,* 22, 4987, 1983.

26. **Chang, Y. C. and Graves, D. J.,** Use of 6-fluoroderivatives of pyridoxal and pyridoxal phosphate in the study of the coenzyme function in glycogen phosphorylase, *J. Biol. Chem.,* 260, 2709, 1985.

27. **Chang, Y. C., Scott, R. D., and Graves, D. J.,** Function of PLP in glycogen phosphorylase: ^{19}F NMR and kinetic studies of phosphorylase reconstituted with 6-fluoropyridoxal and 6-fluoropyridoxal phosphate, *Biochemistry,* 25, 1932, 1986.

28. **Chang, Y. C., Scott, R. D., and Graves, D. J.,** Function of pyridoxal 5′-phosphate in glycogen phosphorylase: a model study using 6-fluoro-5′-pyridoxal- and 5′-deoxy-pyridoxal-reconstituted enzymes, *Biochemistry,* 26, 360, 1987.

29. **Cohen, P.,** The role of cyclic-AMP dependent protein kinase in the regulation of glycogen metabolism in mammalian skeletal muscle, *Curr. Top. Cell. Regul.,* 14, 117, 1978.

30. **Cohen, P.,** Control of Enzyme Activity, 2nd ed., Chapman and Hall, London, 1983a.

31. **Cohen, P.,** Protein phosphorylation and the control of glycogen metabolism in skeletal muscle, *Philos. Trans. R. Soc. London, Ser. B,* 302, 13, 1983b.

32. **Cohen, P.,** Protein phosphorylation and hormone action, *Proc. R. Soc. London, (Biol.) Ser. B,* 234, 115, 1988.

33. **Cohn, M. and Cori, G. T.,** On the mechanism of action of muscle and potato phosphorylase, *J. Biol. Chem.,* 175, 89, 1948.

34. **Cohn, M.,** Mechanism of cleavage of glucose 1-phosphate, *J. Biol. Chem.,* 180, 771, 1949.

35. **Cori, C. F. and Cori, G. T.,** Mechanism of formation of hexosemonophosphate in muscle and isolation of a new phosphate ester, *Proc. Soc. Exp. Biol. Med.,* 34, 702, 1936.

36. **Cori, G. T., Colowick, S. P., and Cori, C. F.,** The action of nucleotides in the disruptive phosphorylation of glycogen, *J. Biol. Chem.,* 123, 381, 1938.

37. **Cori, G. T. and Cori, C. F.,** The kinetics of the enzymic synthesis of glycogen from glucose-1-phosphate, *J. Biol. Chem.,* 135, 733, 1940.

38. **Cori, C. F., Cori, G. T., and Green, A. A.,** Crystalline muscle phosphorylase. III. Kinetics, *J. Biol. Chem.,* 151, 39, 1943.

39. **Cortijo, M., Steinberg, I. Z., and Shaltiel, S.,** Fluorescence of glycogen phosphorylase *b.* Structural transitions and energy transfer, *J. Biol. Chem.,* 246, 933, 1971.

40. **Cortijo, M. and Shaltiel, S.,** On the microenvironment of the pyridoxine 5-phosphate residue in NaBH₄-reduced glycogen phosphorylase *b.* Absorption and fluorescence studies, *Eur. J. Biochem.,* 29, 134, 1972.

41. **De Barsy, T., Stalmans, W., Laloux, M., De Wulf, H., and Hers, H. G.,** The effect of the glucose on the conversion of phosphorylase *a* into phosphorylase *b* or *b′, Biochem. Biophys. Res. Commun.,* 46, 183, 1972.

42. **Detwiler, T. C., Gratecos, D., and Fischer, E. H.,** Rabbit muscle phosphorylase phosphatase-2. Kinetic properties and behavior in glycogen particles, *Biochemistry,* 16, 4818, 1977.

43. **Dombradi, V.,** Structural aspects of the catalytic and regulatory function of glycogen phosphorylase, *Int. J. Biochem.,* 13, 125, 1981.

44. **Dombradi, V., Toth, B., Bot, G., Hajdu, J., and Friedrich, P.,** Interactions of ligands in phosphorylase *a* as monitored by cross-linking and enzymic modifications: synergism of glucose and caffeine manifested in the exposure of the N-terminal segment, *Int. J. Biochem.,* 14, 277, 1982.

45. **Dobradi, V., Gergely, P., and Bot, G.,** Limited proteolysis by subtilisin reveals structural differences between phosphorylase *a* and *b, Int. J. Biochem.,* 15, 1089, 1983a.

46. **Dobradi, V., Toth, B., Gergely, P., and Bot, G.,** Limited proteolysis of glycogen phosphorylase *a* by subtilisin BPN', *Int. J. Biochem.,* 15, 1329, 1983b.

47. **Dreyfus, M., Vandenbunder, B., and Buc, H.,** Stabilization of a phosphorylase *b* active conformation by hydrophobic solvents, *FEBS Lett.,* 95, 185, 1978.

48. **Dreyfus, M., Vandenbunder, B., and Buc, H.,** Mechanism of allosteric activation of glycogen phosphorylase probed by the reactivity of essential arginyl residues. Physicochemical and kinetic studies, *Biochemistry,* 19, 3634, 1980.

49. **Eagles, P. A. M., Iqbal, M., Johnson, L. N., Mosley, J., and Wilson, K. S.,** A tetragonal crystal form of phosphorylase *b., J. Mol. Biol.,* 71, 803, 1972.

50. **Engers, H. D. and Madsen, N. B.,** The effects of anions on the activity of phosphorylase *b, Biochem. Biophys. Res. Commun.,* 33, 49, 1968.

51. **Engers, H. D., Bridger, W. A., and Madsen, N. B.,** Kinetic mechanism of phosphorylase *b, J. Biol. Chem.,* 244, 5936, 1969.

52. **Engers, H. D., Shechosky, S., and Madsen, N. B.,** Kinetic mechanism of phosphorylase *a.* I. Initial velocity studies, *Can. J. Biochem.,* 48, 746, 1970.

53. **Farkas, I., Toth, B., and Gergely, P.,** Regulation of the dephosphorylation of phosphorylase *a* by glucose, AMP and polyamines, *Int. J. Biochem.,* 20, 197, 1988.

54. **Fasold, H., Ortanderl, F., Huber, R., Bartels, K., and Scwager, P.,** Crystallization and crystallographic data of rabbit muscle phosphorylase *a* and *b, FEBS Lett.,* 21, 229, 1972.

55. **Feldmann, K. and Helmreich, E. J. M.,** The pyridoxal 5'-phosphate site in rabbit skeletal muscle glycogen phosphorylase *b:* an ultraviolet and ^{1}H and ^{31}P nuclear magnetic resonance spectroscopic study, *Biochemistry,* 15, 2394, 1976.

56. **Feldmann, K., Jeisel, H. J., and Helmreich, E. J. M.,** Complementation of subunits from glycogen phosphorylases of frog and rabbit skeletal muscle and rabbit liver, *Eur. J. Biochem.,* 65, 285, 1976.

57. **Feldmann, K. and Hull, W. E.,** ^{31}P nuclear magnetic resonance studies of glycogen phosphorylase from rabbit skeletal muscle: Ionization states of pyridoxal 5'-phosphate, *Proc. Natl. Acad. Sci. U.S.A.,* 74, 856, 1977.

58. **Firsov, L. M., Bogacheva, I., and Bresler, S. E.,** Secondary isotope effect in the phosphorylase reaction, *Eur. J. Biochem.,* 42, 605, 1974.

59. **Fischer, E. H. and Krebs, E. G.,** Conversion of phosphorylase *b* to phosphorylase *a* in muscle extracts, *J. Biol. Chem.,* 216, 121, 1955.

60. **Fischer, E. H. and Krebs, E. G.,** The isolation and crystallization of rabbit skeletal muscle phosphorylase *b, J. Biol. Chem.,* 231, 65, 1958.

61. **Fischer, E. H., Kent, A. B., Snyder, E. R., and Krebs, E. G.,** The reaction of sodium borohydride with muscle phosphorylase, *J. Am. Chem. Soc.,* 80, 2906, 1958.

62. **Fischer, E. H., Graves, D. J., Crittenden, E. R. S., and Krebs, E. G.,** Structure of the site phosphorylated in the phosphorylase *b* to *a* reaction, *J. Biol. Chem.,* 234, 1698, 1959.

63. **Fischer, E. H. and Krebs, E. G.,** Muscle phosphorylase *b,* in *Methods in Enzymology,* Vol. 5, Academic Press, New York, 1962, 369.

64. **Fischer, E. H.,** Binding of vitamin B_6-coenzymes and labeling of active site enzymes by sodium borohydride reduction, in *Structure and Activity of Enzymes,* Goodwin, T. W., Harris, J. I., and Hartley, B. S., Eds., 1st Symp. Fed. Eur. Biochem. Soc., Academic Press, London, 1964, 111.

65. **Fletterick, R. J., Sygusch, J., Murray, N., Madsen, N. B., and Johnson, L. N.,** Low-resolution structure of the glycogen phosphorylase *a* monomer and comparison with phosphorylase *b, J. Mol. Biol.,* 103, 1, 1976.

66. **Fletterick, R. J. and Madsen, N. B.,** The structures and related functions of phosphorylase *a, Annu. Rev. Biochem.,* 49, 31, 1980.

67. **Fletterick, R. J. and Sprang, S. R.,** Glycogen phosphorylase structures and function, *Acc. Chem. Res.,* 15, 361, 1982.

68. **Ford, G. C., Eichele, G., and Jansonius, J. N.,** Three-dimensional structure of a pyridoxal-phosphate-dependent enzyme, mitochondrial aspartate aminotransferase, *Proc. Natl. Acad. Sci. U.S.A.,* 77, 2559, 1980.

69. **Fukui, T., Shimomura, S., and Nakano, K.,** Potato and rabbit muscle phosphorylases: comparative studies on the structure, function and regulation of regulatory and nonregulatory enzymes, *Mol. Cell. Biochem.,* 42, 129, 1982.

70. **Fukui, T. and Tagaya, M.,** A Novel Coenzymatic Function of Pyridoxal 5'-phosphate, *BioEssays,* 5, 21, 1986.

71. **Gani, V., Kupfer, A., and Shaltiel, S.,** A micellar model for the pyridoxal 5'-phosphate site of glycogen phosphorylase, *Biochemistry,* 17, 1294, 1978.

72. **Gergely, P., Toth, B., Dombradi, V., Matko, J., and Bot, G.,** Heterotropic interactions of AMP and glucose binding sites in phosphorylase *a* are destroyed by limited proteolysis, *Biochem. Biophys. Res. Commun.,* 113, 825, 1983.

73. **Gold, A. M., Legrand, E., and Sanchez, G. R.,** Inhibition of muscle phosphorylase *a* by 5-Glucono-lactone, *J. Biol. Chem.,* 246, 5700, 1971.

74. **Gold, A. M. and Osber, M. P.,** On the mechanism of glycogen phosphorylase, *Arch. Biochem. Biophys.,* 153, 784, 1972.

75. **Goldsmith, E., Sprang, S., and Fletterick, R. J.,** Structure of maltoheptaose by difference Fourier methods and a model for glycogen, *J. Mol. Biol.,* 156, 411, 1982.

76. **Goldsmith, E. and Fletterick, R. J.,** Oligosaccharide conformation and protein saccharide interactions in solution, *Pure Appl. Chem.,* 55, 577, 1983.

77. **Goldsmith, E. J., Fletterick, R. J., and Withers, S. G.,** The three-dimensional structure of acarbose bound to glycogen phosphorylase, *J. Biol. Chem.,* 262, 1449, 1987.

78. **Goldsmith, E. J., Sprang, S. R., and Fletterick, R. J.,** Alternative binding models for maltopentaose in the activation site of glycogen phosphorylase *a,* in *Protein Carbohydrate Interactions,* Trans. Am. Crystallogr. Assoc., Einspahr, H. M., Ed., 1989a, 87.

79. **Goldsmith, E. J., Sprang, S. R., Hamlin, R., Xuong, N.-H., and Fletterick, R. J.,** Domain separation in the activation of glycogen phosphorylase *a, Science,* 245, 528, 1989b.

80. **Gorenstein, D. G.,** Stereoelectronic effects in biomolecules, *Chem. Rev.,* 87, 1047, 1987.

81. **Graves, D. J., Fischer, E. H., and Krebs, E. G.,** Specificity studies on muscle phosphorylase phosphatase, *J. Biol. Chem.,* 235, 805, 1960.

82. **Graves, J., Mann, S. A. S., Philip, G., and Oliveira, R. J.,** A probe into the catalytic activity and subunit assembly of glycogen phosphorylase. Desensitization of allosteric control by limited tryptic digestion, *J. Biol. Chem.,* 243, 6090, 1968.

83. **Graves, D. J. and Wang, J. H.,** α-glucan phosphorylases-chemical and physical basis of catalysis and regulation, in *The Enzymes,* Vol. 7, 3rd ed., Boyer, P. D., Ed., Academic Press, New York, 1972, 435.

84. **Griffiths, J. R., Dwek, R. A., and Radda, G. K.,** Conformational changes in glycogen phosphorylase studied with a spin-label probe, *Eur. J. Biochem.,* 61, 237, 1976a.

85. **Griffiths, J. R., Dwek, R. A., and Radda, G. K.,** Heterotropic interactions of ligands with phosphorylase *b, Eur. J. Biochem.,* 61, 243, 1976b.

86. **Gusev, N. B., Hajdu, J., and Friedrich, P.,** Mobility of the N-terminal tail of phosphorylase *b* as revealed by cross-linking with bifunctional diimidates, *Biochem. Biophys. Res. Commun.,* 90, 70, 1979.

87. **Hackert, M. L. and Jacobson, R. A.,** The Crystal and Molecular Structure of D-Glucono-(1,5)-lactone, *Acta Crystallogr.,* B27, 203, 1971.

88. **Hajdu, J., Dombradi, V., Bot, G., and Friedrich, P.,** Structural changes in glycogen phosphorylase as revealed by cross-linking with bifunctional diimidates: phosphorylase *b, Biochemistry,* 18, 4037, 1979.

89. **Hajdu, J., Acharya, K. R., Stuart, D. I., McLaughlin, P. J., Barford, D., Oikonomakos, N. G., Klein, H., and Johnson, L. N.,** Catalysis in the crystal: synchrotron radiation studies with glycogen phosphorylase *b, EMBO J.,* 6, 539, 1987.

90. **Hajdu, J., Acharya, K. R., Stuart, D. I., Barford, D., and Johnson, L. N.,** Catalysis in enzyme crystals, *Trends Biochem. Sci.,* 13, 104, 1988.

91. **Hanozet, G., Pircher, H.-P., Vanni, P., Oesch, B., and Semenza, G.,** An example of enzyme hysteresis. The slow and tight interactions of some fully competitive inhibitors with small intestinal sucrase, *J. Biol. Chem.,* 256, 3703, 1981.

92. **Haschke, R. H., Heilmeyer, L. M. G., Meyer, F., and Fischer, E. H.,** Control of phosphorylase activity in a muscle glycogen particle. III. Regulation of phosphorylase phosphate, *J. Biol. Chem.,* 245, 6657, 1970.

93. **Haschke, R. H., Gratz, K. W., and Heilmeyer, L. M. G.,** The role of phosphorylase activity in a muscle glycogen particle, *J. Biol. Chem.,* 247, 5251, 1972.

94. **Hedrick, J. L., Shaltiel, S., and Fischer, E. H.,** On the role of pyridoxal 5'-phosphate in phosphorylase. III. Physicochemical properties and reconstitution of apophosphorylase *b, Biochemistry,* 5, 2117, 1966.

95. **Hehre, E. J., Brewer, C. F., Uchiyama, T., Schlesselmann, P., and Lehmann, J.,** Scope and mechanism of carbohydrase action. Stereospecific hydration of 2,6-anhydro-deoxy-D-gluco-hept-1-enitol catalysed by α- and β-glucosidases and an inverting *exo*-α-glucanase, *Biochemistry,* 19, 3557, 1980.

96. **Heilmeyer, L. M. G., Meyer, F., Haschke, R. H., and Fischer, E. H.,** Control of phosphorylase activity in a muscle glycogen particle. II. Activation by calcium, *J. Biol. Chem.,* 245, 6649, 1970.

97. **Helliwell, J. R., Greenough, T. J., Carr, P. D., Rule, S. A., Moore, P. R., Thompson, A. W., and Worgan, J. S.,** A central data collection facility for protein crystallography, small angle diffraction, and scattering with the Daresbury Laboratory synchrotron radiation source, *Engl. J. Phys. E,* 15, 1363, 1982.

98. **Helliwell, J. R., Papiz, M. Z., Glover, I. D., Habash, J., Thompson, A. W., Moore, P. R., Harris, N., Croft, D., and Pantos, E.,** The wiggler protein crystallography work station at the Daresbury synchrotron radiation source: progress and results, *Nucl. Instrum. Methods Phys. Res.,* Ser. A., 246, 617, 1986.

99. **Helmreich, E. and Cori, C. F.,** The role of adenylic acid in the activation of phosphorylase, *Proc. Natl. Acad. Sci. U.S.A.,* 51, 131, 1964.

100. **Helmreich, E., Michaelides, M. C., and Cori, C. F.,** Effect of substrates and substrate analogs on the binding of 5'-adenylic acid to muscle phosphorylase *a, Biochemistry,* 6, 3695, 1967.

101. **Helmreich, E. J. M. and Klein, H. W.,** The role of pyridoxal phosphate in the catalysis of glycogen phosphorylases, *Angew. Chem. Int. Ed. Engl.,* 19, 441, 1980.

102. **Hers, H. G.,** The control of glycogen metabolism in the liver, *Annu. Rev. Biochem.,* 45, 167, 1976.

103. **Herzfield, J. and Stanley, H. E.,** A general approach to co-operativity and its application to the oxygen equilibrium of hemoglobin and its effectors, *J. Mol. Biol.,* 82, 231, 1974.

104. **Hill, A. V.,** The possible effects of the aggregation of the molecule of haemoglobin on its dissociation curves, *J. Physiol. (London),* 40, 4, 1910.

105. **Ho, H. C. and Wang, J. H.,** Calorimetric studies of the interaction between phosphorylase *b* and its nucleotide activators, *Biochemistry,* 12, 4750, 1973.

106. **Horinishi, N., Tagaya, M., and Fukui, T.,** Base-catalysed reactivation of glycogen phosphorylase reconstituted with a coenzyme-substrate conjugate and its analogues, *FEBS Lett.,* 231, 426, 1988.

107. **Hu, H. Y. and Gold, A. M.,** Kinetics of glycogen phosphorylase *a* with a series of semisynthetic branched saccharides. A model for binding of polysaccharides substrates, *Biochemistry,* 14, 2224, 1975.

108. **Hu, H. Y. and Gold, A. M.,** Inhibition of rabbit muscle glycogen phosphorylase by α-D-glucopyranose 1,2-cyclic phosphate, *Biochim. Biophys. Acta,* 525, 55, 1978.

109. **Huang, C. Y. and Graves, D. J.,** Correlation between subunit interactions and enzymatic activity of phosphorylase *a*. Method for determining equilibrium constants from initial rate measurements, *Biochemistry,* 9, 660, 1970.

110. **Hurley, J. H., Dean, A. M., Thorsness, P. E., Koshland, D. E., Jr., and Stroud, R. M.,** Regulation of isocitrate dehydrogenase by phosphorylation involves no long-range conformational change in the free enzyme, *J. Biol. Chem.,* 265, 3599, 1989.

111. **Hurley, J. H., Dean, A. M., Sohl, J. L., Koshland, D. E., Jr., and Stroud, R. M.,** Regulation of an enzyme by phosphorylation at the active site, *Science,* 249, 1012, 1990.

112. **Hwang, P., Stern, M., and Fletterick, R. J.,** Purification and crystallization of bovine liver phosphorylase, *Biochim. Biophys. Acta,* 791, 252, 1984.

113. **Hwang, P. K., Browner, M., Rath, V., Tugendreich, S., and Fletterick, R. J.,** Design and variation of allosteric regulation in glycogen phosphorylase, Int. Symp. Vitamin B_6 and Carbonyl Catalysis (Abstr.), p. 81, 1990.

114. **Jenkins, J. A., Johnson, L. N., Stuart, D. I., Stura, E. A., Wilson, K. S., and Zanotti, G.,** Phosphorylase: control and activity, *Philos. Trans. R. Soc. London,* Ser. B., 293, 23.

115. **Johnson, L. N., Madsen, N. B., Mosley, J., and Wilson, K. S.,** The crystal structure of phosphorylase *b* at 6 Å resolution, *J. Mol. Biol.,* 90, 703, 1974.

116. **Johnson, L. N., Stura, E. A., Wilson, K. S., Sansom, M. S. P., and Weber, I. T.,** Nucleotide binding to glycogen phosphorylase *b* in the crystal, *J. Mol. Biol.,* 134, 639, 1979.

117. **Johnson, L. N., Jenkins, J. A., Wilson, K. S., Stura, E. A., and Zanotti, G.,** Proposals for the catalytic mechanism of glycogen phosphorylase *b* prompted by crystallographic studies on glucose-1-phosphate binding, *J. Mol. Biol.,* 140, 565, 1980.

118. **Johnson, L. N., Stura, E. A., Sansom, M. S. P., and Babu, Y. S.,** Oligosaccharide binding to glycogen phosphorylase *b, Biochem. Soc. Trans.,* 11, 142, 1983.

119. **Johnson, L. N.,** Protein crystallography, in *Modern Methods in Biochemistry, Part A,* Neuberger/Van Deenen, Eds., 347, 1985.

120. **Johnson, L. N., Acharya, K. R., Stuart, D. I., Barford, D., Oikonomakos, N. G., Hajdu, J., and Varvill, K. M.,** Phosphate-recognition sites in catalysis and control of glycogen phosphorylase *b, Biochem. Soc. Trans.,* 15, 1001, 1987a.

121. **Johnson, L. N., Oikonomakos, N. G., Acharya, K. R., Stuart, D. I., Barford, D., Hajdu, J., and Varvill, K. M.,** The role of the pyridoxal phosphate in glycogen phosphorylase *b*, in *Biochemistry of Vitamin B_6,* Ser. B, Proc. 7th Int. Congr. Chem. Biol., Aspects of Vitamin B6 Catalysis, Korpella, T. P. and Christen, P., Eds., Birkhauser-Verlag, Basel, 1987b, 255.

122. **Johnson, L. N., Cheetham, J., McLaughlin, P. J., Acharya, K. R., Barford, D., and Phillips, D. C.,** Protein-oligosaccharide interactions; lysozyme, phosphorylase, amylases, *Curr. Top. Microbiol. Immunol.,* 139, 81, 1988.

123. **Johnson, L. N. and Hajdu, J.,** Synchrotron studies on enzyme catalysis in crystals, in *Biophysics and Synchrotron Radiation,* Hasnain, S., Ed., Ellis Horwood, Chichester, 1989, 142.

124. **Johnson, L. N., Hajdu, J., Acharya, K. R., Stuart, D. I., McLaughlin, P. J., Oikonomakos, N. G., and Barford, D.,** Glycogen Phosphorylase *b*, in *Allosteric Enzymes,* Herve, G., Ed., CRC Press, Boca Raton, FL, 1989, 81.

125. **Johnson, L. N. and Barford, D.,** Glycogen phosphorylase. The structural basis of the allosteric response and comparison with other allosteric proteins, *J. Biol. Chem.,* 265, 2409, 1990.

126. **Jones, D. C. and Cowgill, R. W.,** Evidence for the binding of pyridoxal 5'-phosphate in a hydrophobic region of glycogen phosphorylase *b* dimer, *Biochemistry,* 10, 4276, 1971.

127. **Johnson, L. N., Acharya, K. R., Jordan, M. D., and McLaughlin, P. J.,** Refined crystal structure of the phosphorylase-heptulose 2-phosphate-oligosaccharide-AMP complex, *J. Mol. Biol.,* 211, 645, 1990.

128. **Illingworth, B., Jansz, H. S., Brown, T. H., and Cori, C. F.,** Observations on the function of pyridoxal 5'-phosphate in phosphorylase, *Proc. Natl. Acad. Sci. U.S.A.,* 44, 1180, 1958.

129. **Kastenschmidt, L. L., Kastenschmidt, J., and Helmreich, E.,** Subunit interactions and their relationship to the allosteric properties of rabbit skeletal muscle phosphorylase *b, Biochemistry,* 7, 3590, 1968a.

130. **Kastenschmidt, L. L., Kastenschmidt, J., and Helmreich, E.,** The effect of temperature on the allosteric transitions of rabbit muscle phosphorylase *b, Biochemistry,* 7, 4543, 1968b.

131. **Kasvinsky, P. J. and Madsen, N. B.,** Activity of glycogen phosphorylase in the crystalline state, *J. Biol. Chem.,* 251, 6852, 1976.

132. **Kasvinsky, P. J., Madsen, N. B., Sygusch, J., and Fletterick, R. J.,** The regulation of glycogen phosphorylase *a* by nucleotide derivatives, *J. Biol. Chem.,* 253, 3343, 1978a.

133. **Kasvinsky, P. J., Shechosky, S., and Fletterick, R. J.,** Synergistic regulation of phosphorylase *a* by glucose and caffeine, *J. Biol. Chem.,* 253, 9102, 1978b.

134. **Kasvinsky, P. J., Madsen, N. B., Fletterick, R. J., and Sygusch, J.,** X-ray crystallographic and kinetic studies of oligosaccharide binding to phosphorylase, *J. Biol. Chem.,* 253, 1290, 1978c.

135. **Kasvinsky, P. J., Fletterick, R. J., and Madsen, N. B.,** Regulation of the dephosphorylation of glycogen phosphorylase *a* and synthase *b* by glucose and caffeine in isolated hepatocytes, *Can. J. Biochem.,* 59, 387, 1981.

136. **Kasvinsky, P. J.,** The effect of AMP on inhibition of muscle phosphorylase *a* by glucose derivatives, *J. Biol. Chem.,* 25, 10805, 1982.

137. **Kent, A. B., Krebs, E. G., and Fischer, E. H.,** Properties of crystalline phosphorylase *b, J. Biol. Chem.,* 232, 549, 1958.

138. **Kipnis, D. M., Helmreich, E., and Cori, C. F.,** Studies of tissue permeability. IV. The distribution of glucose between plasma and muscle, *J. Biol. Chem.,* 234, 165, 1958.

139. **Kirby, A. J.,** Stereoelectronic effects on acetal hydrolysis, *Acc. Chem. Res,* 17, 305, 1984.

140. **Kirsch, J. E., Eichele, G., Ford, G. C., Vincent, M. G., Jansonius, J. N., Gehring, H., and Christen, P.,** Mechanism of action of aspartate aminotransferase proposed on the basis of its spatial structure, *J. Mol. Biol.,* 174, 497, 1984.

141. **Klein, H. W., Schiltz, E., and Helmreich, E. J. M.,** A catalytic role of the dianionic 5'-phosphate of pyridoxal 5'-phosphate in glycogen phosphorylases: formation of a covalent glycosyl intermediate, in *Protein Phosphorylation,* Rosen, O. M. and Krebs, E. G., Eds., Cold Spring Harbor Laboratory, 1981, 305.

142. **Klein, H. W., Palm, D., and Helmreich, E. J. M.,** General acid-base catalysis of α-glucan phosphorylases: stereospecific glucosyl transfer from D-glucal is a pyridoxal 5'-phosphate and orthophosphate (arsenate) dependent reaction, *Biochemistry,* 21, 6675, 1982.

143. **Klein, H. W., Im, M. J., Palm, D., and Helmreich, E. J. M.,** Does pyridoxal 5'-phosphate function in glycogen phosphorylase as an electrophilic or a general acid catalyst?, *Biochemistry,* 23, 5853, 1984.

144. **Klein, H. W., Im, M. J., and Palm, D.,** Mechanism of the phosphorylase reaction. Utilization of D-gluco-hept-1-enitol in the absence of primer, *Eur. J. Biochem.,* 157, 107, 1986.

145. **Kokesh, F. C. and Kakuda, Y.,** Evidence for intermediate formation in the mechanism of potato starch phosphorylase from exchange of the ester and phosphoryl oxygens of α-D-glucopyranosyl 1-phosphate, *Biochemistry,* 16, 2467, 1977.

146. **Koshland, D. E.,** Application of a theory of enzyme specificity to protein synthesis, *Proc. Natl. Acad. Sci. U.S.A.,* 44, 98, 1958.

147. **Koshland, D. E., Nemethy, G., and Filmer, D.,** Comparison of experimental data and theoretical models in proteins containing subunits, *Biochemistry,* 5, 365, 1966.

148. **Krebs, E. G.,** The effect of salmine on the activity of phosphorylase, *Biochim. Biophys. Acta,* 15, 508, 1954.

149. **Krebs, E. G. and Fischer, E. H.,** The phosphorylase *b* to *a* converting enzyme of rabbit skeletal muscle, *Biochim. Biophys. Acta,* 20, 150, 1956.

150. **Krebs, E. G.,** The enzymology of control by phosphorylation, in *The Enzymes,* Vol. 17, 3rd ed., Boyer, P. D. and Krebs, E. G., Eds., Academic Press, New York, 1986, 3.

151. **Ktenas, T. B., Sotiroudis, T. G., Oikonomakos, N. G., and Evangelopoulos, A. E.,** Effect of phenothiazines on the activity of glycogen phosphorylase *b, FEBS Lett.,* 88, 313, 1978.

152. **Ktenas, T. B., Oikonomakos, N. G., Sotiroudis, T. G., Nikolaropoulos, S., and Evangelopoulos, A. E.,** Effect of polymyxins on glycogen phosphorylase, *Biochem. Biophys. Res. Commun.,* 97, 415, 1980.

153. **Ktenas, T. B., Oikonomakos, N. G., Sotiroudis, T. G., Nikolaropoulos, S., and Evangelopoulos, A. E.,** Interaction of aliphatic amines with glycogen phosphorylase, *J. Biochem.,* 92, 2029, 1982.

154. **Lee, Y. M. and Benisek, W. F.,** Inactivation of phosphorylase *b* by potassium ferrate, *J. Biol. Chem.,* 253, 5460, 1978.

155. **Leonidas, D. D., Oikonomakos, N. G., Papageorgiou, A. C., Xenakis, A., Cazianis, C. T., and Bem, F.,** The ammonium sulphate activation of phosphorylase *b, FEBS Lett.,* 261, 23, 1990.

156. **Leonidas, D. D., Oikonomakos, N. G., and Papageorgiou, A. C.,** Sulphate activates phosphorylase *b* by binding to the Ser (P) site, *Biochim. Biophys. Acta,* 1076, 305, 1991.

157. **Lolis, E. and Petsko, G. A.,** Transition-state analogues in protein crystallography: probes of the structural source of enzyme catalysis, *Annu. Rev. Biochem.,* 59, 597, 1990.

158. **Lorek, A., Wilson, K. S., Sansom, M. S. P., Stuart, D. I., Stura, E. A., Jenkins, J. A., Zanotti, G., Hajdu, J., and Johnson, L. N.,** Allosteric interactions of glycogen phosphorylase *b.* A crystallographic study of glucose-6-phosphate and inorganic phosphate binding to di-imidate-cross-linked phosphorylase *b, Biochem. J.,* 218, 45, 1984.

159. **Lutaya, G. and Griffiths, J. R.,** Rapid formation of spermine in skeletal muscle during tetanic stimulation, *FEBS Lett.,* 123, 186, 1981.

160. **Madsen, N. B. and Cori, C. F.,** The interaction of muscle phosphorylase with *p*-chloromercuribenzoate-1. Inhibition of activity and the effect on the molecular weight, *J. Biol. Chem.,* 223, 1055, 1956.

161. **Madsen, N. B.,** Inhibition of glycogen phosphorylase by uridine diphosphate glucose, *Biochem. Biophys. Res. Commun.,* 6, 310, 1961.

162. **Madsen, N. B.,** Allosteric properties of phosphorylase *b, Biochem. Biophys. Res. Commun.,* 15, 390, 1964.

163. **Madsen, N. B. and Shechosky, S.,** Allosteric properties of phosphorylase b. II. Comparison with a kinetic model, *J. Biol. Chem.,* 242, 3301, 1967.

164. **Madsen, N. B.,** Metabolite effects on structure-function relationship in glycogen phosphorylase, in *Molecular Basis of Biological Activity,* Vol. 1, Gaede, K., Horecker, B. L., and Whelan, W. J., Eds., Academic Press, New York, 1972, 13.

165. **Madsen, N. B., Honikel, K. O., and James, M. N. G.,** Studies on glycogen phosphorylase in solution and in the crystalline state, in *Metabolic Interconversion of Enzymes,* Wieland, O., Helmreich, E., and Holzer, H., Eds., Springer, Berlin, 1972, 55.

166. **Madsen, N. B., Kasvinsky, P. J., and Fletterick, R. J.,** Allosteric transitions of phosphorylase *a* and the regulation of glycogen metabolism, *J. Biol. Chem.,* 253, 9097, 1978.

167. Reference deleted.

168. **Madsen, N. B., Shechosky, S., and Fletterick, R. J.,** Site-site interactions in glycogen phosphorylase *b* probed by ligands specific for each site, *Biochemistry,* 22, 4460, 1983.

169. **Madsen, N. B.,** Glycogen phosphorylase, in *The Enzymes,* Vol. 17, 3rd ed., Boyer, P. D. and Krebs, E. G., Eds., Academic Press, New York, 1986, 366.

170. **Madsen, N. B. and Withers, S. G.,** Glycogen phosphorylase, in *Coenzymes and Cofactors,* Vol. 1. Vitamin B_6 Pyridoxal Phosphate, Dolphin, D., Poulson, R., and Avramovic, O., Eds., John Wiley & Sons, New York, 1986, 355.

171. **Madsen, N. B.,** Glycogen phosphorylase and glycogen synthetase, in *A Study of Enzymes,* Vol. 3, Kuby, S. A., Ed., CRC Press, Boca Raton, FL, 1989.

172. **Martensen, T. M., Brotherton, J. E., and Graves, D. J.,** Kinetic studies of the activation of muscle phosphorylase phosphatase, *J. Biol. Chem.,* 248, 8329, 1973.

173. **Martin, J. L.,** Molecular interactions involving glycogen phosphorylase, Ph.D. thesis, University of Oxford, 1990.

174. **Martin, J. L., Johnson, L. N., and Withers, S. G.,** Comparison of the binding of glucose and glucose-1-phosphate derivatives to T-state glycogen phosphorylase *b, Biochemistry,* 29, 10745, 1990.

175. **Martin, J. L., Veluraja, K., Johnson, L. N., Fleet, G. W. J., Ramsden, N., Bruce, I., Oikonomakos, N. G., Papageorgiou, A. C., and Leonidas, D. D.,** The prediction of drug:macromolecule interactions, Proc. 12th Int. Symp. Med. Chem., in press.

176. **Mateo, P. L., Baron, C., Lopez-Mayorga, O., Jimenez, J. S., and Cortijo, M.,** AMP and IMP binding to glycogen phosphorylase *b.* A calorimetric and equilibrium dialysis study, *J. Biol. Chem.,* 259, 9384, 1984.

177. **Mathews, F. S.,** X-ray crystallographic study of glycogen phosphorylase, *Fed. Proc.,* 26, 831, 1967.

178. **McLaughlin, P. J., Stuart, D. I., Klein, H. W., Oikonomakos, N. G., and Johnson, L. N.,** Substrate-cofactor interactions for glycogen phosphorylase *b:* a binding study in the crystal with heptenitol and heptulose-2-phosphate, *Biochemistry,* 23, 5862, 1984.

179. **McLaughlin, P. J.,** Crystallographic studies on glycogen phosphorylase *b,* Ph.D. thesis, University of Oxford, 1985.

180. **McPherson, A.,** Current approaches to macromolecular crystallization, *Eur. J. Biochem.,* 189, 1, 1990.

181. **Melpidou, A. E. and Oikonomakos, N. G.,** Effect of glucose-6-P on the catalytic and structural properties of glycogen phosphorylase *a, FEBS Lett.,* 154, 105, 1983.

182. **Metzger, B., Helmreich, E., and Glaser, L.,** The mechanism of activation of skeletal muscle phosphorylase *a* by glycogen, *Proc. Natl. Acad. Sci. U.S.A.,* 57, 994, 1967.

183. **Meuwly, R. and Vasella, A.,** Synthesis of 1,2-*cis*-configurated glycosylphosphonates, *Helv. Chim. Acta,* 69, 25, 1986.

184. **Meyer, F., Heilmeyer, L. M. G., Haschke, H., and Fischer, E. H.**, Control of phosphorylase activity in a muscle glycogen particle-1. Isolation and characterization of the protein-glycogen complex, *J. Biol. Chem.*, 245, 6642, 1970.

185. **Monanu, M. O. and Madsen, N. B.**, Rabbit liver phosphorylase *a* phosphatase: regulation by glucose and caffeine, *Can. J. Biochem. Cell Biol.*, 63, 115, 1985.

186. **Monod, J., Wyman, J., and Changeux, J-P.**, On the nature of allosteric transitions: a plausible model, *J. Mol. Biol.*, 12, 88, 1965.

187. **Morange, M., Blanco, G., Vandenbunder, B., and Buc, H.**, AMP analogues: their function in the activation of glycogen phosphorylase *b*, *Eur. J. Biochem.*, 65, 553, 1976.

188. **Morange, M., Kolb, A., Buc, H., Chachaty, C., and Langht, G.**, Conformation of purine ribosyl 5′-nucleotides bound to glycogen phosphorylase *b*-Proton T2 relaxation time investigations, *Eur. J. Biochem.*, 74, 99, 1977.

189. **Morgan, H. E. and Parmeggiani, A.**, Regulation of glycogenolysis in muscle. III. Control of muscle glycogen phosphorylase activity, *J. Biol. Chem.*, 239, 2440, 1964.

190. **Mott, D. M. and Bieber, A. L.**, Nucleotide activation of phosphorylase *b* in the presence and absence of salmine, *Biochem. Biophys. Res. Commun.*, 30, 363, 1968.

191. **Mott, D. M. and Bieber, A. L.**, Structural specificity of the adenosine 5′-phosphate site on glycogen phosphorylase *b*, *J. Biol. Chem.*, 245, 4058, 1970.

192. **Murray, A. W. and Atkinson, R. R.**, Adenosine 5′-phosphorothioate as an AMP analog, *Biochemistry*, 7, 4023, 1968.

193. **Nakano, K. and Fukui, T.**, The complete amino acid sequence of potato α-glucan phosphorylase, *J. Biol. Chem.*, 261, 8230, 1986.

194. **Nakano, K., Hwang, P. K., and Fletterick, R. J.**, Complete cDNA sequence for rabbit muscle glycogen phosphorylase, *FEBS Lett.*, 204, 283, 1986.

195. **Najmudin, S.**, The control and activity of phosphorylase: crystallographic binding studies of glucosamine-1-phosphate and adenosine 5′-phosphate to phosphorylase *b*, *Biochemistry*, FHS Part II thesis, University of Oxford, 1984.

196. **Newgard, C. B., Hwang, P. K., and Fletterick, R. J.**, The family of glycogen phosphorylases: structure and function, *Crit. Rev. Biochem. Mol. Biol.*, 24, 69, 1989.

197. **Oikonomakos, N. G., Sotiroudis, T. G., and Evangelopoulos, A. E.**, Interaction of phosphorylase *b* with eosin: influence of substrates and effectors on eosin-enzyme complexes, *Biochem. J.*, 181, 309, 1979.

198. **Oikonomakos, N. G., Melpidou, A. E., and Johnson, L. N.**, Crystallization of pig skeletal phosphorylase *b*. Purification, physical and catalytic characterization, *Biochim. Biophys. Acta*, 832, 248, 1985.

199. **Oikonomakos, N. G., Johnson, L. N., Acharya, K. R., Stuart, D. I., Barford, D., Hajdu, J., Varvill, K. M., Melpidou, A. E., Papageorgiou, A. C., Graves, D. J., and Palm, D.**, Pyridoxal phosphate site in glycogen phosphorylase: structure in native enzyme and in three derivatives with modified cofactors, *Biochemistry*, 26, 8381, 1987.

200. **Oikonomakos, N. G., Acharya, K. R., Stuart, D. I., Melpidou, A. E., McLaughlin, P. J., and Johnson, L. N.**, Uridine (5′)diphospho(1)-α-D-glucose. A binding study to glycogen phosphorylase *b* in the crystal, *Eur. J. Biochem.*, 173, 569, 1988.

201. **Oikonomakos, N. G., Papageorgiou, A. C., Leonidas, D. D., Barford, D., and Johnson, L. N.**, New crystallizations of phosphorylase *b*, Joint CCP4 and ESF-EACBM Newsletter on Protein Crystallography, No. 24, 123, 1989a.

202. **Oikonomakos, N. G., Acharya, K. R., Melpidou, A. E., Stuart, D. I., and Johnson, L. N.**, The binding of glycerol-2-P to glycogen phosphorylase *b* in the crystal, *Arch. Biochem. Biophys.*, 269, 62, 1989b.

203. **Oikonomakos, N. G., Papageorgiou, A. C., Acharya, K. R., Leonidas, D. D., Johnson, L. N., and Evangelopoulos, A. E.**, X-ray and solution studies on glycogen phosphorylase reconstituted with pyridoxal in place of the natural coenzyme, Vitamin B$_6$ and carbonyl catalysis, in press.

204. **Okazaki, T., Nakazawa, A., and Hayaishi, O.**, Studies on the interaction between regulatory enzymes and effectors-II. Effect of adenosine 5′ monophosphate analogues on glycogen phosphorylase *b*, *J. Biol. Chem.*, 243, 5266, 1968.

205. **Palm, D., Schachtele, K. H., Feldmann, K., and Helmreich, E. J. M.**, Is the active form of pyridoxal-*P* in α-glucan phosphorylases a 5′-phosphate dianion?, *FEBS Lett.*, 101, 403, 1979.

206. **Palm, D., Blumenauer, G., Klein, H. W., and Blanc-Muesser, M.**, α-Glucan phosphorylases catalyse the glucosyl transfer from α-D-glucosyl fluoride to oligosaccharides, *Biochem. Biophys. Res. Commun.*, 111, 530, 1983.

207. **Palm, D., Goerl, R., and Burger, K. J.**, Evolution of catalytic and regulatory sites in phosphorylases, *Nature*, 313, 500, 1985.

207a. **Palm, D., Goerl, R., Weidinger, G., Zeier, R., Fischer, B., and Schinzel, R.**, *Escherichia coli* maltodextrin phosphorylase: primary structure and deletion mapping of the C-terminal site, *Z. Naturforsch.*, 42C, 394, 1987.

208. **Palm, D., Klein, H. W., Schinzel, R., Buchner, M., and Helmreich, E. J. M.,** The role of pyridoxal 5'-phosphate in glycogen phosphorylase catalysis, *Biochemistry,* 29, 1099, 1990.

209. **Papageorgiou, A. C., Oikonomakos, N. G., and Leonidas, D. D.,** Inhibition of rabbit muscle glycogen phosphorylase by D-gluconohydroximo-1,5-lactone-N-phenylurethane, *Arch. Biochem. Biophys.,* 272, 376, 1989.

210. **Papageorgiou, A. C., Oikonomakos, N. G., Leonidas, D. D., Bernet, B., Beer, D., and Vasella, A.,** The binding of D-gluconohydroximo-1,5-lactone to glycogen phosphorylase: kinetic, ultracentrifugation and crystallographic studies, *Biochem. J.,* 274, 329, 1991.

211. **Parmeggiani, A. and Morgan, H. E.,** Effect of adenine nucleotides and inorganic phosphate on muscle phosphorylase activity, *Biochem. Biophys. Res. Commun.,* 9, 252, 1962.

212. **Parrish, R. F., Uhing, R. J., and Graves, D. J.,** Effect of phosphate analogues on the activity of pyridoxal reconstituted glycogen phosphorylase, *Biochemistry,* 16, 4824, 1977.

213. **Perutz, M. F.,** Mechanisms of cooperativity and allosteric regulation in proteins, *Q. Rev. Biophys.,* 22, 139, 1989.

214. **Pfeuffer, T., Ehrlich, J., and Helmreich, E.,** Role of pyridoxal 5'-phosphate in glycogen phosphorylase. I. Synthesis of 3'-O-methylpyridoxal 5'-phosphate N-oxide and pyridoxal 5'-phosphate monomethyl ester and the conversion of the N-oxide to pyridoxal 5'-phosphate by apophosphorylase *b* from rabbit skeletal muscle, *Biochemistry,* 11, 2125, 1972a.

215. **Pfeuffer, T., Ehrlich, J., and Helmreich, E.,** Role of pyridoxal 5'-phosphate in glycogen phosphorylase. II. Mode of binding pyridoxal 5'-phosphate to apophosphorylase *b* and the aggregation state of reconstituted phosphorylase proteins, *Biochemistry,* 11, 2136, 1972b.

216. **Phillip, G., Gringel, G., and Palm, D.,** Rabbit muscle phosphorylase derivatives with oligosaccharides covalently bound to the glycogen storage sites, *Biochemistry,* 21, 3043, 1982.

217. **Praly, J.-P. and Lemieux, R. U.,** Influence of solvent on the magnitude of the anomeric effect, *Can. J. Biochem.,* 65, 213, 1987.

218. **Raibaud, O. and Goldberg, M.,** Characterization of two complementary polypeptide chains obtained by proteolysis of rabbit muscle phosphorylase, *Biochemistry,* 12, 5154, 1973.

219. **Romero, P. A., Smith, E. E., and Whelan, W. J.,** Glucosamine as a substitute for glucose in glycogen metabolism, *Biochem. Int.,* 1, 1, 1980.

220. **Rossmann, M. G., Moras, D., and Olsen, K. W.,** Chemical and biological evolution of a nucleotide binding protein, *Nature,* 250, 194, 1974.

221. **Sansom, M. S. P.,** Structural studies of glycogen phosphorylase *b* and of related proteins, Ph.D. thesis, Oxford University, 1983.

222. **Sansom, M. S. P., Babu, Y. S., Hajdu, J., Stuart, D. I., Stura, E. A., and Johnson, L. N.,** The role of pyridoxal phosphate in glycogen phosphorylase *b*. Structure, environment and relationship to catalytic mechanism, in *Chemical and Biological Aspects of Vitamin B₆ Catalysis, Part A,* Evangelopoulos, A. E., Ed., Alan R. Liss, New York, 1984, 127.

223. **Sansom, M. S. P., Stuart, D. I., Acharya, K. R., Hajdu, J., McLaughlin, P. J., and Johnson, L. N.,** Glycogen phosphorylase *b* — the molecular anatomy of a large regulatory enzyme, *J. Mol. Struct.,* 123, 3, 1985.

224. **Schinzel, R. and Palm, D.,** *Escherichia coli* maltodextrin phosphorylase: contribution of active site residues glutamate-637 and tyrosine-538 to the phosphorolytic cleavage of α-glucans, *Biochemistry,* 29, 9956, 1990.

225. **Schnackerz, K. D., Schinzel, R., and Palm, D.,** Pyridoxal 5'-phosphate. A ^{31}P NMR probe reporting small changes in the active site of *E. Coli* phosphorylases obtained by site-directed mutagenesis, Int. Symp. Vitamin B₆ and Carbonyl Catalysis, (Abstr.), 83, 1990.

226. **Schwabe, J. W. R.,** Biochemistry, Part 2 thesis, University of Oxford, 1987.

227. **Sealock, R. W. and Graves, D. J.,** Effect of salt solutions on glycogen phosphorylase. A possible role of the phosphoryl group in phosphorylase *a*, *Biochemistry,* 6, 201, 1967.

228. **Shaltiel, S., Hedrick, J. L., and Fischer, E. H.,** On the role of pyridoxal 5'-phosphate in phosphorylase. II. Resolution of rabbit muscle phosphorylase *b*, *Biochemistry,* 5, 2108, 1966.

229. **Shaltiel, S., Hedrick, J. L., Pocker, A., and Fischer, E. H.,** Reconstitution of apophosphorylase with pyridoxal 5'-phosphate analogs, *Biochemistry,* 8, 5189, 1969.

230. **Shaltiel, S. and Cortijo, M.,** The mode of binding of pyridoxal 5'-phosphate in glycogen phosphorylase, *Biochem. Biophys. Res. Commun.,* 41, 594, 1970.

231. **Shimomura, S. and Fukui, T.,** Characterization of the pyridoxal phosphate site in glycogen phosphorylase *b* from rabbit muscle, *Biochemistry,* 17, 5359, 1978.

232. **Sinnott, M. L., Withers, S. G., and Viratelle, O. M.,** The necessity of magnesium cation for acid assistance of aglycone departure in catalysis by *Escherichia coli* (lacZ) b-galactosidase, *Biochem. J.,* 175, 539, 1978.

233. **Sinnott, M. L.,** *The Chemistry of Enzyme Action,* Elsevier, New York, 1984, 389.

234. **Sinnott, M. L.,** Catalytic mechanisms of enzymic glucosyl transfer, *Chem. Rev.,* 90, 1171, 1990.

235. **Soman, G. and Philip, G.,** The nature of the binding site for aromatic compounds in glycogen phosphorylase, *Biochem. J., 147,* 369, 1975.

236. **Sotiroudis, T. G., Oikonomakos, N. G., and Evangelopoulos, A. E.,** Phosphorylase *b* covalently bound to glycogen, *Eur. J. Biochem.,* 88, 573, 1978.

237. **Sotiroudis, T. G., Oikonomakos, N. G., and Evangelopoulos, A. E.,** Effect of sulfated polysaccharides and sulfate anions on the AMP-dependent activity of phosphorylase *b, Biochem. Biophys. Res. Commun.,* 90, 234, 1979a.

238. **Sotiroudis, T. G., Oikonomakos, N. G., and Evangelopoulos, A. E.,** Effect of polycarboxylates on phosphorylase *b, Biochem. Biophys. Res. Commun.,* 86, 674, 1979b.

239. **Sotiroudis, T. G., Oikonomakos, N. G., and Evangelopoulos, A. E.,** Effect of prostaglandins on phosphorylase *b, Biochim. Biophys. Acta,* 658, 270, 1981.

240. **Sotiroudis, T. G., Cazianis, C. T., Oikonomakos, N. G., and Evangelopoulos, A. E.,** Effect of sodium cholate on the catalytic and structural properties of phosphorylase *b, Eur. J. Biochem.,* 131, 625, 1983.

241. **Sprang, S. R. and Fletterick, R. J.,** The structure of glycogen phosphorylase *a* at 2.5 Å resolution, *J. Mol. Biol.,* 131, 523, 1979.

242. **Sprang, S. R., Goldsmith, E. J., Fletterick, R. J., Withers, S. G., and Madsen, N. B.,** Catalytic site of glycogen phosphorylase: structure of the T state and specificity for α-D-glucose, *Biochemistry,* 21, 5364, 1982.

243. **Sprang, S. R., Goldsmith, E. J., and Fletterick, R. J.,** Structure of the nucleotide activation switch in glycogen phosphorylase *a, Science,* 237, 1012, 1987.

244. **Sprang, S. R., Acharya, K. R., Goldsmith, E. J., Stuart, D. I., Varvill, K. M., Fletterick, R. J., Madsen, N. B., and Johnson, L. N.,** Structural changes in glycogen phosphorylase induced by phosphorylation, *Nature,* 336, 215, 1988.

245. **Stalmans, W., Laloux, M., and Hers, H.-G.,** The interaction of liver phosphorylase *a* with glucose, *Eur. J. Biochem.,* 49, 415, 1974.

246. **Stalmans, W.,** The role of the liver in the homeostasis of blood glucose, *Curr. Top. Cell. Regul.,* 11, 51, 1976.

247. **Steiner, R. F., Greer, L., Bath, R., and Oton, J.,** Structural changes in glycogen phosphorylase *b* induced by the binding of glucose and caffeine, *Biochim. Biophys. Acta,* 611, 269, 1980.

248. **Street, I. P., Armstrong, C. R., and Withers, S. G.,** Hydrogen bonding and specificity. Fluorodeoxy sugars as probes of hydrogen bonding in the glycogen phosphorylase-glucose complex, *Biochemistry,* 25, 6021, 1986.

249. **Street, I. P., Rupitz, K., and Withers, S. G.,** Fluorinated and deoxygenated substrates as probes of transition-state structure in glycogen phosphorylase, *Biochemistry,* 28, 1581, 1989.

250. **Stura, E. A., Zanotti, G., Babu, Y. S., Sansom, M. S. P., Stuart, D. I., Wilson, K. S., Johnson, L. N., and Van de Werve, G.,** Comparison of AMP and NADH binding to glycogen phosphorylase *b, J. Mol. Biol.,* 170, 529, 1983.

251. **Takagi, M., Fukui, T., and Shimomura, S.,** Catalytic mechanism of glycogen phosphorylase: Pyridoxal(5')diphosphol(1)-α-D-glucose as a transition-state analogue, *Proc. Natl. Acad. Sci. U.S.A.,* 79, 3716, 1982.

252. **Tagaya, M. and Fukui, T.,** Catalytic reaction of glycogen phosphorylase reconstituted with a coenzyme-substrate conjugate, *J. Biol. Chem.,* 259, 4860, 1984.

253. **Titani, K., Koide, A., Hermann, J., Ericson, L. H., Kumar, S., Wade, R. D., Walsh, K. A., Neurath, H., and Fischer, E. H.,** Complete amino acid sequence of rabbit skeletal muscle glycogen phosphorylase, *Proc. Natl. Acad. Sci. U.S.A.,* 74, 4762, 1977.

254. **Truscheit, E., Frommer, W., Junge, B., Muller, L., Schmidt, D. D., and Wingender, W.,** Chemistry and biochemistry of microbial α-glucosidase inhibitors, *Angew. Chem. Int. Ed. Engl.,* 20, 744, 1981.

255. **Tu, J. I., Jacobson, G. R., and Graves, D. J.,** Isotopic effects and inhibition of polysaccharide phosphorylase by 1,5-gluconolactone. Relationship to the catalytic mechanism, *Biochemistry,* 10, 1229, 1971.

256. **Uhing, R. J., Janski, A. M., and Graves, D. J.,** The effects of solvents on nucleotide regulation of glycogen phosphorylase, *J. Biol. Chem.,* 254, 3166, 1979.

257. **Uhing, R. J., Lentz, S. R., and Graves, D. J.,** Effects of 1,2-dimethoxyethane on the catalytic and coenzyme properties of glycogen phosphorylase, *Biochemistry,* 20, 2537, 1981.

258. **Vandenbunder, B. and Buc, H.,** The reactivity of arginine residues interacting with glucose-1-phosphate in glycogen phosphorylase, *Eur. J. Biochem.,* 133, 509, 1983.

259. **Varvill, K. M.,** X-ray crystallographic studies on glycogen phosphorylase *b,* Ph.D. thesis, University of Oxford, 1988.

260. **Veinberg, S., Shaltiel, S., and Steinberg, J. Z.,** Factors contributing to absorption and fluorescence characteristics of pyridoxal phosphate in glycogen phosphorylase *b, Isr. J. Chem.,* 12, 421, 1974.

261. **Vereb, G., Fodor, A., and Bot, G.,** Kinetic characterization of rabbit skeletal muscle phosphorylase *ab* hybrid, *Biochim. Biophys. Acta,* 915, 19, 1987.

262. **Vidgoff, J. M., Pocker, A., Hullar, T. L., and Fischer, E. H.,** Interaction of muscle glycogen phosphorylase with pyridoxal 5′-phosphonate, *Biochem. Biophys. Res. Commun.,* 57, 1166, 1974.

263. **Wang, H. H., Shonka, M. L., and Graves, D. J.,** Influence of carbohydrates on phosphorylase structure and activity. I. Activation by preincubation with glycogen, *Biochemistry,* 11, 2296, 1965a.

264. **Wang, J. H., Shonka, M. L., and Graves, D. J.,** The effect of glucose on the sedimentation and catalytic activity of glycogen phosphorylase, *Biochem. Biophys. Res. Commun.,* 18, 131, 1965b.

265. **Wang, J. H., Humniski, P. M., and Black, W. J.,** Effect of polyamines on glycogen phosphorylase. Differential electrostatic interactions and enzymic properties, *Biochemistry,* 7, 2037, 1968.

266. **Wang, J. H., Tu, J. I., and Lo, F. M.,** Effect of glucose-6-phosphate on the nucleotide site of glycogen phosphorylase *b.* A general approach for negative heterotropic interactions, *J. Biol. Chem.,* 245, 3115, 1970a.

267. **Wang, J. H., Kwok, S.-C., Wirch, E., and Suzuki, I.,** Distinct AMP sites in glycogen phosphorylase *b* as revealed by calorimetric studies, *Biochem. Biophys. Res. Commun.,* 40, 1340, 1970b.

268. **Wanson, J. C. and Drochmans, P.,** Rabbit skeletal muscle glycogen, *J. Cell. Biol.,* 38, 130, 1968.

269. **Weber, I. T., Johnson, L. N., Wilson, K. S., Yeates, D. G. R., Wild, D. L., and Jenkins, J. A.,** Crystallographic studies on the activity of glycogen phosphorylase *b, Nature,* 274, 433, 1978.

270. **Willnecker, J., Jahnke, K., and Buehner, M.,** The three-dimensional structure of potato phosphorylase at 2.6 Å resolution, Int. Symp. Vitamin B_6 and Carbonyl Catalysis (Abstr.), 85, 1990.

271. **Withers, S. G., Sykes, B. D., Madsen, N. B., and Kasvinsky, P. J.,** Identical structural changes induced by glycogen phosphorylase by two nonexclusive allosteric inhibitors, *Biochemistry,* 18, 5342, 1979.

272. **Withers, S. G., Madsen, N. B., and Sykes, B. D.,** Active form of pyridoxal phosphate in glycogen phosphorylase. Phosphorous-31 nuclear magnetic resonance investigation, *Biochemistry,* 20, 1748, 1981a.

273. **Withers, S. G., Madsen, N. B., Sykes, B. D., Takagi, M., Shimomura, S., and Fukui, T.,** Evidence for the direct phosphate-phosphate interaction between pyridoxal phosphate and substrate in the glycogen phosphorylase catalytic mechanism, *J. Biol. Chem.,* 256, 10759, 1981b.

274. **Withers, S. G., Madsen, N. B., Sprang, S. R., and Fletterick, R. J.,** Catalytic site of glycogen phosphorylase: structural changes during activation and mechanistic implications, *Biochemistry,* 21, 5372, 1982a.

275. **Withers, S. G., Madsen, N. B., and Sykes, B. D.,** Covalently activated glycogen phosphorylase: a phosphorus-31 NMR and ultracentrifugation analysis, *Biochemistry,* 21, 6716, 1982b.

276. **Withers, S. G., Shechosky, S., and Madsen, N. B.,** Pyridoxal phosphate is not the acid catalyst in the glycogen phosphorylase catalytic mechanism, *Biochem. Biophys. Res. Commun.,* 108, 322, 1982c.

277. **Withers, S. G., MacLennan, D. J., and Street, I. P.,** The synthesis and hydrolysis of a series of deoxyfluoro-D-glucopyranosyl phosphates, *Carbohydr. Res.,* 154, 127, 1986.

278. **Withers, S. G., Percival, M. D., and Street, I. P.,** The synthesis and hydrolysis of a series of deoxyfluoro-α-D-glucopyranosyl and deoxyfluoro-α-D-glucopyranosyl phosphates, *Carbohydr. Res.,* 187, 43, 1989.

279. **Withers, S. G. and Rupitz, K.,** Measurement of active-site homology between potato and rabbit muscle α-glucan phosphorylases through use of a linear free energy relationship, *Biochemistry,* 29, 6405, 1990.

280. **Witters, L. A. and Avruch, J.,** Insulin regulation of hepatic glycogen synthase and phosphorylase, *Biochemistry,* 17, 406, 1978.

281. **Wolfenden, R.,** Analog approaches to the structure of the transition state in enzyme reactions, *Acc. Chem. Res.,* 5, 10, 1972.

282. **Yan, S. C. B., Uhing, R. J., Parrish, R. F., Metzler, D. E., and Graves, D. J.,** A role for pyridoxal phosphate in the control of dephosphorylation of phosphorylase *a, J. Biol. Chem.,* 254, 8263, 1979.

283. **Oikonomakos et al.,** unpublished results.

284. **Johnson et al.,** unpublished results.

285. **Leonidas, D. D., Oikonomakos, N. G., and Papageorgiou, A. C.,** unpublished results.

286. **Vandenbunder, B., Morange, M., and Buc, H.,** 1,N^6-etheno-AMP and 1,N^6-etheno-2′-deoxy-AMP as probes of the activator site of glycogen phosphorylase from rabbit skeletal muscle, *Proc. Natl. Acad. Sci. U.S.A.,* 73, 2696, 1976.

287. **Snape, P., Martin, J. L., and Johnson, L. N.,** unpublished results.

288. **Johnson, L. N.,** unpublished results.

289. **Varvill, K. M., Acharya, K. R., and Johnson, L. N.,** unpublished results.

290. **Oikonomakos, N. G. and Leonidas, D. D.,** unpublished results.

291. **Klein, H. W. and Helmreich, E. J. M.,** The role of pyridoxal 5′-phosphate in phosphorylase catalysis, *Curr. Top. Cell. Regul.,* 26, 281, 1985

292. Reference deleted.

293. **Ford, L. O., Johnson, L. N., Machin, P. A., Phillips, D. C., and Tjian, R.,** Crystal structure of a lysozyme-tetrasaccharide lactone complex, *J. Mol. Biol.,* 88, 349, 1974.

294. **Veluraja et al.,** unpublished results.

295. **Takagi, M., Shimomura, S., and Fukui, T.,** Function of the phosphate group of pyridoxal 5′-phosphate in the glycogen phosphorylase reaction, *J. Biol. Chem.,* 256, 728, 1981.

296. **Hu, S. H. and Johnson, L. N.,** unpublished results.
297. **Oikonomakos, N. G. and Papageorgiou, A. C.,** unpublished results.
298. **Vasella, A.,** personal communication.
299. **Barford, D. and Johnson, L. N.,** unpublished results.
300. **Barford, D., Hu, S. H., and Johnson, L. N.,** unpublished results.
301. **Narendra, N., Seshadri, T. P., and Viswamitra, M. A.,** Structure of the disodium salt of glucose 1-phosphate hydrate $2Na^+ \cdot C_6H_{11}O_9P^{2-} \cdot 3 \cdot 5H_2O$, *Acta Crystallogr.*, C40, 1338, 1984.
302. **Withers, S. G. and Madsen, N. B.,** Nucleotide activation of glycogen phosphorylase *b* occurs only when the nucleotide is in a dianionic form, *Biochem. Biophys. Res. Commun.*, 97, 513, 1980.

Chapter 4

MONO-ADP-RIBOSYLATION OF PROTEINS

Simon van Heyningen

TABLE OF CONTENTS

I. INTRODUCTION

ADP-ribosylation outside the nucleus is a posttranslational modification looking for a function. It was originally discovered as a property of toxins and enzymes secreted by pathogenic bacteria. But, although it is now known to be common in normal eukaryotic cells, it has been studied mostly as a curiosity and because it is experimentally useful; essentially nothing is known about whether it has a real normal biological function, let alone what that function might be.

The reaction being considered is a relatively unusual property of NAD$^+$, quite separate from its properties as a redox coenzyme. It is the transfer of the ADP-ribose moiety of the NAD$^+$ to a protein acceptor with concomitant release of nicotinamide and a proton:

$$NAD^+ + protein = ADP\text{-}ribosyl\text{-}protein + nicotinamide + H^+$$

This reaction can occur nonenzymically, although it is very slow. When it is catalyzed, the range of possible protein substrates is wide and not yet fully characterized or understood. The ADP-ribose can be bound to several different amino acids (e.g., arginine, aspartic acid, cysteine) by different chemical bonds; the substrate proteins can be soluble, insoluble, cytoplasmic, nuclear, membrane-bound, or cytoskeletal. Doubtless many more substrates have still to be discovered; if there is a pattern behind the specificity of these enzymes, it remains to be found.

Research in this area has fallen into two separate areas which have kept remarkably apart: ADP-ribosylation in the nucleus and ADP-ribosylation everywhere else. The reactions and substrates in these two places are rather different and so, in general, are the people doing the research. ADP-ribosylation was first studied in the nucleus, and this remains the only area of biology in which a physiological substrate is relatively well understood. These nuclear reactions frequently involve poly-ADP-ribosylation and differ in many ways from what will be discussed in this chapter.

It is ADP-ribosylation outside the nucleus that is the subject of this chapter. It always involves only one ADP-ribose group (as shown in the equation above), and was first discovered during investigation of the properties of some bacterial protein exotoxins. The bulk of this chapter will be concerned with these reactions. For a long time, extranuclear ADP-ribosylation seemed an esoteric reaction interesting for those concerned with bacterial pathogenicity, but not very important for anyone else. However there is now more and more evidence that ADP-ribosylation is also important in the normal physiology of cells that have never seen a toxin; this point is also discussed below. Several other reviews[1,2] and books[3-5] about ADP-ribosylation have been published recently.

II. DIPHTHERIA AND RELATED TOXINS

A. DIPHTHERIA TOXIN

Diphtheria toxin was the first toxin shown to catalyze ADP-ribosylation more than 20 years ago. Extensive reviews have been published of the toxin in general[6-9] and of its ADP-ribosylation in particular,[10] and so early work will be only briefly summarized in this chapter.

Diphtheria toxin is produced by *Corynebacterium diphtheriae*, growing in the throat of the host. It is a protein that is rapidly passed in the blood stream to other tissues remote from the site of infection (e.g., heart, lung), where it is powerfully cytotoxic.[11] The gene for the toxin is not on the bacterial genome itself, but on a lysogenic beta-phage.[12]

The toxin has been known for many years: it was one of the very first proteins to be crystallized. It has a molecular weight of 58,300, and, although secreted from the bacterium as a single polypeptide chain, is very easily "nicked" by proteolytic enzymes *in vivo* or *in vitro* into two chains A (M_r 21,200) and B (M_r 37,200) that remain joined by a disulphide bond. The whole toxin has been sequenced.[13]

The overall strategy of cytotoxicity is now well understood. Like most toxins, diphtheria toxin has to enter an intoxicated cell before it can exert its action.[14] The B chain of the toxin binds to a receptor on the surface of the cell. This receptor has not been fully identified, but there is good evidence for theories that it is a surface glycoprotein and perhaps part of an anion antiporter. This binding is followed by the translocation of the A chain of the toxin into the cytoplasm of the cell. It is this A chain that carries the ADP-ribosyltransferase activity that is responsible for the cytotoxicity. The A chain is active as an enzyme only when it can fold independently of the B-chain. In a sense, the whole toxin can be thought of as a zymogen of the ADP-ribosyltransferase.

Diphtheria toxin is cytotoxic because the A chain is an inhibitor of protein synthesis, and it was shown in 1964 that NAD^+ had to be present for the inhibition of the incorporation of radioactive amino acid into protein in a cell-free system.[15] Subsequent work showed that this was because the toxin catalyzed the ADP-ribosylation of elongation factor 2 (EF2), a protein essential to protein synthesis, and which catalyzes the transfer of the nascent polypeptide chain from the aminoacyl (A) site to the peptidyl (P) site on the ribosome, with concomitant hydrolysis of GTP. This was the first mono-ADP-ribosylation to be discovered, and that it is the mechanism of intoxication in cells was proved by showing that inhibition of protein synthesis could be reversed by adding an excess of nicotinamide.[16]

Since then, much work has been published on the kinetics and mechanism of the reaction. The equilibrium constant is 0.63 m*M,* and the pH optimum is 8.5 in the direction of ADP-ribosylation, but 5.2 for the reverse (where a proton is one of the reactants).[17] Under physiological conditions, it can be calculated that more than 99.9% of the total EF2 in a cell would be ADP-ribosylated at equilibrium: quite enough to account for the inhibition of protein synthesis, since ADP-ribosyl-EF2, although it binds both GTP and ribosomes, has no useful activity.

Binding of the substrates to the enzyme is relatively tight: Michaelis constant (K_M) values are 5 μM for NAD^+ and 0.5 m*M* for nicotinamide, and direct measurements of the binding of NAD^+, using, for example, equilibrium dialysis and fluorescence, have produced similar values, a K_D of about 8 μM.[18] More detailed kinetic experiments have given evidence that the reaction proceeds through a compulsory order mechanism involving a ternary toxin-EF2-NAD^+ complex.[19] This suggests that NAD^+ binds to the toxin first, and produces some conformational change that allows subsequent binding of EF2: such conformational change has however not been easy to detect in the ultracentrifuge or by fluorescence polarization, although there is evidence that NAD^+ can protect the toxin from proteolytic digestion.

Diphtheria toxin is highly specific for both its substrates. Analogues of NAD bind only weakly if at all (except for thionicotinamide adenine nucleotide, which has a K_M roughly comparable to that of NAD^+). NAD analogues are also rather ineffective competitive inhibitors: the best is adenine (K_i about 40 μM); adenosine is about ten times weaker, and AMP, ADP, ATP, ADP-ribose and $NADP^+$ weaker still.[18]

The specificity for the protein substrate is also marked. EF2 is the only physiological substrate; its prokaryotic analogue, EF-G, is not affected. The other important ADP-ribose acceptor is water; the toxin catalyzes an NAD^+-glycohydrolase activity:

$$NAD^+ + H_2O = \text{nicotinamide} + H^+ + \text{ADP-ribose}$$

This reaction goes much more slowly than the ADP-ribosylation of proteins, but it may be one reason for the apparent nonspecific ADP-ribosylation of other proteins catalyzed by very high concentrations of toxin. Liberated free ADP-ribose can react nonenzymically. At high pH and protein concentration, whole toxin can also catalyze ADP-ribosylation of itself.

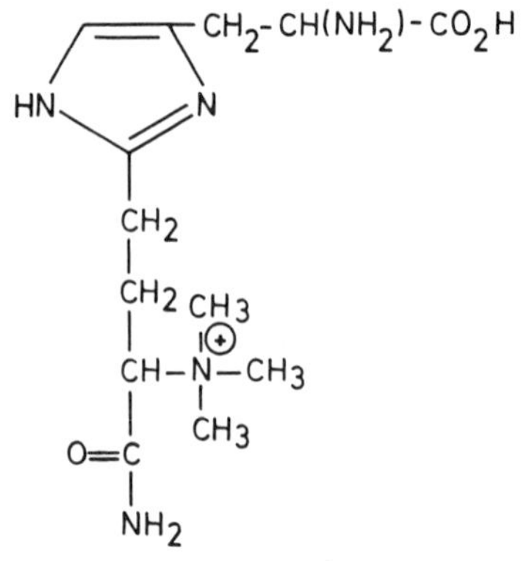

$$\text{CH}_2-\text{CH(NH}_2)-\text{CO}_2\text{H}$$

HN—N

CH$_2$
|
CH$_2$ CH$_3$
| |⊕
CH—N—CH$_3$
| |
O=C   CH$_3$
|
NH$_2$

FIGURE 1.   The structure of diphthamide.[20,21]

The binding site for ADP-ribose on EF2 has been identified and found to be a very strange amino acid: experiments using radioactively labeled NAD$^+$ showed that the residue ADP-ribosylated is a post-translationally modified derivative of histidine properly called 2-(3-carboxyamid-3-[trimethylammonio]-propyl)histidine, but usually called "diphthamide".[20,21] The structure is shown in Figure 1.

ADP-ribose is bound by an alkali-stable alpha glycoside bond from the C-1 of the ribose to the N-1 of the imidazole in the diphthamide.[22]

Although the amino acid sequence around diphthamide is highly conserved over all the very wide range of eukaryotic cells and archaebacteria in which EF2 is found (Figure 2), it is not enough to satisfy the specificity of the toxin. The fact that a tryptic peptide containing diphthamide was not by itself a substrate shows that there must be some other structural feature of the intact protein that is important for the specificity, presumably its three-dimensional conformation.[23] Diphthamide has never been demonstrated in any protein except EF2.

Some cell lines are resistant to diphtheria toxin because they have abnormal EF2.[24] (Others, more commonly, are resistant because they do not take up toxin from the outside,[25] but this is irrelevant to this discussion.) There is evidence that at least some types of cell, e.g., CHO-K1 cells, contain two copies of the gene for EF2, and there are mutants in which one has become non-ADPR-ribosylatable, while the other has remained normal.[26] Some modification-deficient mutants can be divided into at least three complementation groups, consistent with ideas based on the chemistry of diphthamide, that there are at least three posttranslational modification reactions between histidine and diphthamide. Cloning experiments[27] have shown similarity in sequence between EF2 and some other GTP-binding proteins as described below. They have also shown that at least one mutation conferring resistance to toxin is caused by a glycine to arginine change; what effect this has on the properties of EF2 is not clear.[28]

B. STRUCTURE OF THE TOXIN

Crystal structure determinations of diphtheria toxin are under way in several laboratories, but although some preliminary results and identification of secondary structure in a dimer of the toxin have been published,[29] a full structure is not yet available. The structure of the rather similar exotoxin of *Pseudomonas aeruginosa* has been determined (Section II.C), and some experiments with diphtheria toxin should be interpreted in the light of this.

The A chain has 193 residues, and must have a binding site for all the substrates of ADP-ribosylation. Curiously, it also has another highly specific binding site for nucleotides: the K_D for the dinucleotide ApUp is about 9 pM, among the tightest ligand binding constants observed in any system.[30] The function of this site, if any, is unclear: competition studies

Rat liver	phe	asp	val	his	asp	val	thr	leu	his	ala	asp	ala	ile	DIP	arg
Beef liver	phe	asp	val	his	asp	val	thr	leu	his	ala	asp	ala	ile	DIP	arg
Yeast	val	asn	ile	leu	asp	val	thr	leu	his	ala	asp	ala	ile	DIP	arg
Wheat germ	phe	glu	val	ser	asp	val	val	leu	his	thr	asp	ala	ile	DIP	arg

FIGURE 2. Conserved sequence around the diphthamide (DIP) residue in EF2 from four different eukaryotic sources.[22a]

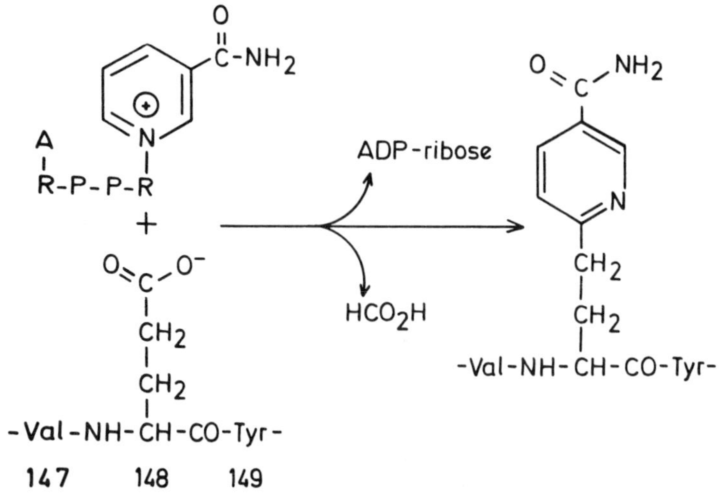

FIGURE 3. Photolabeling of a glutamic acid residue in diphtheria toxin by NAD$^+$. (From Carroll, S. F., McCloskey, J. A., Crain, P. F., Oppenheimer, N. J., Marschner, T. M., and Collier, R. M., *Proc. Natl. Acad. Sci. U.S.A.*, 82, 7273, 1985.)

show that there is overlap between it and the NAD$^+$-binding site, and also perhaps interaction with the "P site", a region on the B chain that binds phosphorylated ligands.[31]

The most direct evidence for the nature of the active site of the toxin has come from the elegant affinity labeling experiments of Collier and his group.[32,33] They showed that when NAD$^+$ labeled with ^{14}C was incubated under ultra-violet radiation with whole toxin or isolated A chain (but not with inactive mutants of the A chain), the label became attached to essentially only one residue in the protein, glutamic acid 148, and that this led to the loss of enzymic activity. Further labeling and mass-spectroscopic experiments showed that the reaction shown in Figure 3 had occurred. The product is an α-amino-γ-(6-nicotinami-dyl)butyric acid residue, and the reaction, involving decarboxylation and formation of a new carbon-carbon bond is a remarkable one, clearly caused by free radical formation under the ultraviolet radiation. The exact mechanism is not known, but the existence of the reaction does imply that the methylene carbon of glu 148 and the C-6 of nicotinamide are in close contact in the toxin-NAD complex, and this might clearly be relevant to the mechanism of catalysis.

This inactivation suggests that the glutamic acid residue is involved with binding of NAD$^+$, and this conclusion has been confirmed by site-directed mutagenesis experiments showing that, when glutamic acid was changed to aspartic acid, more than 99% of the activity was lost although the binding affinity for NAD$^+$ was very little altered, and that total deletion of this residue also abolished activity.[34]

The other residue that has been implicated in the action of the A chain is trp 153 (obviously close to glu 148), whose chemical modification abolishes activity,[35] and whose fluorescence spectrum is greatly affected by NAD$^+$-binding.[18]

C. *PSEUDOMONAS* EXOTOXIN A

Pseudomonas aeruginosa is an opportunistic pathogen, for example, of wounds, that secretes many different proteins. Among the most active, and certainly the best studied, is the exotoxin A (PTA, M_r 66,600) which catalyzes the ADP-ribosylation of EF2 just like diphtheria toxin, but whose role in pathogenicity is not clear.[36]

Diphtheria and *Pseudomonas* toxins catalyze the same reaction with diphthamide: indeed

MEMBRANE
TRANSLOCATION

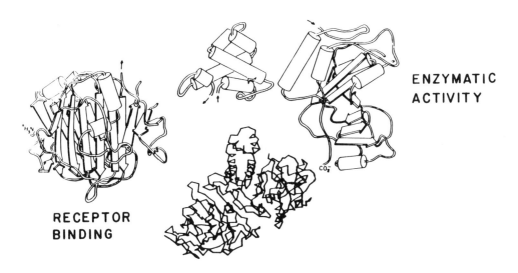

ENZYMATIC
ACTIVITY

RECEPTOR
BINDING

FIGURE 4.   Three separate domains in the structure of *Pseudomonas* exotoxin A: domain I is involved with receptor binding, domain II with translocation across the membrane, and domain III with the enzymatic activity. Figure kindly supplied by Dr. D. B. McKay.

either can reverse the reaction catalyzed by the other. But they are not particularly similar molecules in other ways. PTA does not have the same two-chain structure, although it is secreted as a proenzyme that does not catalyze ADP-ribosylation: it is taken up by cells (through different receptors) and proteolytically activated intracellularly.

Although less research has been done on PTA than on diphtheria toxin, more is now known about its three-dimensional structure because of the work of McKay and his group who successfully solved the structure of crystals of the PTA proenzyme in a remarkably short time.[37] The most important observation is that the molecule has three distinct structural domains that seem to be almost separately folded in the whole protein (see Figure 4). Their functions were originally proposed following inspection of the model and a knowledge of the chemistry of the activation reaction, and were subsequently confirmed principally by deletion analysis of the gene expressed in *Escherichia coli*. The major part of domain I (Ia, residues 1 to 252) carries the site that binds to a receptor on the surface of cells; domain II (residues 253 to 364) is the part of the protein that helps translocation through the plasma membrane.[38] The ADP-ribosyltransferase activity is chiefly a property of domain III (residues 365 to 404) which has a very prominent cleft, but a small portion of domain I (Ib, residues 365 to 404) is also involved. Although the proenzyme whose structure was determined cannot actually catalyze the reaction with diphthamide in EF2 without activation, it does have weak NAD^+-glycohydrolase activity. This suggests that the conformation of domain III in the proenzyme must be much the same as it is in the active fragment, perhaps constrained sterically by parts of the other domains which are proteolytically removed during activation.

Several experiments have given further information about the binding domain. First, Carroll and Collier[39] showed by photoaffinity labeling experiments similar to those they did with diphtheria toxin (see above, Section II.B) that glutamic acid 553 was involved in the binding of NAD^+, and this residue is known to be in the binding cleft of domain III. Site-directed mutagenesis of the toxin cloned in *E. coli* showed that deletion of this residue produced a stable immunoreactive product with less than 10^{-6} times the activity of wild type.[40] Mutation to aspartic acid also almost completely destroyed ADP-ribosyltransferase activity.[41] Aspartic acid is shorter than glutamic acid: to test the effect of introducing a longer residue, Lukac and Collier[42] first changed glutamic acid to cysteine by site-directed mutagenesis; this derivative had less than 1/10,000 the enzymic activity of wild type. They then introduced a long chain carboxyl group by specific reaction of the cysteine with io-doacetic acid, making an analogue of glutamic acid with one more methylene carbon. This restored most of the activity, giving a protein with about one sixth of the activity of wild type, and showing both that a free carboxyl group is essential for activity, and that the active site of the protein is flexible enough to accommodate a longer residue. It is not known what the function of this carboxyl group is; it might act as a general base catalyst, or perhaps as a stabilizer for some positively charged transition state.

Further conclusions about the mechanism of PTA came from comparison of its structure and that of diphtheria toxin. Although originally amino acid similarities between the two toxins had not been detected, once a crystal structure was available and it was clear where to look, such similarities became clear.[43-45] There is one particular stretch of about 60 residues within the enzymatic domains where there is particular similarity: 24 of these residues are either identical or highly conserved (Figure 5). Furthermore almost all of this region is in the enzymatic cleft seen in the three-dimensional structure, where adenosine (an effective competitive inhibitor of NAD^+) is known to bind. It seems almost certain that the NAD^+-binding sites of diphtheria toxin and PTA are very similar in structure.

Brandhuber et al.[44] have put forward a model for NAD^+ binding which fits with their structure and with crystallographic difference maps of the binding site. A conserved pen-tapeptide, trp 466 to tyr 470, interacts extensively with NAD^+, perhaps accounting for the changes in tryptophan fluorescence found with diphtheria toxin. Positively charged conserved residues may also interact with negatively charged phosphates on the NAD^+; they are arginines 458 and 467 in PTA, and lysine 39 (known to be involved with NAD^+ binding[45]) and 51 in diphtheria toxin. There are also two tyrosine residues in this neighborhood that might be important. Directed mutagenesis of tyr 470 to phenylalanine does not alter any of the enzymic activity, so there seems little chance that it is important.[46] Tyrosine 481, on the other hand, is implicated. Its iodination abolishes enzyme activity[44] and its mutagenesis to phenylalanine[46] produces a protein with unchanged NAD^+-glycohydrolase activity (and K_M for NAD^+), but reduced ADP-ribosyltransferase activity, suggesting that this residue might be involved with the interaction with EF2.

Another residue that seems to be critical for activity is histidine 426, as shown by Wozniak et al.[47] who showed that its mutation to tyrosine resulted in a loss of activity which could be restored by complementation back to histidine, and furthermore that monoclonal antibodies that inhibit enzymic activity of wild type will not react with this mutant. However, both the crystal structure and preliminary spectroscopic experiments suggest that the 426 residue is not in the binding site; perhaps it also is involved with the reaction with EF2.

III. CHOLERA AND SIMILAR TOXINS

Cholera is an overwhelming diarrhoea caused by *Vibrio cholerae* growing in the gut and secreting a protein toxin. A similar diarrhoea is also caused by some strains of *E. coli*, which secrete a protein very similar in action and in structure to cholera toxin (over 70% identical in amino acid sequence). The mechanism by which these toxins cause the diarrhoea is complex, but at root is caused by their ability to activate adenylate cyclase in the intestinal brush-border cells. Adenylate cyclase is an intracellular enzyme which catalyzes the formation of the second messanger cyclic AMP from ATP, an important method of signal transduction for many hormones. Cholera toxin differs from hormones in that it has essentially no tissue specificity. A very wide range of eukaryotic cells, from dictyostelium to drosophila, from rat enterocytes to pigeon erythrocytes, has been shown to be affected by the toxin.

```
Diphtheria   gly   ala   asp   asp   val   val   asp   ser   ser   lys   ser   phe   val   met   glu   asn   PHE
             (1)
Pseudomonas  glu   arg   leu   leu   gln   ala   his   arg   gln   leu   glu   glu   arg   gly   tyr   val   PHE
             (420)

ser   ser   TYR   HIS   GLY   THR   lys   pro   gly   tyr   val   asp   SER   ILE   gln   lys   GLY   ile   gln   LYS
val   gly   TYR   HIS   GLY   THR   phe   leu   glu   ala   ala   gln   SER   ILE   val   phe   GLY   gly   val   ARG

pro   LYS   SER   gly   thr   gln   gly   asn   tyr   ASP   asp   asp   ile   TRP   LYS   GLY   PHE   TYR   ser   thr   asp
ala   ARG   SER   ....(gln,asp,leu)..........   ASP   ala   ile   TRP   ARG   GLY   PHE   TYR   ile   ala   gly

asn   lys   tyr   asp   ALA   ala   GLY   TYR   ser   val   ASP   asn   GLU   asn   pro   leu   ser   GLY   LYS
                                                                                                      (76)
asp   pro   ala   leu   ALA   tyr   GLY   TYR   TYR   ala   ASP   gln   ASP   pro   asp   ala   arg   GLY   ARG
                                                                                                      (492)
```

FIGURE 5. Similarities in the amino acid sequence of diphtheria toxin and Pseudomonas exotoxin A.[44]

A. STRUCTURE AND STRATEGY OF CHOLERA TOXIN

Cholera toxin (which has been sequenced at the DNA level, and whose genetics[12,48,49] and mechanism of assembly[50] are quite well understood) has a complex subunit structure (for general reviews of the toxin see References 1 and 51 to 53). A three-dimensional structure has not yet been published, but diffraction measurements on crystals of both cholera and *E. coli* toxins are under way.[54] The toxin has one A subunit, itself synthesized as a single polypeptide chain but normally found "nicked" by proteolysis into two chains linked by a disulphide bond, A1 (M_r 21,000) and A2 (7 000). The A subunit is joined to a complex of five B subunits (each M_r 11,500). Each of these subunits has a different action. Subunit B binds ganglioside GM1 on the outer plasma membrane of intoxicated cells.[55] It is because this ganglioside is so widespread among eukaryotes that cholera toxin has no tissue specificity. Subunit B has no other activity, but this initial binding leads in some way, still not fully understood, to the entry into the cell of the A1 chain. (The function of the A2 chain seems to be mainly concerned with the interaction of the A and B subunits.) It is the A1 chain that carries the ADP-ribosyltransferase activity. It is either inactive or very weakly active with intact whole cells, but when the cells have been lysed, the A1 chain can activate cyclase by itself, showing that the function of the B subunits had been to deliver the A1 peptide to its site of action. Indeed B subunits can also protect intact cells from the action of the toxin, presumably by occupying all the available ganglioside receptors. This strategy in which an otherwise inactive binding component ushers an active component into the cell is common to several different toxins including diphtheria toxin.[52]

B. ACTION OF THE A1 CHAIN

The A1 chain catalyzes the ADP-ribosylation of a G protein, analogously to the action of diphtheria toxin. G proteins are a complex family of regulatory proteins that bind guanine nucleotides, and have several different functions. For general reviews see References 56 to 58. Functional aspects of G proteins will be considered only in outline in this chapter, which concentrates on the actual ADP-ribosylation reactions.

The main G-protein substrate of cholera toxin is a part of the intracellular adenylate cyclase complex (sometimes also called *N* protein, because it binds nucleotides, or even G/F protein, because it can be activated by guanine nucleotides or fluoride). There are at least two different types of G protein that are involved with adenylate cyclase: G_s, which stimulates the catalytic subunit of the cyclase, and G_i, which inhibits it. These two proteins are one of the most important routes by which hormones affect the cyclase complex. They have been purified and shown to be similar in structure and action. Both have three different subunits α, β, and γ in a total M_r of about 80,000 to 90,000. The α subunits bind guanine nucleotides, and are different in the two forms ($G_s\alpha$ about 42,000, and $G_i\alpha$ about 41,000). When $G_s\alpha$ binds GTP and Mg^{2+}, it is activated, and this leads to activation of the cyclase probably following dissociation of the whole complex into separate α and $\beta + \gamma$ components. However G_s also has intrinsic GTPase activity, so that the bound nucleotide is rapidly hydrolysed to GDP; in this state, it no longer activates the cyclase.

The other two subunits (β about 35,000, and γ about 5000 to 10,000) can be the same in both G_i and G_s, although they may also exist in several different forms. Some observed heterogeneity in the β,γ subunits may be caused by stable combination of similar β subunits with different γ subunits. There are certainly at least two different, although closely related forms of both of the subunits in nonretinal cells, and this diversity means that there must be a minimum of four forms of the β,γ combination. The function of the β,γ subunits is probably concerned with anchoring into the plasma membrane and perhaps communication with rest of the cyclase complex.

It is the $G_s\alpha$ subunit that is the substrate for ADP-ribosylation by cholera toxin. ($G_i\alpha$ is the substrate for pertussis toxin, see below, section IV.A.) This ADP-ribosylation has several physiologically important effects. First of all, and probably most important, it inhibits

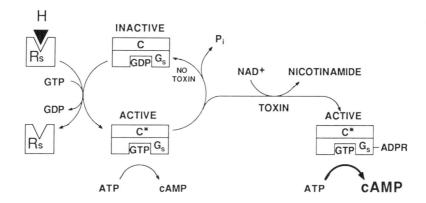

FIGURE 6. A simplified idea of the mechanism of activation of adenylate cyclase by cholera toxin. H represents a hormone that binds to its receptor R_s; C and C* are the inactive and active forms respectively of the catalytic component of adenylate cyclase; G_s is the regulatory component. For details, see text.

the intrinsic GTPase activity of the α subunit. This keeps it in an active state. This is the underlying mechanism of activation by cholera toxin, and explains the early observations that activation was long lasting, and could be mimicked by GTP analogues that cannot be hydrolyzed (e.g., guanosine $5'$-[β,γ-imodo]triphosphate, Gpp(NH)p). In several systems (e.g., rat liver plasma membranes[59]) there is evidence that the α subunits are released from the β,γ complex and the cell membrane following ADP-ribosylation by toxin.

Figure 6 shows a simplified summary of some of the action of cholera toxin and of the interaction of the subunits of G_s with the catalytic subunit of cyclase.

Experiments using NAD^+ labeled with ^{32}P have shown that cholera toxin can bring about ADP-ribosylation of a protein of M_r about 42,000 in many different tissues, of which the first was avian erythrocytes. In many of these tissues, several other bands are also labeled; it is not always clear whether all these bands are derivatives of the $G_{s\alpha}$ protein. It should be remembered that in the strictest sense, there is still relatively little evidence about how the diarrhoea of cholera can be caused by activation of adenylate cyclase; there is more than a formal possibility that ADP-ribosylation of some other protein might be involved. Several systems have been reported in which some effect of the toxin is not accompanied by any increase in cAMP concentration (e.g., chemotaxis in mouse macrophages,[60] inhibition of receptor-mediated increases in inositol triphosphate concentration).[61]

C. COFACTORS FOR ADP-RIBOSYLATION

Relatively recent experiments have shown that various cytosolic proteins are required as cofactors for maximum efficiency of toxin action. ADP-ribosylation factor (ARF) is a protein (M_r about 21,000) purified from detergent extracts of rabbit liver membranes, which has a high-affinity binding site for guanine nucleotides and is required for ADP-ribosylation of purified G_s by cholera toxin.[62,63] It is active only when binding GTP or analogues. This observation led to a proposal that the actual substrate of the toxin might be a $G_{s\alpha} \cdot$ ARF complex, but it has been shown that ARF can also increase the activity of the toxin with other substrates, e.g., water, guanidine analogues such as agmatine, other protein substrates, and the auto-ADP-ribosylation of the A1 peptide itself (see below, Section III.F).[64] These reactions are also stimulated by GTP and analogues and clearly suggest a direct GTP-dependent interaction of ARF with toxin (possibly increasing its affinity for NAD^+, see below Section III.G). Bovine brain also contains two different ARF proteins (sARF1 and sARF2) that are soluble, and which also enhance ADP-ribosylation of G_s and other substrates. They have a similar M_r to the membrane-bound ARF and also require GTP.[65] Gill and coworkers[66-68] have prepared a membrane GTP-binding protein named S that is required for toxin activity and a soluble protein termed CF (cytosolic factor) needed for the action of S. It seems likely that S and ARF are the same.

The amino acid sequences of the three ARF proteins have been determined at the DNA level:[69,70] they all have similarities in the regions for guanine nucleotide binding between each other and with G_s and other similar G proteins, although the region of G_s that is ADP-ribosylated by toxin is absent, showing that although they are all in a sense substrates for the toxin, they must interact with it in different ways. The N terminus of ARF is blocked by myristylation.[71] Whatever the normal biological function of the ARFs might be, it must be common to most types of cell, since specific anti-ARF antibodies detected its presence in a very wide range of eukaryotic cells, including yeast, *Dictyostelium,* and man.[71]

D. OTHER PROTEIN SUBSTRATES OF THE TOXIN

The A1 chain can catalyze its own auto-ADP-ribosylation, although this is probably not physiologically important. Apart from this and its "usual" substrate, G_s, cholera toxin can ADP-ribosylate several other proteins, most of which are also G proteins. The most widely studied of these is the alpha subunit of transducin, a peripheral membrane protein of rod outer segments in the eye, which is very similar to G_s and G_i in structure and function. This ADP-ribosylation is rather inefficient, but seems to have a similar action to that in G_s; for example, transducin is a better substrate when it is binding GTP, and its intrinsic GTPase activity is inhibited by the modification. There is evidence that the actual substrate is a transient complex containing the GTP-binding α subunit, the β and γ subunits, and a photoinduced form of rhodopsin.[72] (The site of modification in transducin is discussed below, Section III.F.)

Cholera toxin has also been used to identify a possible new G-protein involved in insulin action; the hormone can inhibit ADP-ribosylation by the toxin of a protein in plasma membranes of M_r about 25,000.[73] This protein is found in many different tissues and has been tentatively called G_{ins}.[74]

In many other experiments, several other sometimes unidentified proteins have been found to be ADP-ribosylated, e.g., tubulin[75] and a major erythrocyte membrane protein.[76] Some soluble proteins, even globin, are sometimes affected, and it has been suggested that at high enough toxin concentration almost every protein possessing a suitable arginine could be ADP-ribosylated, although perhaps not by the normal mechanism, and certainly not significantly. The toxin can even affect ADP-ribosylation in the nucleus,[77] where there are known to be toxin-binding sites.[78]

E. OPTIMUM CONDITIONS FOR ADP-RIBOSYLATION

It is not always experimentally easy to obtain ADP-ribosylation catalyzed by any of the toxins in the laboratory; a complex cocktail of components needs to be added.[79] A particular problem is that many cells contain high activities of NAD glycohydrolase so that [^{32}P]-NAD$^+$ added at high activity is rapidly hydrolyzed. Several inhibitors of these enzymes are usually added, e.g., isonicotinic acid hydrazide and 3-acetylpyridine nucleotide.[80] An alternative approach has been to add NADP$^+$ which competes for enzymes that degrade NAD$^+$, but does not itself affect ADP-ribosylation.[81,82] Thymidine is a useful inhibitor of poly-ADP-ribosylation.

F. ENZYMOLOGY AND SUBSTRATE SPECIFICITY OF THE TOXIN

The residue modified by cholera toxin is arginine. This was first suggested as a result of experiments using low-molecular-weight artificial substrates such as arginine (or, better, arginine ethyl ester) and guanidine,[83] but was confirmed when arginine in the peptide ser-arg-val-lys was identified as the ADP-ribosylated residue in transducin.[72] Little is known of the chemistry of the reaction itself, but stereoscopic evidence suggests a direct displacement of the nicotinamide of NAD^+ by the guanidino group.[84]

Like most similar enzymes, cholera toxin can also use water as a substrate, i.e., show an NAD^+ glycohydrolase activity, although there is always a potential problem with this measurement as *V. cholerae* secretes large amounts of an entirely different NAD^+ glycohydrolase enzyme, and it is difficult to be quite certain that preparations of toxin have been entirely freed from this as a contaminant.[85,86] There are other useful low-molecular-weight substrates that have been found. For example, Tait and Nassau[87] carried out a systematic test of many arginine derivatives, and found that the side chain was important in determining whether or not the compound was ADP-ribosylated. The best substrates, e.g., (12-[{phenylmethyl}amino]dodecyl) guanidine (compound A599), tended to have a large hydrophobic domain with no charge centres near the guanidine. This agreed with a previous finding[88] that (^{125}I)-guanyl tyramine was a better substrate than guanyl tyramine alone, presumably because halogenation of the aromatic ring decreased the dipole moment of the phenolic hydroxyl, so increasing hydrophobicity. Perhaps a hydrophobic group mimics the lipid environment of G_s. Agmatine ([4-aminobutyl]guanidine)[89] and certain (benzylidineamino)guanidines[90] were also good substrates. The absorption spectrum of the latter is changed by ADP-ribosylation, making them potentially useful, although unfortunately rather insensitive, substrates for assay of the toxin.[91] Investigating the action of several different such derivatives suggested an S_N2 reaction mechanism in which the deprotonated guanidino group makes a nucleophilic attack on the C-1 carbon of the ribose.

Although cholera toxin is relatively nonspecific at high concentrations, it does not ADP-ribosylate all arginine residues: there must be some extra specificity requirements in the reaction. There is some evidence for more than one site of ADP-ribosylation even on G_s.[92,93] Kharadia and Graves[94] investigated ADP-ribosylation by toxin of a heptapeptide which was a good substrate for cAMP-dependent protein kinase, and which had two arginine residues. One of these was preferentially modified, and there was an inverse correlation between phosphorylation and ADP-ribosylation.

G. MECHANISM OF THE ADP-RIBOSYLATION REACTION

The kinetic mechanism of ADP-ribosylation by the toxin has been studied by two different groups with different results. Mekalonos et al.,[88] who used (^{125}I)-guanyl tyramine as a substrate, found evidence for an ordered sequential mechanism in which NAD^+ was the first substrate to bind. However later experiments with agmatine[89] supported a sequential random order mechanism in which either substrate can bind to the toxin first. Such kinetic experiments are always hard to interpret, especially when one of the substrates is so different from the substrate *in vivo*.

Whatever the mechanism, the binding of NAD^+ seems to be remarkably weak. The Michaelis constant, K_M, is about 5 mM when the enzyme is working as an NAD^+-glycohydrolase.[88,89,95] Surprisingly, this value is increased to about 50 mM in the presence of the artificial substrate A599 (see above),[95] which itself had a K_M of 0.4 mM. Only molecules having both the adenine and nicotinamide moieties of NAD^+, but with minor alteration in the nicotinamide ring, can be competitive inhibitors of this reaction, suggesting that the binding site must recognize the whole molecule.

The direct binding constant for NAD^+ (rather than kinetic catalytic constants) is hard to measure because it is so high. However it turns out that saturating NAD^+ concentrations protect subunit A essentially completely from digestion with trypsin, and measurement of the extent of that digestion at different NAD^+ concentrations gave a K_D for NAD^+ of about 4 mM.[95] This protective effect was also shown by some of the competitive inhibitors of the ADP-ribosyltransferase activity.

This low affinity for NAD^+ is closer to that found with dehydrogenases than with other ADP-ribosyltransferases, and much lower than the intracellular concentration of NAD^+, thought to be around 70 μM. There is no obvious explanation for why it is so low; perhaps it is altered *in vivo*, either by the hydrophobic environment of G_s, or by the ARFs that activate the toxin (see above, section III.C.).

Even less is known about the nature of the binding site for either substrate on the toxin. Photoaffinity labeling experiments with NAD^+ showed specific binding to cholera toxin although the whole NAD molecule became bound whereas in the similar earlier experiments with diphtheria and PTA (sections II.B. and II.C.) the nicotinamide ring was bound.[96] However it was not possible to identify a particular binding site. Fluorometric titration of the A subunit shows some conformational change affecting the environment of at least one tryptophan residue when it is activated by cleavage of the disulphide bond; this may be relevant to the structure of the active site.[97]

Some authors[98] have used secondary structure prediction methods on the primary sequence to show evidence for a "Rossman fold" in cholera toxin similar to the NAD^+-binding site of many dehydrogenases. But there is no evidence of any such fold in the known structure of PTA (section II.C.), and no real reason to believe that it exists.

IV. PERTUSSIS TOXIN

Bordetella pertussis, the pathogen of whooping cough, secretes many different proteins, and it is only relatively recently that one of them has been recognized as the most important pathogenic agent. This protein is usually called pertussis toxin, but also pertussigen or islet-activating protein. It is an ADP-ribosyltransferase, and has many interesting similarities to the other toxins already discussed. For a recent book covering all aspects, see Wardlaw and Parton,[99] and for a review of the toxin, see Burns,[100] and of its genetics, see Gross et al.[101]

A. STRUCTURE AND STRATEGY OF THE TOXIN

Pertussis toxin has a subunit structure that is almost uniquely complex for a soluble protein.[102] There are five different subunits, named for their mobility on electrophoresis in SDS: S1 (M_r 28,000), S2 (23,000), S3 (22,000), S4 (11,700), and S5 (9300). The molecular weight of the whole toxin is 117,000. The topology of the subunits is not clear; indeed even their stoichiometry is uncertain. It is thought that there is one copy of each of them except S4, of which there are two. The whole protein has been sequenced at the DNA level,[103,104] and since there is significant similarity between S2 and S3 it has been suggested that an S2-S4 dimer is joined to a rather similar S3-S4 dimer by the S5. Clear evidence on this will have to wait for the completion of structural and crystallographic studies.

Pertussis toxin has a two-component structure much like the cholera and diphtheria families. In this case, it is S1 that is the A component, and all the rest of the subunits that form the B component. The B component binds to a receptor on the cell surface and this leads to incorporation of some of the toxin into the membrane and subsequent translocation of the A component into the cell. Little is known of the nature of this receptor, but there is evidence that it is a type of glycoprotein with branched mannose residues. The B component of pertussis toxin has some intrinsic activity of its own, e.g., as a mitogen of T cells.[105]

The S1 subunit A component is the ADP-ribosyltransferase. It catalyzes the same sort of reaction as cholera toxin, but the most important native substrate is the alpha subunit of G_i, the guanine-nucleotide-binding regulatory protein of adenylate cyclase described above, section III.B. This protein normally mediates the interaction between cyclase and inhibitory hormones (or other stimuli). When it is ADP-ribosylated, it is effectively uncoupled from the receptors of these hormones, and so can no longer inhibit, leading to a virtual activation. In general, G-proteins that have been ADP-ribosylated by the toxin lose their transduction activity.[106] This can happen in several tissues in which the toxin can ADP-ribosylate a variety of G-proteins. All these proteins have the α,β,γ subunit structure described above, and in all the α subunit is only affected when tightly bound to β and γ. Examples other than G_i

include transducin in retinal cells, which can no longer mediate visual stimuli when modified by toxin;[107] the G_o protein of unknown function in nervous tissue[108] and in chromaffin cells;[109] G_p proteins involved in coupling receptors and phospholipase C[110] (of which only some seem to be substrates for toxin); G proteins coupled to ion channels, but unaffected by cAMP;[111] as well as some apparently unrelated proteins such as tubulin. Many other diverse effects in different cell types have been explained as ADP-ribosylation of such proteins, e.g., stimulation of the evoked release of catecholamines from cultured adrenal chromaffin cells by toxin seems to be dependent on a putative G protein of M_r about 40,000.[112,113]

It is possible that pertussis toxin can affect G proteins other than by ADP-ribosylation; toxin-induced conformational changes in G have been detected even in the absence of NAD^+.[114]

B. SUBSTRATE OF THE REACTION

The residue modified by pertussis toxin was first identified in transducin. Although an original report suggested that it was asparagine[114] subsequent experiments have made it clear that the true substrate is a cysteine residue (number 351 in the sequence).[115] Low-molecular-weight sulphydryl-containing compounds can also act as substrates, albeit rather inefficiently,[116] (K_M for cysteine is about 105 mM). The product of ADP-ribosylation of cysteine was purified and presumably has a covalent bond between the C-1 atom of ribose and the sulphur atom of cysteine. Such ADP-ribosylated cysteine residues are also identifiable by their ready and specific cleavage with mercury ions,[117] but not with hydroxylamine.[118] The low-molecular-weight guanidine derivatives that are substrates for cholera toxin (above Section III.F) do not react with pertussis toxin.

C. MECHANISM OF THE REACTION

Few kinetic experiments have been done with pertussis toxin. The K_M for NAD^+ is about 25 μM with water or cysteine as the substrate,[116,119] comparable with that of diphtheria toxin, and much lower than that of cholera toxin. The binding constant, K_D, for NAD^+ is about the same.[120]

The A subunit can be activated in a number of ways, of which the most important is by thiol reagents. Moss et al.[119] showed that dithiothreitol increased NAD^+-glycohydrolase activity, and that very high concentrations (250 mM) were required for maximum activity. It has subsequently become clear[116] that this is because the thiol has two roles: as an activator at low concentrations, but also as a substrate at high concentrations. Thiol reagents activate the toxin by reducing the single disulphide bond in the subunit when it is secreted by the bacterium. This reduction liberates two free sulphydryl groups, and alters the conformation of the subunit so that it can no longer readily associate with the B component. When the reduced toxin is alkylated with iodoacetate, the enzymic activities are lost, suggesting that one sulphydryl group is necessary for activity.[121] NAD^+ can partially protect from alkylation, suggesting that it interacts with at least one of the cysteine residues.

The critical residue is cysteine 41. A tryptic fragment of the A subunit containing only this one of the two cysteines is still active, and its activity is greatly reduced after reaction with N-ethylmaleimide.[122] However mutagenesis experiments[123,124] showed that its role is more complicated. Although deleting cys 41 reduced enzymic activity very substantially, its replacement by several other residues, e.g., serine, glycine, proline, or asparagine, produced a protein with only slightly lower maximum enzymic activities but no longer activated by thiol or inhibited by alkylation. However these derivatives bound NAD^+ much less efficiently, suggesting that the sulphydryl group of cys 41 is not involved in the catalytic site of the enzyme, but is involved in the NAD^+-binding site. If the cysteine is replaced by a negatively charged residue (aspartic or glutamic acids), then the enzymic activity is greatly reduced, presumably because the NAD^+ is then actually repulsed.

A number of other treatments also activate the A subunit, probably mainly by making it more easily reduced.[121,125] For example, nucleotides such as ATP bind very tightly to the toxin and promote its dissociation; they also activate the enzymic activity of whole toxin, perhaps by promoting the release of free subunit A.[126,127] Detergents (especially zwitterions such as CHAPS, 3-[{3-cholamidopropyl}dimethylammonio]-1-lysophosphatidylcholine) and phospholipids also activate the toxin (provided thiol is present),[128] and increase the affinity for nucleotides (e.g., the activation constant for ATP is about 0.2 μM in the presence of 1% CHAPS and 10 mM dithiothreitol). This very powerful activating effect of compounds analogous to what might be found in the eukaryotic cell suggests that pertussis toxin may not need activating by other factors in the way that cholera toxin does (Section III.C).[125]

Clues about other residues involved in the enzymic activity came from the observation[129] that tryptic peptides from the A subunit lacking the C-terminal 20% of residues (about 40 to 50) still had enzymic activity, but could not associate with the B component. This suggested that the enzymic site lay in the N-terminal portion, and that the C-terminal portion was involved with interaction between subunits. These results was confirmed by several genetic experiments, mostly carried out with the long-term aim of producing a suitable toxoid. The subunits can be expressed in *E. coli*.[130,131] For example,[132] a truncated A subunit expressed in *E. coli* and lacking the last 48 C-terminal residues (including cysteine 200) retained NAD$^+$-glycohydrolase and ADP-ribosyltransferase activity. Similarly, Pizza et al.[133] constructed a number of deletion and substitution mutants of subunit A and expressed them in *E. coli*. The shortest peptide that was still enzymically active had residues 2 to 179 present. Barbieri and Cortina[134] showed that mutants lacking either residues 2 to 22 or 153 to 180 also lacked enzymic activity.

The A subunit has two regions, each about eight residues long, that show near identity to parts of the A1 peptide of cholera and *E. coli* toxins,[103,104,135] as well as some cross-immunoreactivity. These are as follows:

Pertussis	(8)tyr arg tyr asp ser arg pro pro(15)
Cholera	(6)tyr arg ala asp ser arg pro pro(13)
E. coli	(6)tyr arg ala asp ser arg pro pro(13)
Pertussis	(51)val ser thr ser ser ser arg arg(58)
Cholera	(60)val ser thr ser ile ser leu arg(67)
E. coli	(60)val ser thr ser leu ser leu arg(67)

Two double mutations in these regions abolished enzymic activity: residues 8 and 9 (tyrosine and arginine to aspartic acid and serine), residues 50 and 53 (phenylalanine and serine to glutamic acid and isoleucine).[135] Single mutations of aspartic acid 11 to serine and arginine 13 to leucine also greatly decreased activity.[134] In another experiment, six truncated toxin subunits were expressed in *E. coli;* of these, three had all of the first region present (residues 8 to 15), and were enzymically active; another three, which lacked most of these eight residues, also lacked activity.[136] One particular mutant in this region, in which arginine 9 was changed to a lysine, a relatively conservative change, had only 0.02% of the native enzyme activity, suggesting an important role for this residue.[137]

Another approach is to look for residues that might be analogous to those shown to be important in diphtheria and *Pseudomonas* toxins, for example glutamic acid. There is some disagreement about which of several fairly close glutamic acids this might be. Deletion of glu 106 has no effect on enzymic activity,[138] but changing glu 139 to serine abolished most of the activity, pointing to a function for that residue.[134] Deletion of glu 129 or its mutation to glycine or aspartic acid also abolished activity.[133,138] Perhaps both glutamates are involved. Recent photoaffinity labeling experiments like those done with the other toxins[140] showed glu 129 to be the residue that interacts with NAD$^+$.

Analogies with the other toxins also suggest that a tryptophan residue might be important, and indeed deletion or replacement of trp 26 in pertussis toxin (which lies between the two regions of similar sequence) also abolishes enzymic activity.[138]

A mutant strain of *B. pertussis* producing whole toxin lacking ADP-ribosyltransferase activity in this sort of way was found also to be low in most other signs of pathogenicity, confirming the view that ADP-ribosylation is important in pathogenesis.[141] Many of the mutant toxins discussed above could therefore not be pathogenic, but since they retain their immunogenicity and antigenicity, might be excellent candidates as vaccines.

V. TOXINS THAT ADP-RIBOSYLATE ACTIN

During the search for a mechanism of botulinum toxin, it became clear a few years ago that one particular type of botulinum toxin, known as C2, is an ADP-ribosyltransferase using actin as a substrate.[142-144] This form of botulinum toxin is quite different from the other botulinum toxins and tetanus toxin discussed in section VI, below; it is highly toxic in a variety of ways, but not neurotoxic.[145] It has a two-component strategy similar in some ways to that of diphtheria, cholera, or pertussis toxins, but in this case the two components are distinct proteins, I (M_r about 50,000) and II (about 100,000), not linked covalently or noncovalently. It is component II (after proteolytic processing) that binds to the surface of eukaryotic cells, leading to the introduction into the cell of component I, which has by itself both ADP-ribosyltransferase and NAD$^+$ glycohydrolase activities.

Recently several other toxins that also ADP-ribosylate actin, and have a similar but not identical structure have been identified: the iota toxin from *Clostridium perfringens*,[146,147] and toxins from *C. difficile*,[148] and *C. spiroforme*.[149,150]

The main substrate protein for all these toxins both *in vitro* and in intact cells is the monomeric G-actin of microfilaments. Skeletal muscle actin is not a substrate for C2 toxin but it is a substrate for the *C. perfringens* toxin. The residue affected has been identified in skeletal muscle and cytoplasmic actin as arginine 177 (which is thought to be in the nucleotide-binding region).[151,152] The K_M for NAD$^+$ is 4 μM, comparable with that for most of the other ADP-ribosylating toxins.[153]

There is good evidence that this modification of actin as a component of the microfilaments of affected cells is physiologically significant. It largely prevents the polymerisation of G-actin monomers to F-actin filaments, leading to the destruction of the microfilament network and a rounding up of the cells.[154] The mechanism appears to be two fold. ADP-ribosylated G-actin binds tightly (K_{eq} about 10^8 M^{-1}) to the barbed ends of preformed filaments so preventing further addition of monomers.[155,156] ADP-ribosylation also decreases the ADPase activity of the actin (essential for its polymerisation) and increases the rate of exchange of bound ATP.[157] This effect is presumably one reason for many biological effects of the toxin, and may make it a useful probe for the function of the cytoskeleton, for example in the evoked exocytosis of catecholamines from cultured adrenal cells,[158] and in O_2-production and secretion in activated human neutrophils.[159,160]

VI. BOTULINUM NEUROTOXINS

Different strains of *Clostridium botulinum* produce a neurotoxin, commonly stated to be the most toxic material known (for general reviews see references 161 to 165). They are most active in the peripheral nervous system; all block the release of neurotransmitters at presynaptic nerve terminals, and have a generally similar structure and a molecular weight of around 150,000. At least seven different types have been distinguished, mainly serologically: A, B, C1, D, E, F, and G. (Strains C1 and D usually also secrete the entirely different C2 toxin referred to above, section V). Tetanus toxin, secreted by *Clostridium tetani*, is also rather similar in structure,[166-168] and is now generally thought to have the same action, at least at a molecular level. Tetanus toxin has been fully sequenced.[169,170]

They all have two chains: a heavy chain (M_r about 100,000) and a light chain (about 50,000), joined by a disulphide bond. The heavy chain binds to a cell-surface receptor (probably a ganglioside at least for tetanus toxin, but perhaps also a protein).[71] The light chain of tetanus toxin has also recently been reported to have an activity of its own.[172] The analogies with the two-component strategy of other toxins have proved hard to resist, but have not been verified, and indeed there is evidence that things are rather more complicated.[173,174]

Remarkably little is known about the molecular action of these toxins even though they have been studied for decades; this probably reflects our generally slight knowledge of the mechanism of neurosecretion. A report that botulinum C and D toxins catalyze ADP-ribosylation of a protein having M_r about 21,000 in bovine adrenal cells was therefore most interesting.[175,176]

Subsequently, this activity has been studied in some detail, and the general opinion is that it is not a property of the neurotoxins themselves, but rather of an entirely different exoenzyme secreted by these strains, and often contaminating preparations of neurotoxin. This enzyme called botulinum C3 can be distinguished from the neurotoxin (C1 and D) immunologically and enzymically.[177-179] The enzyme has been purified to homogeneity and shown to have a M_r of about 25,000, and a K_M for NAD$^+$ of 2 μM.[180] Its substrate in many types of cell (e.g., platelets, brain membranes, cultured adrenal cells, etc.) is a previously unrecognized G-protein[181,182] with a M_r of about 21,000 and definitely not one of the ARFs associated with cholera toxin.[183] This molecular weight initially suggested that the protein might be the p21 product of the *ras* oncogene and its ADP-ribosylation in several types of cells produced morphological and maturational changes similar to those produced by microinjected *ras*. However, later experiments[179,184] showed immunologically that the two were similar but not identical. More recently, it has been suggested that at least one substrate is the family of guanine-nucleotide-binding *rho* proteins, which are related to *ras*, may also be involved in the regulation of cell proliferation and differentiation, and of which many different forms have been found in animal tissues.[185] A substrate protein has been purified from several different sources,[186-188] and shown to have an amino acid sequence very similar to that of *rho*,[189] and an asparagine residue that could be ADP-ribosylated.[190] Furthermore, the C3 enzyme could directly ADP-ribosylate the *rho* protein when it was expressed in *E. coli*. Treatment of Vero cells with C3 brings about morphological changes and the disappearance of microfilaments, presumably because of ADP-ribosylation of a *rho* protein.[188]

C3 may turn out to be a very useful protein for investigating this complex system, although one must remember that its specificity has not yet been established and there could be a variety of proteins that can act as substrates, not all of which are related to *rho*, or physiologically relevant.[182,190]

It is still not universally agreed that there is an ADP-ribosyltransferase distinct from the neurotoxin; there have been several reports implying that the toxins from strains C and D can themselves catalyze ADP-ribosylation.[176,184,191,192] But the separate existence of a C3 transferase has now been accepted by most authors.[189]

Botulinum toxin has several different biological activities. For example, it can inhibit the evoked release of catecholamines from primary cultures of bovine adrenal chromaffin cells,[193,194] and it affects the intracellular calcium concentration in neutrophils. But it is clear that these activities are separate from the ADP-ribosylation catalyzed by some preparations of the toxin in both these systems.[195,196] Analogy with other toxins has been a useful tool in determining the mechanism of action of both cholera and pertussis toxins, but it does look as if it is misleading when considering botulinum. This conclusion is reinforced by studying tetanus toxin, which is generally thought to have a very similar mechanism to botulinum toxin at the molecular level. A small, but possibly significant similarity in the amino acid sequence of tetanus toxin and the NAD^+-binding domains of diphtheria and *Pseudomonas* toxins has been noticed.[197] Yet strenuous (and largely unpublished) efforts in many laboratories, including the author's, have failed to find any evidence that any preparation of tetanus toxin can catalyze ADP-ribosylation.

VII. ADP-RIBOSYLATION BY OTHER TOXINS

There are many other ADP-ribosylating extracellular proteins secreted by a variety of bacteria and mostly little characterized. Cholera-like toxins, for example, are produced by *Salmonella typhimurium, Campylobacter jejuni, Plesiomonas shigelloides,*[198] and *Aeromonas hydrophila.*[199,200] Even some strains of *E. coli* produce another ADP-ribosylating and cyclase-activating toxin, different from the heat-labile cholera-like toxin, but in many ways similar to it;[201] other strains also have a protein that will ADP-ribosylate some of the bacterium's own proteins. *Ps aeruginosa* secretes an ADP-ribosylating exoenzyme S as well as exotoxin A.[202] The list is growing and likely to prove endless, but conclusions about the function of these proteins are still awaited.

Some bacteriophage also produce ADP-ribosyltransferases: the main substrate of two proteins produced by T4 phage, for example, is arginine residues on the alpha subunit of *E. coli* RNA polymerase, and N4 phage modifies several other *E. coli* proteins.[203] No doubt there would be many more ADP-ribosylating viruses if they were really searched for.

VIII. ENDOGENOUS ADP-RIBOSYLATION

Although most of the work done on ADP-ribosylation has concentrated on bacterial toxins, it is also an important activity in normal cells. There are two main ways of showing this: indirectly, by measuring the amount of ADP-ribosylated protein present *in vivo,* and directly, by looking for ADP-ribosyltransferases.

Several methods[204,205] have been used to measure the total amount of ADP-ribose linked to proteins in normal cells, and it is clear that there are many such residues: 31.8 pmol mono-ADP-ribose linked to arginine residues per mg protein in rat liver for example.[205] Most of these are linked to proteins having M_r about 40,000 to 50,000,[206] similar to those of the regulatory G-proteins. Such experiments, however, can give little idea what the functional significance of the ADP-ribosylation might be, or how and when the modification takes place. Many of these ADP-ribose groups are of course in the nucleus, but there is more than 100-fold more ADP-ribose present in the mono- rather than in the poly-form.

Enzymes catalyzing mono-ADP-ribosylation outside the nucleus (EC 2.4.2.30) have been isolated from a wide variety of sources, starting with turkey erythrocytes,[207] and such activities are clearly widespread,[208] and can be purified and characterized,[209] although the protein substrates have not usually been identified. The systems are often complex: there are four different transferases in turkey erythrocytes[210] with different kinetics and specificity; A and B are soluble, but C and D are membrane-associated. Transferase A differs from almost all the other ADPR-ribosyltransferases mentioned in this chapter in that $NADP^+$ can be a substrate as well as NAD^+. There are probably transferases in most tissues of almost all species:[8] animals, birds, *Xenopus,*[211] molluscs *(Helix pomatia),*[212] *Drosophila,*[213] and so on. These mono transferases can be distinguished from the poly(ADP-ribose) transferases more often found in the nucleus by their ability to use 2′-deoxy-NAD^+ as a substrate.[214]

In most cases, the function of these enzymes remains unknown. But there are some examples that have been published recently where a possible function has been suggested in the control of a wide variety of physiological processes. Phosphorylase kinase, for example, is an important enzyme in the regulation of glycogen breakdown in muscle, and is itself regulated by phosphorylation by protein kinase. When the enzyme, purified from rabbit skeletal muscle, was ADP-ribosylated at arginine residues (using a purified transferase from the nucleus), there was a marked reduction in the extent of this phosphorylation-dependent activation.[215] Since muscle is known to contain transferases,[208] this could be a physiologically important mechanism.

Other experiments have looked at control points more obviously suggested by the action of the bacterial toxins. For example, the GTPase activity of transducin (known to be a substrate for cholera and pertussis toxins, Sections III.D and IV.B) was inhibited when it was ADP-ribosylated using the turkey erythrocyte enzyme,[216] although probably not by the same mechanism as cholera toxin uses. Tanuma et al.[217] purified a specific ADP-ribosyl-transferase of cysteine residues from human erythrocytes, and showed that it could use the alpha subunit of G_i as a substrate, just like pertussis toxin. An endogenous inhibitor of pertussis-toxin-catalyzed ADP-ribosylation has been purified from bovine brain.[218]

However there still remains little direct evidence for an involvement of ADP-ribosylation in physiological control of G-proteins. It is very difficult to look for ADP-ribosylation in intact cells directly mainly because of the problem in achieving high internal radioactivity in $[^{32}P]NAD^+$. Jacquemin et al.[219] got round this by measuring the extent to which G_s in plasma membrane preparation from rat adipocytes could be ADP-ribosylated by cholera toxin. They showed that sustained stimulation of G_i by adenosine in the intact cells decreased significantly the toxin-catalyzed incorporation of ^{32}P in membranes prepared from these cells, presumably because they were already ADP-ribosylated. This is indirect, but persuasive, evidence that the adenosine is ultimately affecting the endogenous ADP-ribosylation of G_s.

The substrate on EF-2 for diphtheria toxin is a specific residue, diphthamide, produced by three posttranslational modifications, and that is found in essentially all eukaryotes (Section II.A); it must play some very important role in protein synthesis. An enzyme that catalyzes the specific ADP-ribosylation of the same diphthamide residues has been purified from a range of cells.[220,221] Recently, it has been shown that this enzyme is active *in vivo:*[222] up to 35% of the EF-2 of transformed baby hamster kidney cells was shown to be ADP-ribosylated when these cells were grown in culture, and this ADP-ribosylation was readily reversible, as would be expected for a process important in control. Since there are cell lines that lack diphthamide (and are therefore resistant to diphtheria toxin), this control process cannot be essential for the life of a cultured cell. But it might be more important when the cell is in its natural state as part of a larger organism.

Most of the mono-ADP-ribosylated proteins in cells are in the mitochondria,[223] and these organelles also have high ADP-ribosyltransferase and NAD^+ glycohydrolase activities.[224] The soluble fraction from rat liver mitochondria has an enzyme which ADP-ribosylates a specific protein, and is made up of two subunits each of M_r about 100,000.[225] There is also what seems to be a different enzyme in sub-mitochondrial particles (presumably the inner mitochondrial membrane); its acceptor has M_r about 30,000, and may also be ADP-ribo-sylated nonenzymically.[225] It has also been shown indirectly that in intact mitochondria there are at least three different classes of proteins ADP-ribosylated *in vivo;* one of these has the properties of an ADP-ribosyl-cysteine such as is formed by pertussis toxin.[226] A possible function of this ADP-ribosylation in the control of calcium release from mitochondria has been suggested, based on the observed properties of the modified proteins and the link between NAD^+ hydrolysis in mitochondria and the release of calcium.[5] Calcium transport in the endoplasmic reticulum may also be controlled by ADP-ribosylation: a Ca^{2+}-dependent ATPase from the endoplasmic reticulum of rabbit skeletal muscle was partially inhibited by ADP-ribosylation.[227] Another possible function for ADP-ribosylation was suggested by the observation[228] that nicotinamide (an inhibitor of the reaction) prevented the activation of oxidative phosphorylation by triiodothyronine in hyperthyroid rat liver mitochondria.

Finally, there is one prokaryotic system in which ADP-ribosylation is clearly an important control mechanism: dinitrogenase reductase from the photosynthetic bacterium *Rhodospirillum rubrum*. This is one of the two enzymes that catalyze nitrogen fixation. It is inactivated by ADP-ribosylation of arginine 101 using the enzyme dinitrogenase-reductase ADP-ribo-syltransferase (DRAT), and activated by glycohydrolysis using a specific enzyme dinitro-genase reductase-activating glycohydrolase (DRAG). DRAG and DRAT have both been purified to homogeneity and fully characterized, and their role in the control of this system confirmed.[229] DRAT has a K_M for NAD$^+$ for 2 mM, which is rather high, like the K_M of cholera toxin but unlike that of most other transferases. It will ADP-ribosylate enzymes from several different nitrogen-fixing bacteria, but, unlike the bacterial toxins, it will not use water as a substrate.[230] Of all systems in which ADP-ribosylation has been implicated, this is the only one where an unequivocal role for it has been established, although the mechanism by which it works remains unknown. Whatever it is, ADP-ribosylation of the enzyme from one species can affect the nitrogenase of another.[231] An enzyme with a similar activity to DRAG, catalyzing the hydrolysis of ADP-ribosylarginine bonds, has been isolated and purified from turkey erythrocytes,[232] suggesting that a similar cycle in which different enzymes are used for the modification and demodification reactions might be used in eu-karyotic cells.

IX. CONCLUSION

It is a remarkable fact that so many bacterial toxins work by catalyzing ADP-ribosylation, a reaction essentially unknown before the mechanism of diphtheria toxin was established. Whooping cough, cholera, diphtheria, and the rest are diseases with essentially nothing in common produced by entirely different bacteria, yet, at root, they have the same molecular mechanism. What can the explanation for this be?

The first idea was always that ADP-ribosylation would turn out to be a control mechanism used physiologically as well as pathologically, and that was subverted by opportunistic toxins. It seemed inconceivable, for example, that the complex series of modifications required to produce diphthamide could have arisen solely so that it could be a substrate for diphtheria toxin. Since this idea was first advanced, evidence (as summarized above) has accumulated that ADP-ribosylation is indeed physiologically important, although we still know very little about any details. It must be that the toxins are interfering with normal control processes, but it is still not known how physiological ADP-ribosylation is switched on or off. Hormones, such as the glycoprotein hormones like follicle stimulating hormone or thyrotropin, that have some analogies in their action to cholera toxin and might possibly work in the same sort of way certainly do not themselves catalyze ADP-ribosylation. The obvious suggestion is that they might activate the endogenous transferases, but there is no evidence for that idea as yet.

Finally the question arises as to how the bacteria could have acquired such a complex activity that affects eukaryotes even assuming that their ability to produce toxins does give them some selective advantage. If similar transferase activities are indeed likely to have been present in the hosts of the ancestors of modern bacteria, then it is tempting to postulate, though impossible to prove, that some kind of gene transfer between host and parasite must have taken place. This would imply that there should be structural similarities between the toxins and some eukaryotic transferases or control proteins. Such similarities have been searched for, but never found, although there are some secreted proteins from bacteria that appear to have a structure similar to that of a eukaryotic protein, e.g., the calmodulin-activated adenylate cyclase of *Bordetella pertussis*.[233] It is possible that the original ''uptake'' of a eukaryotic gene might have occurred in only one species. As discussed above, there are sequence similarities, particularly in the NAD$^+$-binding regions between diphtheria and *Pseudomonas* toxins, and between pertussis toxin and the cholera and cholera like toxins. There are also some similarities in sequence between these two otherwise rather different families;[234] perhaps they all evolved from a common ancestor.

Whatever the origins of these toxins, their study has led us to a greater understanding of control mechanisms in normal cells. It is interesting, perhaps even gratifying, that investigation of toxins that have killed so many people has produced such unexpected results.

ACKNOWLEDGMENTS

I am grateful to Dr. D. Longbottom and Dr. M. Lobban for their helpful discussions, and to the Medical Research Council and the Wellcome Trust for grants towards my own research.

NOTE ADDED IN PROOF

This chapter includes coverage of work published up to August 1989. Since then more detail has appeared of several aspects of ADP-ribosylation, but there have been few major advances. The conclusions of the chapter remain true.

There has been a great deal of work on the active S1 subunit of pertussis toxin and localization of activity to particular residues,[235-237] as well as the binding of NAD^+.[238] Domenighini et al. have published a useful paper summarizing what is known about the NAD^+-binding site of several toxins and advancing some suggestions about structure.[239] There has been more information about the cellular function of the ADP-ribosylation factors that activate cholera toxin.[240] Additional evidence has been published that cholera toxin can ADP-ribosylate more than one arginine residue in $G_{s\alpha}$, certainly in a tumor in which the usual substrate arginine residue had been replaced.[241]

Probably the most important result, although one whose implications have not yet had time to be fully appreciated, is the three-dimensional structure of *E. coli* heat-labile toxin determined by Sixma et al.,[242] using X-ray diffraction. The active A subunit (presumably very similar in structure to the A subunit of cholera toxin) shows some interesting similarities to the ADP-ribosylating domain of *Pseudomonas* exotoxin A. It has been reported that a structure of cholera toxin is about to be announced.[243]

REFERENCES

1. **Lai, C. Y.,** Bacterial protein toxins with latent ADP-ribosyl transferases activities, *Adv. Enzymol.,* 58, 110, 1986.
2. **Ueda, K. and Hayaishi, O.,** ADP-ribosylation, *Annu. Rev. Biochem.,* 54, 73, 1985.
3. **Hayaishi, O. and Ueda, K., Eds.,** *ADP-Ribosylation Reactions,* Academic Press, New York, 1982.
4. **Althaus, F. R., Hilz, H., and Shall, S., Eds.,** *ADP-Ribosylation of Proteins,* Springer-Verlag, Berlin, 1985.
5. **Althaus, F. R. and Richter, C.,** *ADP-Ribosylation of Proteins,* Springer-Verlag, Berlin, 1987.
6. **Collier, R. J.,** Diphtheria toxin: mode of action and structure, *Bacteriol. Rev.,* 39, 54, 1975.
7. **Pappenheimer, A. M.,** Diphtheria toxin, *Annu. Rev. Biochem.,* 46, 69, 1977.
8. **Uchida, T.,** Diphtheria toxin: biological activity, in *Molecular Action of Toxins and Viruses,* Cohen, P. and van Heyningen, S., Eds., Elsevier, Amsterdam, 1982, chap. 1.
9. **Ward, W. H. J.,** Diphtheria toxin: a novel cytocidal enzyme, *Trends Biochem. Sci.,* 12, 28, 1987.
10. **van Heyningen, S.,** ADP-ribosylation by bacterial toxins, in *The Enzymology of Post-Translational Modification of Proteins,* Freedman, R. B. and Hawkins, H. C., Eds., Academic Press, London, 1980, 388.
11. **Saelinger, C. B.,** Use of exotoxin A to inhibit protein synthesis, *Meth. Enzymol. Relat. Areas Mol. Biol.,* 165, 226, 1988.
12. **Betley, M. J., Miller, V. L., and Mekalanos, J. J.,** Genetics of bacterial enterotoxins, *Annu. Rev. Microbiol.,* 40, 577, 1986.
13. **Kaczorek, M., Delpeyroux, F., Chenciner, N., Streek, R. E., Murphy, J. R., Boquet, P., and Tiollais, P.,** Nucleotide sequence and expression of the diphtheria tox 228 gene in *Escherichia coli, Science,* 221, 855, 1983.

14. **Olsnes, S., Moskaug, J. O., Stenmark, H., and Sandvig, K.,** Diphtheria toxin entry: protein translocation in the reverse direction, *Trends Biochem. Sci.,* 13, 348, 1988.

15. **Collier, R. J. and Pappenheimer, A. M.,** Studies on the mode of action of diphtheria toxin. II. Effect of toxin on amino acid incorporation in cell-free systems, *J. Exp. Med.,* 120, 1019, 1964.

16. **Gill, D. M., Pappenheimer, A. M., Brown, R., and Kurnick, J. J.,** Studies on the mode of action of diphtheria toxin. VII. Toxin-stimulated hydrolysis of nicotinamide adenine dinucleotide in mammalian cell extracts, *J. Exp. Med.,* 129, 1, 1969.

17. **Honjo, T., Nishizuka, Y., Kato, I., and Hayaishi, O.,** Adenosine diphosphate ribosylation of aminoacyl transferase II and inhibition of protein synthesis by diphtheria toxin, *J. Biol. Chem.,* 246, 4251, 1971.

18. **Kandel, J., Collier, R. J., and Chung, D. W.,** Interaction of fragment A from diphtheria toxin with nicotinamide adenine dinucleotide, *J. Biol. Chem.,* 249, 2088, 1974.

19. **Chung, D. W. and Collier, R. J.,** The mechanism of ADP-ribosylation of elongation factor 2 catalyzed by fragment A from diphtheria toxin, *Biochim. Biophys. Acta,* 483, 248, 1977.

20. **van Ness, B. G., Howard, J. B., and Bodley, J. W.,** ADP-ribosylation of elongation factor 2 by diphtheria toxin. NMR spectra and proposed structures of ribosyl-diphthamide and its hydrolysis products, *J. Biol. Chem.,* 255, 10710, 1980.

21. **van Ness, B. G., Howard, J. B., and Bodley, J. W.,** ADP-ribosylation of elongation factor 2 by diphtheria toxin. Isolation and properties of the novel ribosyl-amino acid and its hydrolysis products, *J. Biol. Chem.,* 255, 10717, 1980.

22. **Oppenheimer, N. J. and Bodley, J. W.,** Diphtheria toxin. Site and configuration of ADP-ribosylation of diphthamide in elongation factor 2, *J. Biol. Chem.,* 256, 8579, 1981.

22a. **Brown, B. A. and Bodley, J. W.,** Primary structure at the site in beef and wheat elongation factor 2 of ADP-ribosylation by diphtheria toxin, *FEBS Lett.,* 103, 253, 1979.

23. **van Ness, B. G., Barrowclough, B., and Bodley, J. W.,** Recognition of elongation factor 2 by diphtheria toxin is not solely defined by the presence of diphthamide, *FEBS Lett.,* 120, 4, 1980.

24. **Moehring, J. M., Moehring, T. J., and Danley, D.,** Posttranslational modification of elongation factor 2 in diphtheria toxin resistant mutants of CHO-K1 cells, *Proc. Natl. Acad. Sci. U.S.A.,* 77, 1010, 1980.

25. **Moehring, J. T. and Moehring, J. M.,** Selection and characterization of cells resistant to diphtheria toxin and Pseudomonas exotoxin A: presumptive translational mutants, *Cell,* 11, 447, 1977.

26. **Kohno, K., Uchida, T., Mekada, E., and Okada, Y.,** Characterization of diphtheria-toxin-resistant mutants lacking receptor function or containing nonribosylatable elongation factor 2, *Somatic Cell Mol. Gen.,* 11, 421, 1985.

27. **Kohno, K., Uchida, T., Ohkubo, H., Nakanishi, S., Nakanishi, T., Fukui, T., Ohtsuka, E., Ikehara, M., and Okada, Y.,** Amino acid sequence of mammalian elongation factor 2 predicted from the cDNA sequence: homology with GTP-binding proteins, *Proc. Natl. Acad. Sci. U.S.A.,* 83, 4978, 1986.

28. **Kohno, K., Nakanishi, T., Ohmura, F., and Uchida, T.,** Studies on structure and functions of elongation factor 2 in relation to ADP ribosylation using its cDNA, in *Bacterial Protein Toxins,* Fehrenbach, F. J., Alouf, J. E., Falmagne, P., Goebel, W., Jeljaszewicz, J., Jurgens, D., and Rappuoli, R., Eds., Gustav Fischer, Stuttgart, 1988, 207.

29. **Kantardjieff, K., Dijkstra, B., Westbrook, E. M., Barbieri, J. T., Carroll, S. F., Collier, R. J., and Eisenberg, D.,** Structural studies of Diphtheria toxin, *Protein Structure Folding and Design,* 2, 187, 1987.

30. **Collins, C. M., Barbieri, J. T., and Collier, R. J.,** Interaction of diphtheria toxin with adenylyl-(3',5')-uridine 3'-monophosphate. I. Equilibrium and kinetic measurements, *J. Biol. Chem.,* 259, 15154, 1984.

31. **Lory, S. and Collier, R. J.,** Diphtheria toxin: nucleotide binding and toxin heterogeneity, *Proc. Natl. Acad. Sci. U.S.A.,* 77, 267, 1980.

32. **Carroll, S. F. and Collier, R. J.,** NAD binding site of diphtheria toxin: identification of a residue within the nicotinamide subsite by photochemical modification with NAD, *Proc. Natl. Acad. Sci. U.S.A.,* 81, 3307, 1984.

33. **Carroll, S. F., McCloskey, J. A., Crain, P. F., Oppenheimer, N. J., Marschner, T. M., and Collier, R. J.,** Photoaffinity labeling of diphtheria toxin fragment A with NAD: Structure of the photoproduct at position 148, *Proc. Natl. Acad. Sci. U.S.A.,* 82, 7273, 1985.

34. **Tweten, R. K., Barbieri, J. T., and Collier, R. J.,** Diphtheria toxin. Effect of substituting aspartic acid for glutamic acid 148 on ADP-ribosyltransferase activity, *J. Biol. Chem.,* 260, 10392, 1985.

35. **Michel, A. and Dirkx, J.,** Occurrence of tryptophan in the enzymatically active site of diphtheria toxin fragment A, *Biochim. Biophys. Acta,* 491, 286, 1977.

36. **Iglewski, B. H., Liu, P. V., and Kabat, D.,** Mechanism of action of *Pseudomonas aeruginosa* exotoxin A: adenosine diphosphate-ribosylation of mammalian elongation factor 2 *in vitro* and *in vivo, Infect. Immun.,* 15, 138, 1977.

37. **Allured, V. S., Collier, R. J., Carroll, S. F., and McKay, D. B.,** Structure of exotoxin A of *Pseudomonas aeruginosa* at 3.0 Angstrom resolution, *Proc. Natl. Acad. Sci. U.S.A.,* 83, 1320, 1986.

38. **Chaudhary, V. K., Xu, Y., FitzGerald, D., Adhya, S., and Pastan, I.,** Role of domain II of Pseudomonas exotoxin in the secretion of proteins into the periplasm and medium by *Escherichia coli, Proc. Natl. Acad. Sci. U.S.A.,* 85, 2939, 1988.

39. **Carroll, S. F. and Collier, R. J.,** Active site of *Pseudomonas aeruginosa* exotoxin A. Glutamic acid 553 is photolabeled by NAD and shows functional homology with glutamic acid 148 of diphtheria toxin, *J. Biol. Chem.,* 262, 8707, 1987.

40. **Lukac, M., Pier, G. B., and Collier, R. J.,** Toxoid of *Pseudomonas aeruginosa* exotoxin A generated by deletion of an active-site residue, *Infect. Immun.,* 56, 3095, 1988.

41. **Douglas, C. M. and Collier, R. J.,** Exotoxin A of *Pseudomonas aeruginosa*: substitution of glutamic acid 553 with aspartic acid drastically reduces toxicity and enzymic activity, *Infect. Immun.,* 169, 4967, 1987.

42. **Lukac, M. and Collier, R. J.,** Restoration of enzymic activity and cytotoxicity of mutant, E553C, *Pseudomonas aeruginosa* exotoxin A by reaction with iodoacetic acid, *J. Biol. Chem.,* 263, 6146, 1988.

43. **Carroll, S. F. and Collier, R. J.,** Amino acid sequence homology between the enzymic domains of diphtheria toxin and *Pseudomons aeruginosa* exotoxin A, *Mol. Microbiol.,* 2, 293, 1988.

44. **Brandhuber, B. J., Allured, V. S., Falbel, T. G., and McKay, D. B.,** Mapping the enzymatic active site of *Pseudomonas aeruginosa* exotoxin A, *Proteins,* 3, 146, 1988.

45. **Zhao, J. M. and London, E.,** Localization of the active site of diphtheria toxin, *Biochemistry,* 27, 3398, 1988.

46. **Lukac, M. and Collier, R. J.,** *Pseudomonas aeruginosa* exotoxin A: effects of mutating tyrosine-470 and tyrosine-481 to phenylalanine, *Biochemistry,* 27, 7629, 1988.

47. **Wozniak, D. J., Hsu, L. Y., and Galloway, D. R.,** His-426 of the *Pseudomonas aeruginosa* exotoxin A is required for ADP-ribosylation of elongation factor II, *Proc. Natl. Acad. Sci. U.S.A.,* 85, 8880, 1988.

48. **Guidolin, A. and Manning, P. A.,** Genetics of *Vibrio cholerae* and its bateriophages, *Microbiol. Rev.,* 51, 285, 1987.

49. **Manning, P. A.,** Genetic approaches to the study of *Vibrio cholerae, Microbiol. Sci.,* 5, 196, 1988.

50. **Hardy, S. J. S., Holmgren, J., Johansson, S., Sanchez, J., and Hirst, T. R.,** Coordinate assembly of multisubunit proteins: oligomerization of bacterial enterotoxins *in vivo* and *in vitro, Proc. Natl. Acad. Sci. U.S.A.,* 85, 7109, 1988.

51. **van Heyningen, S.,** Cholera toxin, *Biosci. Rep.,* 2, 135, 1982.

52. **van Heyningen, S.,** Similarities in the action of different toxins, in *Molecular Action of Toxins and Viruses,* Cohen, P. and van Heyningen, S., Eds., Elsevier, Amsterdam, 1982, 169.

53. **van Heyningen, S.,** Cholera and related toxins, in *Molecular Medicine,* Vol. 1, Malcolm, A. D. B., Ed., IRL Press, Oxford, 1984, 1.

54. **Spangler, B. D. and Westbrook, E. M.,** Crystallization of isoelectrically homogeneous cholera toxin, *Biochemistry,* 28, 1333, 1989.

55. **Ribi, H. O., Ludwig, D. S., Mercer, K. L., Schoolnik, G. K., and Kornberg, R. D.,** Three-dimensional structure of cholera toxin penetrating a lipid membrane, *Science,* 239, 1272, 1988.

56. **Gilman, A. G.,** G proteins and dual control of adenylate cyclase, *Cell,* 36, 577, 1984.

57. **Levitzki, A.,** From epinephrine to cyclic AMP, *Science,* 241, 800, 1988.

58. **Weiss, E. R., Kelleher, D. J., Woon, C. W., Soparkar, S., Osawa, S., Heasley, L. E., and Johnson, G. L.,** Receptor activation of G proteins, *FASEB J.,* 2, 2841, 1988.

59. **Lynch, C. J., Morbach, L., Blackmore, P. F., and Exton, J. H.,** Alpha subunits of Ns are released from the plasma membrane following cholera toxin activation, *FEBS Lett.,* 200, 333, 1986.

60. **Askamit, R. R., Backlund, P. S., and Cantoni, G. L.,** Cholera toxin inhibits chemotaxis by a cAMP-independent mechanism, *Proc. Natl. Acad. Sci. U.S.A.,* 82, 7475, 1985.

61. **Imboden, J. B., Shoback, D. M., Pattison, G., and Stobo, J. D.,** Cholera toxin inhibits the T-cell antigen receptor-mediated increases in inositol triphosphate and cytoplasmic free calcium, *Proc. Natl. Acad. Sci. U.S.A.,* 83, 5673, 1986.

62. **Kahn, R. A. and Gilman, A. G.,** Purification of a protein cofactor required for ADP-ribosylation of the stimulatory regulatory component of adenylate cyclase by cholera toxin, *J. Biol. Chem.,* 259, 6228, 1984.

63. **Kahn, R. A. and Gilman, A. G.,** The protein cofactor necessary for ADP-ribosylation of Gs by cholera toxin is itself a GTP binding protein, *J. Biol. Chem.,* 261, 7906, 1986.

64. **Tsai, S. U., Noda, M., Adamik, R., Moss, J., and Vaughan, M.,** Enhancement of choleragen ADP-ribosyltransferase activities by guanyl nucleotides and a 19-kDa membrane protein, *Proc. Natl. Acad. Sci. U.S.A.,* 84, 5139, 1987.

65. **Tsai, S. C., Noda, M., Adamik, R., Chang, P. P., Chen, H. C., Moss, J., and Vaughan, M.,** Stimulation of choleragen enzymatic activites by GTP and two soluble proteins purified from bovine brain, *J. Biol. Chem.,* 263, 1768, 1988.

66. **Neer, E. J., Wolf, L. G., and Gill, D. M.,** The stimulatory guanine-nucleotide regulatory unit of adenylate cyclase from bovine cerebral cortex. ADP-ribosylation and purification, *Biochem. J.,* 241, 325, 1987.

67. **Gill, D. M. and Coburn, J.,** ADP-ribosylation by cholera toxin: functional analysis of a cellular system that stimulates the enzymic activity of cholera toxin fragment A1, *Biochemistry,* 26, 6364, 1987.

68. **Woolkalis, M., Gill, D. M., and Coburn, J.,** Assay and purification of cytosolic factor required for cholera toxin activity, *Meth. Enzymol. Relat. Areas Mol. Biol.,* 165, 246, 1988.

69. **Price, S. R., Nightingale, M., Tsai, S. C., Williamson, K. C., Adamik, R., Chen, H. C., Moss, J., and Vaughan, M.,** Guanine nucleotide-binding proteins that enhance choleragen ADP-ribosyltransferase activity: nucleotide and deduced amino acid sequence of an ADP-ribosylation factor cDNA, *Proc. Natl. Acad. Sci. U.S.A.,* 85, 5488, 1988.

70. **Sewell, J. L. and Kahn, R. A.,** Sequences of the bovine and yeast ADP-ribosylation factor and comparison to other GTP-binding proteins, *Proc. Natl. Acad. Sci. U.S.A.,* 85, 4620, 1988.

71. **Kahn, R. A., Goddard, C., and Newkirk, M.,** Chemical and immunological characterization of the 21-kDa ADP-ribosylation factor of adenylate cyclase, *J. Biol. Chem.,* 263, 8282, 1988.

72. **van Dop, C., Tsubokawa, M., Bourne, H. R., and Ramachandran, J.,** Amino acid sequence of retinal transducin at the site ADP-ribosylated by cholera toxin, *J. Biol. Chem.,* 259, 696, 1984.

73. **Heyworth, C. M., Whetton, A. D., Wong, S., Martin, B. R., and Houslay, M. D.,** Insulin inhibits the cholera toxin-catalysed ribosylation of a M_r-25,000 protein, *Biochem. J.,* 228, 593, 1985.

74. **Waldo, G. L., Evans, T., Fraser, E. D., Northup, J. K., Martin, M. W., and Harden, T. K.,** Identification and purification from bovine brain of a guanine nucleotide binding protein distinct from Gs, Gi, and Go, *Biochem. J.,* 246, 431, 1987.

75. **Hawkins, D. J. and Browning, E. T.,** Tubulin adenosine diphosphate ribosylation is catalyzed by cholera toxin, *Biochemistry,* 21, 4474, 1982.

76. **Gill, D. M. and Wookalis, M.,** Toxins which activate adenylate cyclase, *Ciba Found. Symp.,* 112, 57, 1985.

77. **Aranda, A., Pascual, A., Copp, R., and Samuels, H.,** Cholera toxin affects nuclear ADP-ribosylation in GH1 cells, *Biochem. Biophys. Res. Commun.,* 150, 323, 1988.

78. **Parkinson, M. E., Smith, C. G., Garland, P. B., and van Heyningen, S.,** Identification of cholera toxin-binding sites in the nucleus of intestinal epithelial cells, *FEBS Lett.,* 242, 309, 1989.

79. **Gill, D. M. and Woolkalis, M.,** [^{32}P]ADP-ribosylation of proteins catalysed by cholera toxin and related heat-labile enterotoxins, *Meth. Enzymol. Relat. Areas Mol. Biol.,* 165, 235, 1988.

80. **Gill, D. M. and Coburn, J.,** ADP-ribosylation of membrane proteins by bacterial toxins in the presence of NAD glycohydrolase, *Biochim. Biophys. Acta,* 954, 65, 1988.

81. **Longabaugh, J. P., Vatner, D. E., Graham, R. M., and Homcy, C. E.,** NADP improves the efficiency of cholera toxin catalyzed ADP-ribosylation in liver and heart membranes, *Biochem. Biophys. Res. Commun.,* 137, 328, 1986.

82. **Kawai, Y., Whitsel, C., and Arinze, I. J.,** NADP enhances cholera and pertussis toxin-catalysed ADP-ribosylation of membrane proteins, *J. Cycl. Nucl. Prot. Phos. Res.,* 11, 265, 1986.

83. **Moss, J. and Vaughan, M.,** Mechanism of action of choleragen: evidence for ADP-ribosyltransferase activity with arginine as acceptor, *J. Biol. Chem.,* 252, 2455, 1977.

84. **Oppenheimer, N. J.,** Structural determination and stereospecificity of the choleragen-catalyzed reaction of NAD with guanidines, *J. Biol. Chem.,* 253, 4907, 1978.

85. **Tait, R. M. and van Heyningen, S.,** The adenylate cyclase-activating activity of cholera toxin is not associated with a nicotinamide-adenine dinucleotide glycohydrolase activity, *Biochem. J.,* 175, 1059, 1978.

86. **Moss, J., Stanley, S. J., and Lin, M. C.,** NAD glycohydrolase and ADP-ribosyltransferase activities are intrinsic to the A1 peptide of cholera toxin, *J. Biol. Chem.,* 254, 11993, 1979.

87. **Tait, R. M. and Nassau, P. M.,** Artificial low-molecular-mass substrates of cholera toxin, *Europ. J. Biochem.,* 143, 213, 1984.

88. **Mekalanos, J. J., Collier, R. J., and Romig, W. R.,** Enzymic activity of cholera toxin, *J. Biol. Chem.,* 254, 5855, 1979.

89. **Osborne, J. C., Stanley, S. J., and Moss, J.,** Kinetic mechanisms of two NAD:arginine ADP-ribosyl-transferases: the soluble, salt-stimulated transferase from turkey erythrocytes and choleragen, a toxin from *Vibrio cholerae, Biochemistry,* 24, 5235, 1985.

90. **Soman, G., Narayanan, J., Martin, B. L., and Graves, D. J.,** Use of substituted (benzylidineamino) guanidines in the study of guanidino group specific ADP-ribosyltransferases, *Biochemistry,* 25, 4113, 1986.

91. **Soman, G., Tomer, K. B., and Graves, D. J.,** Assay of mono ADP-ribosyltransferase activity by using guanylhydrazones, *Anal. Biochem.,* 134, 101, 1983.

92. **Ward, W. H. J. and van Heyningen, S.,** Structural studies on the polypeptide substrates of cholera toxin, *Biochem. Biophys. Res. Commun.,* 108, 1181, 1982.

93. **Ward, W. H. J. and van Heyningen, S.,** The sites on the regulatory component of adenylate cyclase which are ADP-ribosylated by cholera toxin, *Biochem. Biophys. Res. Commun.,* 105, 928, 1982.

94. **Kharadi, S. V. and Graves, D. J.,** Relationship of phosphorylation and ADP-ribosylation using a synthetic peptide as a model substrate, *J. Biol. Chem.,* 262, 17379, 1987.

95. **Galloway, T. S. and van Heyningen, S.,** Binding of NAD$^+$ by cholera toxin, *Biochem. J.,* 244, 225, 1987.

96. **Galloway, T. S., Tait, R. M., and van Heyningen, S.,** Photolabeling of cholera toxin by NAD$^+$, *Biochem. J.,* 242, 927, 1987.

97. **de Wolf, M., van Dessel, G., Lagrou, A., Hilderson, H. J., and Dierick, W.,** Structural features of the binding site of cholera toxin inferred from fluorescence measurements, *Biochim. Biophys. Acta,* 8832, 165, 1985.

98. **Duffy, L. K., Kurosky, A., and Lai, C. Y.,** Cholera toxin A subunit: functional sites correlated with regions of secondary structure, *Arch. Biochem. Biophys.,* 239, 549, 1985.

99. **Wardlaw, A. C. and Parton, R.,** *Pathogenesis and Immunity in Pertussis,* Wiley, Chichester, 1988.

100. **Burns, D. L.,** Subunit structure and enzymic activity of pertussis toxin, *Microbiol. Sci.,* 5, 285, 1988.

101. **Gross, R., Arico, B., and Rappuoli, R.,** Genetics of pertussis toxin, *Mol. Microbiol.,* 3, 119, 1989.

102. **Tamura, M., Nogimori, K., Murai, S., Yajima, M., Ito, K., Katada, T. K., Ui, M., and Ishii, S.,** Subunit structure of the islet-activating protein, pertussis toxin, in conformity with the A-B model, *Biochemistry,* 21, 5516, 1982.

103. **Locht, C. and Keith, J. M.,** Pertussis toxin gene: nucleotide sequence and genetic organization, *Science,* 232, 1258, 1986.

104. **Nicosia, A., Perugini, M., Franzini, C., Casagli, M. C., Borri, M. G., Antoni, G., Almoni, M., Neri, P., Ratti, G., and Rappuoli, R.,** Cloning and sequencing of the pertussis toxin genes: operon structure and gene duplication, *Proc. Natl. Acad. Sci. U.S.A.,* 83, 4631, 1986.

105. **Tamura, M., Nogimori, K., Yajima, M., Ase, K., and Ui, M.,** A role of the B-oligomer moiety of islet-activating protein, pertussis toxin, in development of biological effects on intact cells, *J. Biol. Chem.,* 258, 6756, 1983.

106. **Marayama, T. and Ui, M.,** [³H]GDP release from rat and hamster adipocyte membranes independently linked to receptors involved in activation or inhibition of adenylate cyclase. Differential susceptibility to two bacterial toxins, *J. Biol. Chem.,* 258, 3319, 1984.

107. **van Dop, C., Yamanaka, G., Steinberg, F., Sekura, R. D., Manclark, C. R., Stryer, L., and Bourne, H. R.,** ADP-ribosylation of transducin by pertussis toxin blocks the light-stimulated hydrolysis of GTP and cGMP in retinal photoreceptors, *J. Biol. Chem.,* 254, 8582, 1984.

108. **Katada, T., Oinuma, M., and Ui, M.,** Two guanine nucleotide-binding proteins in rat brain serving as the specific substrate of islet-activating protein, pertussis toxin. Interaction of the alpha subunits with beta subunits in development of their biological activities, *J. Biol. Chem.,* 261, 8182, 1986.

109. **Toutant, M., Aunis, D., Bockaert, J., Homburger, V., and Rouot, B.,** Presence of three pertussis toxin substrates and G$_{o\alpha}$ immunoreactivity in both plasma and granule membranes of chromaffin cells, *FEBS Lett.,* 215, 339, 1987.

110. **Ui, M., Okajima, F., Murayama, T., Nakamura, T., Kurose, H., Itoh, H., and Ohta, H.,** A role of the inhibitory guanine nucleotide-binding regulatory protein in signal transduction via Ca^{2+}-mobilizing receptors, in *Adrenergic Receptors: molecular properties and therapeutic implications,* Lefkowitz, R. J. and Lindenlaub, E., Eds., Schattauer, Stuttgart, 1985, 209.

111. **Ui, M., Katada, T., and Kurose, H.,** Islet-activating protein, pertussis toxin, as a probe for the biosignaling system, in *Natural Products and Biological Activities,* Imura, H., Goto, T., Murachi, T., and Nakajima, T., Eds., University of Tokyo Press, Tokyo, 1986, 247.

112. **Tanaka, T., Yokohama, H., Negishi, M., Hayashi, H., Ito, S., and Hayaishi, O.,** Pertussis toxin facilitates secretagogue-induced catecholamine release from cultured bovine adrenal chromaffin cells, *Biochem. Biophys. Res. Commun.,* 144, 907, 1987.

113. **Brockelhurst, K. W. and Pollard, H. B.,** Pertussis toxin stimulates delayed-onset, Ca^{2+}-dependent catecholamine release and the ADP-ribosylation of a 40 kDa protein in bovine adrenal chromaffin cells, *FEBS Lett.,* 234, 439, 1988.

114. **Manning, D. R., Fraser, B. A., Kahn, R. A., and Gilman, A. G.,** ADP-ribosylation of transducin by islet-activating protein. Identification of asparagine as the site of ADP-ribosylation, *J. Biol. Chem.,* 259, 749, 1984.

114. **Ribeiro-Neto, A. P. and Rodbell, M.,** Pertussis toxin induces structural changes in G$_\alpha$ proteins independently of ADP-ribosylation, *Proc. Natl. Acad. Sci. U.S.A.,* 86, 2577, 1989.

115. **West, R. E., Moss, J., Vaughan, J. M., Liu, T., and Liu, T. Y.,** Pertussis toxin-catalyzed ADP-ribosylation of transducin. Cysteine 347 is the ADP-ribose receptor, *J. Biol. Chem.,* 260, 14428, 1985.

116. **Lobban, M. D. and van Heyningen, S.,** Thiol reagents are substrates for the ADP-ribosyltransferase activity of pertussis toxin, *FEBS Lett.,* 233, 229, 1988.

117. **Meyer, T., Koch, R., Fanick, W., and Hilz, H.,** ADP-ribosyl proteins formed by pertussis toxin are specifically cleaved by mercury ions, *Biol. Chem. Hoppe-Seyler,* 369, 579, 1988.

118. **Aktories, K., Just, I., and Rosenthal, W.,** Different types of ADP-ribose protein bonds formed by botulinum C2 toxin, botulinum ADP-ribosyltransferase C3 and pertussis toxin, *Biochem. Biophys. Res. Commun.,* 156, 361, 1988.

119. **Moss, J., Stanley, S. J., Burns, D. L., Hsias, J. A., Yost, D. A., Myers, G. A., and Hewlett, E. L.,** Activation by thiol of the latent NAD glycohydrolase and ADP-ribosyltransferase activities of *Bordetella pertussis* toxin (islet-activating protein), *J. Biol. Chem.,* 258, 11879, 1983.

120. **Lobban, M. D., Irons, L. I., and van Heyningen, S.,** Binding of NAD to pertussis toxin, *Biochim. Biophys. Acta,* 1078, 155, 1991.

121. **Kaslow, H. R. and Lesikar, D. D.,** Sulfhydryl-alkylating reagents inactivate the NAD glycohydrolase activity of pertussis toxin, *Biochem.*, 26, 4397, 1987.

122. **Burns, D. L. and Manclark, C. R.,** Role of cysteine 41 of the A subunit of pertussis toxin, *J. Biol. Chem.*, 264, 564, 1989.

123. **Kaslow, H. R., Schlotterbeck, J. D., Mar, V. L., and Burnette, W. N.,** Alkylation of cysteine 41, but not cysteine 200, decreases the ADP-ribosyltransferase activity of the S1 subunit of pertussis toxin, *J. Biol. Chem.*, 264, 6386, 1989.

124. **Locht, C., Lobet, Y., Feron, C., Cieplak, W., and Keith, J. M.,** The role of cysteine 41 in the enzymatic activities of the pertussis toxin S1 subunit as investigated by site-directed mutagenesis, *J. Biol. Chem.*, 265, 455, 1990.

125. **Kaslow, H. R., Lim, L.-K., Moss, J., and Lesikar, D. D.,** Structure-activity analysis of the activation of pertussis toxin, *Biochemistry*, 26, 123, 1987.

126. **Burns, D. L. and Manclark, C. R.,** Adenine nucleotides promote dissociation of pertussis toxin, *J. Biol. Chem.*, 261, 4324, 1986.

127. **Mattera, R., Codina, J., Sekura, R. D., and Birnbaumer, L.,** The interaction of nucleotides with pertussis toxin, *J. Biol. Chem.*, 261, 11173, 1986.

128. **Moss, J., Stanley, S. J., Watkins, P. W., Burns, D. L., Manclark, C. R., Kaslow, H. R., and Hewlett, R. L.,** Stimulation of the thiol-dependent ADP-ribosyltransferase and NAD glycohydrolase activities of *Bordetella pertussis* toxin by adenine nucleotides, phospholipids, and detergents, *Biochemistry*, 25, 2720, 1986.

129. **Burns, D. L., Hausman, S. Z., Lindner, W., Robey, F. A., and Manclark, C. R.,** Structural characterization of pertussis toxin A subunit, *J. Biol. Chem.*, 262, 17677, 1987.

130. **Barbieri, J. T., Rappuoli, R., and Collier, R. J.,** Expression of the S-1 catalytic subunit of pertussis toxin in *Escherichia coli*, *Infect. Immun.*, 55, 1321, 1987.

131. **Burnette, W. N., Mar, V. L., Ceiplak, W., Morris, C. F., Kaljot, K. T., Marchitto, K. S., Sachdev, R. K., Locht, C., and Keith, J. M.,** Direct expression of *Bordetella-pertussis* toxin subunits to high levels in *Escherichia coli*, *Bio/technology*, 6, 699, 1988.

132. **Locht, C., Cieplak, W., Marchitto, K. S., Sato, H., and Keith, J. M.,** Activities of the complete and truncated forms of pertussis toxin subunits S1 and S2 synthesized in *Escherichia coli*, *Infect. Immun.*, 55, 2546, 1987.

133. **Pizza, M., Bartoloni, A., Prugnola, A., Silvstri, S., and Rappuoli, R.,** Subunit S1 of pertussis toxin: mapping of the regions essential for ADP-ribosyltransferase activity, *Proc. Natl. Acad. Sci. U.S.A.*, 85, 7521, 1988.

134. **Barbieri, J. T. and Cortina, G.,** ADP-ribosyltransferase mutations in the catalytic S-1 subunit of pertussis toxin, *Infect. Immun.*, 56, 1934, 1988.

135. **Capiau, C., Petre, J., Van Damme, J., Puype, M., and Vandekerchove, J.,** Protein-chemical analysis of pertussis toxin reveals homology between the subunits S2 and S3, between S1 and the A chains of enterotoxins of *Vibrio cholerae* and *Escherichia coli* and identifies S2 as the haptoglobin-binding subunit, *FEBS Lett.*, 204, 336, 1986.

136. **Cieplak, W., Burnette, W. N., Mar, V. L., Kaljot, K. T., Morris, C. F., Chen, K. K., Sato, H., and Keith, J. M.,** Identification of a region in the S1 subunit of pertussis toxin that is required for enzymatic activity and that contributes to the formation of a neutralizing antigenic determinant, *Proc. Natl. Acad. Sci. U.S.A.*, 85, 4667, 1988.

137. **Burnette, W. N., Cieplak, W., Mar, V. L., Kaljot, K. T., Sato, H., and Keith, J. M.,** Pertussis toxin S1 mutant with reduced enzyme activity and a conserved protective epitope, *Science*, 242, 72, 1988.

138. **Locht, C., Capiau, C., and Feron, C.,** Identification of amino acid residues essential for the enzymatic activities of pertussis toxin, *Proc. Natl. Acad. Sci. U.S.A.*, 86, 3075, 1989.

140. **Cockle, S. A.,** Identification of an active site residue in subunit S1 of pertussis toxin by photocrosslinking to NAD, *FEBS Lett.*, 249, 329, 1989.

141. **Black, W. J., Munoz, J. J., Peacock, M. G., Schad, P. A., Cowell, J. L., Burchall, J. J., Lim, M., Kent, A., Steinman, L., and Falkow, S.,** ADP-ribosyltransferase activity of pertussis toxin and immunomodulation by *Bordetella pertussis*, *Science*, 240, 656, 1988.

142. **Aktories, K., Ankenbauer, T., Scherin, P., and Jakobs, K. H.,** ADP-ribosylation of platelet actin by botulinum C2 toxin, *Europ. J. Biochem.*, 161, 155, 1986.

143. **Aktories, K., Barmann, M., Ohishi, I., Tsuyama, S., Jakobs, K. H., and Habermann, E.,** Botulinum C2 toxin ADP-ribosylates actin, *Nature, London*, 322, 390, 1986.

144. **Ohishi, I. and Tsuyama, S.,** ADP-ribosylation of nonmuscle actin with component 1 of C2 toxin, *Biochem. Biophys. Res. Commun.*, 136, 802, 1986.

145. **Aktories, K., Barmann, M., Chhatwal, G. S., and Presek, P.,** New class of microbial toxins ADP-ribosylates actin, *Trends Pharmacol. Sci.*, 8, 158, 1987.

146. **Simpson, L. L., Stiles, B. G., Zepeda, H. H., and Wilkins, T. D.,** Molecular basis for the pathological actions of *Clostridium perfringens* iota toxin, *Infect. Immun.*, 55, 118, 1987.

147. **Schering, B., Barmann, M., Chhatwal, G. S., Geipel, U., and Aktories, K.,** ADP-ribosylation of skeletal muscle and non-muscle actin by *Clostridium perfringens* iota toxin, *Eur. J. Biochem.,* 171, 225, 1988.

148. **Popoff, M. R., Rubin, E. J., Gill, D. M., and Boquet, P.,** Actin-specific ADP-ribosyltransferase produced by a *Clostridium difficile* strain, *Infect. Immun.,* 56, 2299, 1988.

149. **Popoff, M. R. and Boquet, P.,** *Clostridium spiroforme* toxin is a binary toxin which ADP-ribosylates cellular actin, *Biochem. Biophys. Res. Commun.,* 152, 1361, 1988.

150. **Simpson, L. L., Stiles, B. G., Zepeda, H., and Wilkins, T. D.,** Production by *Clostridium spiroforme* of an iotalike toxin that possesses mono(ADP-ribosyl)transferase activity: identification of a novel class of ADP-ribosyltransferases, *Infect. Immun.,* 57, 255, 1989.

151. **Vandekerckhove, J., Schering, B., Barmann, M., and Aktories, K.,** *Clostridium perfringens* iota toxin ADP-ribosylates skeletal muscle actin in Arg-177, *FEBS Lett.,* 225, 48, 1987.

152. **Vanderkerckhove, J., Schering, B., Barmann, M., and Aktories, K.,** Botulinum C2 toxin ADP-ribosylates cytoplasmic beta/gamma actin in arginine 177, *J. Biol. Chem.,* 263, 696, 1988.

153. **Simpson, L. L., Zepeda, H., and Ohishi, I.,** Partial characterization of the enzymatic activity associated with the binary toxin (type C2) produced by *Clostridium botulinum, Infect. Immun.,* 56, 24, 1988.

154. **Reuner, K. H., Presek, P., Boschek, C. B., and Aktories, K.,** Botulinum C2 toxin ADP-ribosylates actin and disorganizes the microfilament network in intact cells, *Europ. J. Cell Biol.,* 43, 134, 1987.

155. **Wegner, A. and Aktories, K.,** ADP-ribosylated actin caps the barbed ends of actin filaments, *J. Biol. Chem.,* 263, 13739, 1988.

156. **Weigt, C., Just, I., Wegner, A., and Aktories, K.,** Nonmuscle actin ADP-ribosylated by botulinum C2 toxin caps actin filaments, *FEBS Lett.,* 246, 181, 1989.

157. **Giepel, U., Just, I., Schering, B., Haas, D., and Aktories, K.,** ADP-ribosylation of actin causes increase the rate of ATP exchange and inhibition of ATP hydrolysis, *Eur. J. Biochem.,* 179, 229, 1989.

158. **Matter, K., Dreyer, F., and Aktories, K.,** Actin involvement in exocytosis from PC12 cells: studies on the influence of botulinum C2 toxin on stimulated noradrenaline release, *J. Neurochem.,* 52, 370, 1989.

159. **Norgauer, J., Kownatzki, E., Seifert, R., and Aktories, K.,** Botulinum C2-toxin ADP-ribosylates actin and enhances O_2 production and secretion but inhibits migration of activated human neutrophils, *J. Clin. Invest.,* 82, 1376, 1988.

160. **Al-Mohanna, F. A., Ohishi, I., and Hallett, M. B.,** Botulinum C2 toxin potentiates activation of the neutrophil oxidase: Further evidence of a role for actin polymerization, *FEBS Lett.,* 219, 40, 1987.

161. **Bleck, T. P.,** Pharmacology of tetanus, *Clin. Neuropharmacol.,* 9, 103, 1986.

162. **Habermann, E. and Dryer, F.,** Clostridial neurotoxins: handling and action at the molecular and cellullar levels, *Curr. Top. Microbiol. Immunol.,* 129, 93, 1986.

163. **Simpson, L. L.,** Molecular pharmacology of botulinum toxin and tetanus toxin, *Annu. Rev. Pharmacol. Toxicol.,* 26, 427, 1986.

164. **van Heyningen, S.,** Tetanus Toxin, in *International Encyclopedia of Pharmacology and Toxicology,* Vol. 119, Dorner, F. and Drews, J., Eds., Pergamon Press, Oxford, 1986, 549.

165. **Mellanby, J.,** Recent work on the mechanism of action of tetanus toxin, in *Anaerobes Today,* Hardie, J. M. and Borriello, S. P., Eds., Wiley, Chichester, 1988, 69.

166. **DasGupta, B. R. and Datta, A.,** Botulinum neurotoxin type B (strain 657): partial sequence and similarity with tetanus toxin, *Biochimie,* 70, 811, 1988.

167. **Tsuzuki, K., Yokosawa, N., Syuto, B., Ohishi, I., Fuji, N., Kimura, K., and Oguma, K.,** Establishment of a monoclonal antibody recognizing an antigenic site common to *Clostridium botulinum* type B, C1, D and E toxins and tetanus toxin, *Infect. Immun.,* 56, 898, 1988.

168. **Halpern, J. L., Smith, L. A., Seamon, K. B., Groover, K. A., and Habig, W. H.,** Sequence homology between tetanus and botulinum toxins detected by antipeptide antibody, *Infect. Immun.,* 57, 18, 1989.

169. **Eisel, U., Jarausch, W., Goretzki, K., Henschen, A., Engels, J., Weller, U., Hudel, M., Habermann, E., and Niemann, H.,** Tetanus toxin: primary structure, expression in *E. coli,* and homology with botulinum toxins, *EMBO J.,* 5, 2495, 1986.

170. **Fairweather, N. F. and Lyness, V. A.,** The complete nucleotide sequence of tetanus toxin, *Nucl. Acid Res.,* 14, 7809, 1986.

171. **Montecucco, C.,** How do tetanus and botulinum toxins bind to neuronal membranes?, *Trends Biochem. Sci.,* 11, 314, 1986.

172. **Ahnert-Hilger, G., Weller, U., Dauzenroth, M. E., Habermann, E., and Gratzl, M.,** The tetanus toxin light chain inhibits exocytosis, *FEBS Lett.,* 242, 245, 1989.

173. **Maisey, E. A., Wadsworth, J. D. F., Poulain, B., Shone, C. C., Melling, J., Gibbs, P., Tauc, L., and Dolly, J. O.,** Involvement of the constituent chains of botulinum neurotoxins A and B in the blockade of neurotransmitter release, *Eur. J. Biochem.,* 177, 683, 1988.

174. **Poulain, B., Tauc, L., Maisey, E. A., Wadsworth, J. D. F., Mohan, P. M., and Dolly, J. O.,** Neurotransmitter release is blocked intracellularly by botulinum neurotoxin, and this requires uptake of both toxin polypeptides by a process mediated by the larger chain, *Proc. Natl. Acad. Sci. U.S.A.,* 85, 4090, 1988.

175. **Ohashi, Y. and Narumiya, S.,** ADP-ribosylation of a relative mol wt 21,000 membrane protein by type D botulinum toxin, *J. Biol. Chem.,* 262, 1430, 1987.

176. **Ohashi, Y., Kamiya, T., Fujiwara, M., and Narumiya, S.,** ADP-ribosylation by type C1 and D botulinum neurotoxins: stimulation by guanine nucleotides and inhibition by guanidino-containing compounds, *Biochem. Biophys. Res. Commun.,* 142, 1032, 1987.

177. **Rosener, S., Chhatwal, G. S., and Aktories, K.,** Botulinum ADP-ribosyltransferase C3 but not botulinum neurotoxins C1 and D ADP-ribosylates low molecular mass GTP-binding proteins, *FEBS Lett.,* 224, 38, 1987.

178. **Aktories, K., Weller, U., and Chhatwal, G. S.,** *Clostridium botulinum* type C produces a novel ADP-ribosyltransferase distinct from botulinum C2 toxin, *FEBS Lett.,* 212, 109, 1987.

179. **Rubin, E. J., Gill, D. M., Boquet, P., and Popoff, M. R.,** Functional modification of a 21-kilodalton G protein when ADP-ribosylated by exoenzyme C3 of *Clostridium botulinum, Mol. Cell Biol.,* 8, 418, 1988.

180. **Aktories, K., Rosener, S., Blaschke, U., and Chhatwal, G. S.,** Botulinum ADP-ribosyltransferase C3. Purification of the enzyme and characterization of the ADP-ribosylation reaction in platelet membranes, *Europ. J. Biochem.,* 172, 445, 1988.

181. **Aktories, K. and Frevert, J.,** ADP-ribosylation of a 21-24 kDa eukaryotic protein(s) by C3, a novel botulinum ADP-ribosyltransferase, is regulated by guanine nucleotide, *Biochem. J.,* 247, 363, 1987.

182. **Quilliam, L. A., Brown, J. H., and Buss, J. E.,** A 22 kDa ras-related G-protein is the substrate for an ADP-ribosyltransferase from *Clostridium botulinum, FEBS Lett.,* 238, 22, 1988.

183. **Tsai, S. C., Adamik, R., Moss, J., and Aktories, K.,** Separation of the 24 kDa substrate for botulinum C3 ADP-ribosyltransferase and the cholera toxin ADP-ribosylation factor, *Biochem. Biophys. Res. Commun.,* 152, 957, 1988.

184. **Banga, H. S., Gupta, S. K., and Feinstein, M. B.,** Botulinum toxin D ADP-ribosylates a 22-24 kDa membrane protein in platelets and HL-60 cells that is distinct from p21N-RAS, *Biochem. Biophys. Res. Commun.,* 155, 263, 1988.

185. **Aktories, K., Braun, U., Rosener, S., Just, I., and Hall, A.,** The rho gene product expressed in *E. coli* is a substrate of botulinum ADP-ribosyltransferase C3, *Biochem. Biophys. Res. Commun.,* 158, 209, 1989.

186. **Morii, N., Sekine, A., Ohashi, Y., Nakao, K., Imura, H., Fujiwara, M., and Narumiya, S.,** Purification and properties of the cytosolic substrate for Botulinum ADP-ribosyltransferase, *J. Biol. Chem.,* 263, 12420, 1988.

187. **Braun, U., Habermann, B., Just, I., Aktories, K., and Vandekerchove, J.,** Purification of the 22 kDa protein substrate of botulinum ADP-ribosyltransferase C3 from bovine brain cytosol and its characterization as a GTP-binding protein highly homologous to the rho gene product, *FEBS Lett.,* 243, 70, 1989.

188. **Chardin, P., Boquet, P., Madaule, P., Popoff, M. R., Rubin, E. J., and Gill, D. M.,** The mammalian G protein rhoC is ADP-ribosylated by *Clostridium botulinum* exoenzyme C3 and affects actin microfilaments in Vero cells, *EMBO J.,* 8, 1087, 1989.

189. **Narumiya, S., Sekine, A., and Fujiwara, M.,** Substrate for botulinum ADP-ribosyltransferase, Gb, has an amino acid sequence homologous to a putative rho gene product, *J. Biol. Chem.,* 263, 17255, 1988.

190. **Bococh, G. M., Parkos, C. A., and Mumby, S. M.,** Purification and characterization of the 22,000-Dalton GTP-binding protein substrate for ADP-ribosylation by botulinum toxin, G22K, *J. Biol. Chem.,* 263, 16774, 1988.

190. **Sekine, A., Fujiwara, M. A., and Narumiya, S.,** Asparagine residue in the rho gene product is the modification site for botulinum ADP-ribosyltransferase, *J. Biol. Chem.,* 264, 8602, 1989.

191. **Narumiya, S., Morii, N., Ohno, K., Ohashi, Y., and Fujiwara, M.,** Subcellular distribution and isoelectric heterogeneity of the substrate for ADP-ribosyl transferase from *Clostridium botulinum, Biochem. Biophys. Res. Commun.,* 150, 1122, 1988.

192. **Matsuoka, I., Sakuma, H., Syuto, B., Moriishi, K., Kubo, S., and Kurihara, K.,** ADP-ribosylation of 24-26-kDa GTP-binding proteins localized in neuronal and non-neuronal cells by botulinum neurotoxin D, *J. Biol. Chem.,* 264, 706, 1989.

193. **Knight, D. E., Tonge, D. A., and Baker, P. F.,** Inhibition of exocytosis in bovine adrenal medullary cells by botulinum toxin type D, *Nature (London),* 317, 719, 1985.

194. **Knight, D. E.,** Botulinum toxin types A, B and D inhibit catecholamine secretion from bovine adrenal medullary cells, *FEBS Lett.,* 207, 222, 1986.

195. **Adam-Vizi, V., Rosener, S., Aktories, K., and Knight, D. E.,** Botulinum toxin-induced ADP-ribosylation and inhibition of exocytosis are unrelated events, *FEBS Lett.,* 238, 277, 1988.

196. **Mege, J. L., Volpi, M., Becker, E. L., and Sha'afi, R. I.,** Effect of botulinum D toxin on neutrophils, *Biochem. Biophys. Res. Commun.,* 152, 926, 1988.

197. **Montecucco, C.,** Some theoretical considerations on the cellular mechanism of action of clostridial neurotoxins, in *Proc. 8th Int. Conf. Tetanus,* Nistico, G., Bizzini, B., Bytchenko, B., and Triau, R., Eds., Pythagora, Rome, 1988, chap. 11.

198. **Gardner, S. E., Fowlston, S. E., and George, W. L.,** *In vitro* production of cholera toxin-like activity by *Plesiomonas shigelloides, J. Infect. Dis.,* 156, 720, 1987.

199. **Schultz, A. J. and McCardell, B. A.,** DNA homology and immunological cross-reactivity between *Aeromonas hydrophila* cytotonic toxin and cholera toxin, *J. Clin. Microbiol.,* 26, 57, 1988.

200. **Rose, J. M., Houston, C. W., Coppenhaver, D. H., Dixon, J. D., and Kurosky, A.,** Purification and chemical characterization of a cholera toxin-cross-reactive cytolytic enterotoxin produced by a human isolate of *Aeromonas hydrophila, Infect. Immun.,* 57, 1165, 1989.

201. **Chang, P. P., Moss, J., Twiddy, E. M., and Holmes, R. K.,** Type II heat-labile enterotoxin of *Escherichia coli* activates adenylate cyclase in human fibroblasts by ADP ribosylation, *Infect. Immun.,* 55, 1854, 1987.

202. **Iglewski, B. H., Sadoff, J., Bjorn, M. J., and Maxwell, E. S.,** *Pseudomonas aeruginosa* exoenzyme S: an adenosine diphosphate ribosyltransferase distinct from toxin A, *Proc. Natl. Acad. Sci. U.S.A.,* 75, 3211, 1978.

203. **Skoro, R.,** T4 and N4 phage-encoded ADP-ribosyltransferases, in *ADP-Ribosylation Reactions,* Hayaishi, O. and Ueda, K., Eds., Academic Press, London, 1982, 647.

204. **Bredhorst, R., Wielckens, K., Gartemann, A., Lengyel, H., Klapproth, K., and Hilz, H.,** Two different types of bonds linking single ADP-ribose residues covalently to proteins. Quantification in eukaryotic cells, *Eur. J. Biochem.,* 92, 129, 1978.

205. **Payne, D. M., Jacobson, E. L., Moss, J., and Jacobson, M. K.,** Modification of proteins by mono(ADP-ribosylation) *in vivo, Biochem.,* 24, 7540, 1985.

206. **Jacobson, M. K., Payne, D. M., Smith, K. P., Cardenas, M. E., Moss, J., and Jacobson, E. L.,** Mono ADP-ribosylation of proteins at arginine *in vivo,* in *ADP-Ribosylation of Proteins,* Althaus, F. R., Hilz, H., and Shall, S., Eds., Springer, Berlin, 1985, 526.

207. **Moss, J. and Vaughan, M.,** Isolation of an avian erythrocyte protein possessing ADP-ribosyltransferase activity and capable of activating adenylate cyclase, *Proc. Natl. Acad. Sci. U.S.A.,* 75, 3621, 1978.

208. **Soman, G. and Graves, D. J.,** Endogenous ADP-ribosylation in skeletal muscle membranes, *Arch. Biochem. Biophys.,* 260, 56, 1988.

209. **Burtscher, H. J., Auer, B., Klocker, H., Schweiger, M., and Hirsch-Kauffmann, M.,** Isolation of ADP-ribosyltransferase by affinity chromatography, *Anal. Biochem.,* 152, 285, 1986.

210. **Moss, J., West, R. E., Osborne, J. C., and Levine, R. L.,** Characterization of NAD:arginine mono-ADP-ribosyltransferase in turkey erythrocytes: determination of substrate specificity, in *ADP-Ribosylation of Proteins,* Althaus, F. R., Hilz, H., and Shall, S., Eds., Springer, Berlin, 1985, 512.

211. **Godeau, F., Belin, D., and Koide, S. S.,** Mono(adenosine diphosphate ribosyl)transferase in Xenopus tissues. Direct demonstration by a zymographic localization in sodium dodecyl sulfate-polyacrylamide gels, *Anal. Biochem.,* 137, 287, 1984.

212. **Burtscher, H. J., Klocher, H., Schneider, R., Auer, B., Hirsch-Kaufmann, M., and Schweiger, M.,** ADP-ribosyltransferase from *Helix pomatia, Biochem. J.,* 248, 859, 1987.

213. **Hopkins, R. S., Stamnes, M. A., Simon, M. I., and Hurley, J.,** Cholera toxin and pertussis toxin substrates and endogenous ADP-ribosyltransferase activity in *Drosophila melanogaster, Biochim. Biophys. Acta,* 970, 355, 1988.

214. **Alvarez-Gonzalez, R., Moss, J., Niedergang, C., and Althaus, F. R.,** Selective probing of ADP-ribosylation reactions with oxidized 2′-deoxy-nicotinamide adenine dinucleotide, *Biochemistry,* 27, 5378, 1988.

215. **Tsuchiya, M., Tanigawa, Y., Ushiroyama, T., Matsuura, R., and Shimoyama, M.,** ADP-ribosylation of phosphorylase kinase and block of phosphate incorporation into the enzyme, *Eur. J. Biochem.,* 147, 33, 1985.

216. **Watkins, P. A., Kanaho, Y., and Moss, J.,** Inhibition of the GTPase activity of transducin by an NAD^+:arginine ADP-ribosyltransferase from turkey erythrocytes, *Biochem. J.,* 248, 749, 1987.

217. **Tanuma, S., Kohtaro, K., and Endo, H.,** Eukaryotic mono(ADP-ribosyl)transferase that ADP-ribosylates GTP-binding regulatory Gi protein, *J. Biol. Chem.,* 263, 5485, 1988.

218. **Hara-Yokoyama, M. and Furuyama, S.,** An endogenous inhibitor of the ADP-ribosylation of GTP-binding proteins by pertussis toxin is present in bovine brain, *Biochem. Biophys. Res. Commun.,* 160, 67, 1989.

219. **Jacquemin, C., Thibout, H., Lambert, B., and Correze, C.,** Endogenous ADP-ribosylation of G$_s$ subunit and autonomous regulation of adenylate cyclase, *Nature (London),* 323, 182, 1986.

220. **Lee, H. and Iglewski, W. J.,** Cellular ADP-ribosyltransferase with the same mechanism of action as diphtheria toxin and *Pseudomonas aeruginosa* toxin A, *Proc. Natl. Acad. Sci. U.S.A.,* 81, 2703, 1984.

221. **Iglewski, W. J., Lee, H., and Muller, P.,** ADP-ribosyltransferase from beef liver which ADP-ribosylates elongation factor-2, *FEBS Lett.,* 173, 113, 1984.

222. **Fendrick, J. L. and Iglewski, W. J.,** Endogenous ADP-ribosylation of elongation factor 2 in polyoma virus-transformed baby hamster kidney cells, *Proc. Natl. Acad. Sci. U.S.A.,* 86, 554, 1989.

223. **Adamietz, P., Wielckens, K., Bredhorst, R., Lengyel, H., and Hilz, H.,** Subcellular distribution of mono(ADP-ribose) protein conjugates in rat liver, *Biochem. Biophys. Res. Commun.,* 101, 96, 1981.

224. **Masmoudi, A. and Mandel, P.,** ADP-ribosyl transferase and NAD glycohydrolase activities in rat liver mitochondria, *Biochemistry,* 26, 1965, 1987.

225. **Hilz, H., Koch, R., Fanick, W., Klapproth, K., and Adamietz, P.,** Nonenzymic ADP-ribosylation of specific mitochondrial polypeptides, *Proc. Natl. Acad. Sci. U.S.A.,* 81, 3929, 1984.

226. **Frei, B. and Richter, C.,** Mono(ADP-ribosylation) in rat liver mitochondria, *Biochemistry,* 27, 529, 1988.

227. **Hara, N., Tsuchiya, M., Mishima, K., Tanigawa, Y., and Shimoyama, M.,** ADP-ribosylation of Ca^{2+}-dependent ATPase *in vitro* suppresses the enzyme activity, *Biochem. Biophys. Res. Commun.,* 148, 989, 1987.

228. **Thomas, W. E. and Mowbray, J.,** Evidence for ADP-ribosylation in the mechanism of rapid thyroid hormone control of mitochondria, *FEBS Lett.,* 223, 279, 1987.

229. **Saari, L. L., Pope, R. M., Murrell, S. A., and Ludden, P. W.,** Studies on the activating enzyme for iron protein of nitrogenase from *Rhodosprillum rubrum, J. Biol. Chem.,* 4973, 4977, 1986.

230. **Lowery, R. G. and Ludden, P. W.,** Purification and properties of dinitrogenase reductase ADP-ribosyl-transferase from the photosynthetic bacterium *Rhodospirillum rubrum, J. Biol. Chem.,* 263, 16714, 1988.

231. **Murrell, S. A., Lowery, R. G., and Ludden, P. W.,** ADP-ribosylation of dinitrogenase reductase from *Clostridium pasteurianum* prevents its inhibtion of nitrogenase from *Azotobacter vinelandii, Biochem. J.,* 251, 609, 1988.

232. **Moss, J., Tsai, S. C., Adamik, E., Chen, H. C., and Stanley, S. J.,** Purification and characterization of ADP-ribosylarginine hydrolysase from turkey erythrocytes, *Biochem,* 27, 5819, 1988.

233. **Goyard, S., Orlando, C., Sabatier, S. M., Labruyere, E., d'Alayer, J., Fontan, G., van Rietschoten, J., Mock, M., Danchin, A., Ullmann, A., and Monneron, A.,** Identification of a common domain in calmodulin-activated eukaryotic and bacterial adenylate cyclases, *Biochemistry,* 28, 1964, 1989.

234. **Gill, D. M.,** Sequence homologies among the enzymically active portions of ADP-ribosylating toxins, in *Bacterial Protein Toxins,* Fehrenbach, F. J., Alouf, J. E., Falmagne, P., Goebel, W., Jeljaszewicz, J., Jurgens, D., and Rappuoli, R., Eds., Fischer, Stuttgart, 1988, 315.

235. **Antoine, R. and Locht, C.,** Roles of the disulfide bond and the carboxy-terminal region of the S1 subunit in the assembly and biosynthesis of pertussis toxin, *Infect. Immun.,* 58, 1518, 1990.

236. **Cortina, G. and Barbieri, J. T.,** Localization of a region of the S1 subunit of pertussis toxin required for efficient ADP-ribosyltransferase activity, *J. Biol. Chem.,* 266, 3022, 1991.

237. **Krueger, K. M., Mendemueller, L. M., and Barbieri, J. T.,** Protease treatment of persussis toxin identifies the preferential cleavage of the S1 subunit, *J. Biol. Chem.,* 266, 8122, 1991.

238. **Lobban, M. D., Irons, L. I., and van Heyningen, S.,** Binding of NAD^+ to pertussis toxin, *Biochim. Biophys. Acta,* 1078, 155, 1991.

239. **Domenighini, M., Montecucco, C., Ripka, W. C., and Rappuoli, R.,** Computer modelling of the NAD binding site of ADP-ribosylating toxins—active site structure and mechanism of NAD binding, *Mol. Microbiol.,* 5, 23, 1991.

240. **Stearns, T., Willingham, M. C., Botstein, D., and Kahn, R. A.,** ADP-Ribosylation factor is functionally and physically associated with the Golgi complex, *Proc. Natl. Acad. Sci. U.S.A.,* 87, 1238, 1990.

241. **Landis, C. A., Master, S. B., Spada, A., Pace, A. M., Bourne, H. R., and Vallar, L.,** GTPase inhibiting mutations activate the alpha chain of G_s and stimulate adenylyl cyclase in human pituitary tumours, *Nature (London),* 340, 692, 1989.

242. **Sixma, T. K., Pronk, S. E., Kalk, K. H., Wartna, E. S., van Zanten, B. A. M., Witholt, B., and Hol, W. G. J.,** Crystal structure of a cholera toxin-related heat-labile enterotoxin from *E. coli, Nature (London),* 351, 371, 1991.

243. **Gibbons, A.,** New 3-d protein structure revealed. The shape of cholera, *Science,* 253, 382, 1991.

Chapter 5

GLYCAN STRUCTURES IN GLYCOPROTEINS AND THEIR ANALYSIS BY FAST ATOM BOMBARDMENT MASS SPECTROMETRY

Anne Dell, Roy A. McDowell, and Mark E. Rogers

TABLE OF CONTENTS

I. INTRODUCTION

The main objective of this chapter is to provide biopolymer scientists with a guide to the glycan structural problems that can be solved in the glycoprotein area using the powerful technique of Fast Atom Bombardment Mass Spectrometry (FAB-MS). Many of the strategies employed in glycoprotein FAB-MS exploit the characteristic structural features of glycoprotein glycans that are dictated by their biosynthesis. The first part of this chapter, therefore, contains an overview of current understanding of the structure and biosynthesis of oligosaccharide moieties in glycoproteins. This survey is not intended to be a comprehensive review of biosynthesis (several excellent and in-depth reviews of aspects of biosynthesis are already available, see, e.g., References 1 to 5) but serves as a source of background information to facilitate understanding of the FAB-MS discussion which follows.

The term glycoprotein will be used in this chapter to describe enzymically glycosylated proteins containing carbohydrate chains attached via either an *N*-glycosidic linkage to asparagine or via an *O*-glycosidic linkage to a hydroxylated amino acid, most commonly serine or threonine. Also included in our discussion are the proteoglycans, e.g., chondroitin sulphate and heparin, which constitute a special sub-class of glycoproteins in which long linear polysaccharides containing tandem repeats of sulphated disaccharides are attached to a protein core. In addition we will cover glycan anchors which link the C terminal carboxyl group of certain proteins and glygoproteins to a lipid moiety embedded in the outer leaflet of the plasma membrane.

Glycoproteins have the following characteristic features:[6]

1.   A limited set of core structures occur at the protein attachment sites. These (with a few exceptions) are common to all glycoproteins irrespective of the species of origin.
2.   A great variety of structural motifs decorate these core moieties.
3.   Each protein attachment site normally carries a family of closely related glycans, i.e., glycoproteins are heterogeneous in their glycan content.
4.   The pattern of oligosaccharide heterogeneity at each glycosylation site appears to be nonrandom but may vary from one cell type to another.

With one important exception (see Section II.C below), the carbohydrate chains of eukaryotic glycoproteins are biosynthesised in the lumenal compartments of the endoplasmic reticulum and the Golgi apparatus. The mature glycoproteins are either secreted from the cell or remain membrane bound with the carbohydrate chains occurring on the part of the protein which is exterior to the cell (or on the lumenal side in the case of intracellular compartments). Prokaryotic cells, which lack the specialised intracellular compartments which characterise eukaryotic cells, do not biosynthesise glycoproteins within the cell, although some bacteria (notably thermophilic organisms) produce a limited number of glycoproteins whose carbohydrate chains are believed to be synthesised on the outside of the cell wall.

II. OVERVIEW OF STRUCTURE AND BIOSYNTHESIS OF GLYCOPROTEIN GLYCANS

A. *N*-GLYCANS
1. Overview[1,5]

All asparagine-linked oligosaccharides share the same pentasaccharide core (Figure 1) which is derived from a common biosynthetic precursor, the structure of which is shown in Figure 2.* This lipid-linked precursor is synthesised in the endoplasmic reticulum (ER) by

---

*   All figures appear at end of this chapter.

the stepwise addition of two *N*-acetylglucosamine (GlcNAc), nine mannose (Man) and three glucose (Glc) residues onto dolichol phosphate. Biosynthesis of the precursor is initiated in the lumen of the ER, but the overall topography of biosynthesis is complicated by the fact that the nucleotide donor of mannose (GDP-Man), which is used in the first five mannose additions, is unable to enter the lumen of the ER. This problem is overcome by "flipping" of the $GlcNAc_2$-P-P-dolichol precursor to the cytoplasmic surface where the first five mannose residues are added, followed by "flopping" back to the lumenal side where the remaining four mannosyl residues are derived from dolichol-linked donors. Addition of three glucose residues completes the precursor biosynthesis.

Proteins destined for glycosylation begin their synthesis on ribosomes attached to the cytoplasmic face of the ER and are extruded through the lipid bilayer into the lumenal compartment where glycosylation occurs co-translationally via transfer en bloc of the $Glc_3Man_9GlcNAc_2$ moiety to specific asparagine residues. A necessary prerequisite for glycosylation is the occurrence of the acceptor asparagine in the tri-peptide sequence -Asn-X-Thr or Ser- where X may be any amino acid with the exception of proline. Not all Asn-X-Thr/Ser sites in eukaryotic proteins are glycosylated however, and additional factors that control glycosylation remain to be determined. Hence it is not yet possible to predict which consensus sequences in a newly determined protein sequence are likely to be glycosylated.

The oligosaccharide precursor is processed in the ER and Golgi via pathways which, at least in part, appear to be common to all plant and animal cell types. Processing involves stepwise trimming by specific exoglycosidases and stepwise addition of new sugar residues, the latter reactions being catalyzed by specific glycosyl transferases. The major pathways are summarised below. Of necessity this discussion is oversimplified and the reader seeking more detailed accounts of *N*-glycan biosynthesis should consult specialist reviews[1-7] and recent literature pertaining to research on the relevant exoglycosidases and glycosyl transferases where there has been an explosion of fruitful activity facilitated by the technologies of modern molecular biology and cell biology (see, for example References 8 to 12).

2. Trimming in the ER

Two alpha glucosidases (I and II) remove the three glucose residues from the $Glc_3Man_9GlcNAc_2$ precursor and an alpha mannosidase trims a single mannose residue from the nonreducing end of the inner branch to give the $Man_8GlcNAc_2$ structure shown in Figure 3. The trimmed glycoprotein then migrates to the Golgi. This transfer occurs via vesicles which bud from the ER and fuse with the Golgi membrane.

3. Golgi Processing Events

In the cis Golgi the $Man_8GlcNAc_2$ precursor is directed into one of four major biosynthetic pathways the nature of which dictates the type of glycan which is ultimately formed:

Pathway (a): formation of mannose-6-phosphate — Glycan moieties of enzymes destined for the lysosome are phosphorylated as shown in Figure 4. Sequential actions of *N*-acetylglucosaminylphospho-transferase and *N*-acetylglucosamine-1-phosphodiester alpha-*N*-acetylglucosaminidase yield $Man_8GlcNAc_2$ structures containing one or more mannose-6-phosphate residues at the penultimate nonreducing positions of the two outer branches. These act as recognition markers for translocation to the lysosome.[13]

Pathway (b): high mannose structures — Trimming by alpha mannosidases without any subsequent glycosyl additions results in glycans having the composition $Man_xGlcNAc_2$ where x is most commonly 5 to 8 although truncation to the tri-mannosyl core is possible. These structures are called "high mannose". Truncation does not appear to be directed along a unique path; hence mixtures of isomers are formed depending on the order of trimming. Examples are given in Figure 5.

Pathway (c): yeast glycoproteins — Yeast glycoproteins constitute a unique class of molecules whose glycan structures differ significantly from those found in other eukaryotic glycoproteins.[7] The Asn-linked oligosaccharides are exclusively of the high mannose type but are very much larger than those derived from Pathway (b). The number of mannosyl residues can be as high as 150 or more and some side-chain terminal mannosyl residues may be attached via phosphodiester instead of normal glycosidic linkages (Figure 6). The long mannan chains are biosynthesised from the common $Man_8GlcNAc_2$ precursor by step-wise addition of mannosyl residues.

Pathway (d): complex and hybrid structures — The majority of eukaryotic glyco-proteins (except for yeast) are directed into this pathway which begins as a single path and then divides into two branches, one leading to a multitude of complex-type structures, the other leading to a more limited array of hybrid structures.

Alpha mannosidase I trims three mannosyl residues from the $Man_8GlcNAc_2$ precursor to give the $Man_5GlcNAc_2$ glycan (1) shown in Figure 7. This is the substrate for GlcNAc transferase I which transfers a GlcNAc residue from UDP-GlcNAc to the 3-linked mannose of the tri-mannosyl core to give the $Man_5GlcNAc_3$ structure (2) shown in Figure 7. This is an important intermediate which is a substrate for two key enzymes viz GlcNAc transferase III and alpha mannosidase II. Action of alpha mannosidase II directs the glycoprotein down the route leading to complex structures; action of GlcNAc transferase III blocks subsequent processing by alpha mannosidase II and results in hybrid molecules being formed.

Pathway (d)-(i): formation of hybrid structures — The route to hybrid structures is shown in Figure 8. GlcNAc transferase III transfers a GlcNAc residue from UDP-GlcNAc to the 4-position of the central mannose in the tri-mannosyl core of the $Man_5GlcNAc_3$ precursor. Presence of this "bisecting" GlcNAc prevents the removal of the two mannose residues which remain attached to the 6-arm of the tri-mannosyl core. Subsequent biosyn-thesis involves elaboration of the 3-linked branch of the trimannosyl core to give typical complex-type moieties which are produced by the glycosyl transferase reactions described below for complex structures. A typical mature hybrid structure (3) is shown in Figure 8.

Pathway (d)-(ii): complex-type core structures[4] — The route to complex-type struc-tures begins with the alpha mannosidase II-catalyzed removal of the two nonreducing man-nosyl residues from the key $Man_5GlcNAc_3$ intermediate, and is followed by GlcNAc trans-ferase II-catalyzed attachment of GlcNAc to the upper mannose of the trimannosyl core to give the biantennary structure 4 (Figure 9). This is one of the major core structures present in complex glycans. The other major core structures are shown in Figure 9. There are two triantennary structures (5 and 6) produced by further action of GlcNAc transferases IV and V respectively and the tetraantennary core (7) is the result of action of both these transferases. Action of GlcNAc transferase III on the above cores yields bisected structures, e.g., 9 (Figure 10) (cf hybrid structures above). Once the bisecting GlcNAc is added further branch-ing is inhibited. A pentaantennary core (8) has been found in glycoproteins of the hen oviduct and this is the most highly branched *N*-glycan described to date. GlcNAc transferase VI, which catalyzes the addition of the bisecting GlcNAc to form the pentaantennary core, has recently been characterized[14] and it appears to be present in a number of tissues in the hen indicating that the unusual pentaantennary structure may not be restricted to oviduct gly-coproteins. Similar structures have not, however, been described for mammalian systems.

Fucosylation of the terminal GlcNAc residue of the trimannosyl chitobiose core (i.e., the GlcNAc attached to Asn) is observed in many glycoproteins to give, for example, core structure 10 (Figure 10). In mammalian glycoproteins the fucose is attached at position 6 of the GlcNAc residue; in plant glycoproteins fucosylation occurs at position 3. In plant glycoproteins,[15] core regions are further modified by the attachment of xylose to the 2-position of the beta mannose residue (structure 11, Figure 10). This modification appears to be unique to plants.

Pathway (d)-(iii): processing of complex cores in the trans Golgi — The GlcNAc and fucosyl transferases which produce the core oligosaccharides described in (d)-(ii) are located in the medial Golgi. Further processing in the trans Golgi converts the small pool of core structures into an extensive array of mature oligosaccharides which differ in branch length and in the disposition of common structural motifs.[1] Extension of the branches is normally initiated by a galactosyl transferase that transfers galactose from UDP-Gal to the 4-position of a nonreducing GlcNAc residue to produce the *N*-acetyllactosamine moiety, Galβ1-4GlcNAc, which is the building block for the branches of most complex structures. Many complex-type glycans contain only a single *N*-acetyllactosamine moiety in each branch but there are some in which tandem repeats of Galβ 1-4GlcNAcβ1-3 are present. Branches may be elongated by only one or two of these repeats but some membrane bound glycoproteins carry polysaccharide chains composed of many tandem repeats of this structural motif. These polysaccharides are called polylactosaminoglycans.[16] A GlcNAc transferase involved in catalysing the GlcNAcβ1-3 Gal reaction in polylactosaminoglycan biosynthesis has recently been purified[17] and *in vitro* studies implicate it in both the initiation and elongation of the polylactosaminoglycan chains. The enzyme appears to prefer substrates which carry a 6-linked antenna on the alpha mannose attached to the 6-position of the beta mannose of the trimannosyl core (see Figure 11). Polylactosaminoglycans from non-foetal cells usually contain branched antenna of the structure shown in Figure 12.[18] Branching is initiated by a GlcNAc transferase that specifically adds GlcNAc to the 6-positions of in-chain galactose residues. The branches are then elongated by galactosyl transferase in an analogous manner to antenna biosynthesis. Normally the branches are short, containing only one repeat of *N*-acetyllactosamine.

Biosynthesis of all complex structures is completed by a variety of "capping" reactions, the most important being sialylation and fucosylation.[1,4] Sites of sialylation and fucosylation include the backbone as well as the terminal residue of each branch and consequently an enormous number of potential structures is possible. However, substrate specificities for the various sialyl and fucosyl transferases limit the number of structures actually formed. The main sites for sialylation are the 3- and 6-positions of terminal galactose and at the 6-position of GlcNAc residues adjacent to NeuAcα2-3Gal moieties. Fucose is commonly attached to the 3- or 6-positions of GlcNAc and the 2-position of Gal. In Figure 13 examples of complex glycans are given which illustrate some of the potential structural diversity afforded by variations in the numbers and lengths of antennae and in sites and extent of sialylation and/or fucosylation.

Other capping groups for terminal galactose in complex glycans include alpha-linked Gal, alpha-linked GalNAc and sulphate. The best characterised example of the last is thyroglobulin[19] where terminal Gal-3-SO$_4$ has been observed on biantennary as well as on some more highly branched carbohydrate structures. Modification with sulphate is, in fact, a common feature of a wide range of oligosaccharides in glycoproteins. Sulphated residues that have been observed in *N*-glycans include GlcNAc[20] (in complex structures), Man[21] (in high mannose and hybrid structures) and, importantly, GalNAc.[22] Sulphated GalNAc is found at the nonreducing termini of complex glycans present in the pituitary glycoprotein hormones. In the sulphated branches of these glycans the *N*-acetyllactosamine moiety is absent and is replaced by GalNAcβ1-4GlcNAc-. Terminal GalNAc residues are specifically recognised by a pituitary sulphotransferase which catalyses sulphation at the 4-position. The structures shown in Figure 14 are typical examples of GalNAc-4-SO$_4$-containing *N*-glycans.

B. *O*-GLYCANS

The biosynthesis of *O*-linked oligosaccharides has not been as well defined as that of *N*-linked structures. It is, nevertheless, well established that, in contrast to *N*-glycans, the initial glycosylation event involves transfer of a single sugar residue to the acceptor hydroxylated amino acid. In mammals the sugar residue attached to Ser and Thr is GalNAc (except for cytoplasmic glycoproteins, see Section II.C) and the transfer from UDP-GalNAc is catalyzed by a specific GalNAc transferase.[23] The specificity of this enzyme for protein acceptor sites is not known and there does not appear to be a consensus sequence for *O*-glycosylation analogous to that for *N*-glycosylation. The subcellular location of GalNAc addition is still the subject of some dispute but most of the available evidence points to the cis Golgi or a compartment between the ER and the cis Golgi.

Following GalNAc addition to the protein, *O*-glycans are further biosynthesized by stepwise addition of glycosyl residues. There appear to be six core structures[24,25] common to mammalian *O*-glycans and these are shown in Figure 15. These core structures are then substituted with moieties similar to those described earlier for complex glycans, e.g., *N*-acetyllactosamine, fucose, sialic acid, sulphate, etc. Figure 16 contains some examples of *O*-glycans which illustrate the structural diversity that arises from variable substitution of the core oligosaccharides. Many mucin-type glycoproteins, e.g., the constituents of human respiratory mucus, which are molecules containing multiple glycosylation sites, are very highly heterogeneous in their *O*-glycan content.[24] It is likely that several hundred different carbohydrate chains are present in some mucins. In contrast, there are some glycoproteins which have only a limited number of glycosylation sites and they are usually far less heterogeneous. For example, erythropoietin, the kidney derived hormone that stimulates the proliferation of red blood cells, has one attachment site for *O*-glycans (in addition to 3 sites for *N*-glycosylation) at which one major oligosaccharide (structure 1, Figure 16) and four minor oligosaccharides reside.[26]

The *O*-glycans of nonmammalian glycoproteins are very different from their mammalian counterparts. In plants, for example, oligosaccharides are linked through a galactose residue to Ser and through either arabinose or galactose to hydroxyproline (Figure 17).[27] In yeast the linkage sugar is mannose[7] and this sugar has also been observed at the reducing end of *O*-glycans obtained from glycoproteins of coral[28] (Figure 17). Interestingly there appear[29] to be *O*-glycosidic mannose-linked oligosaccharides in the chondroitin sulphate proteoglycan of brain which have structural features similar to "normal" mammalian *O*-glycans, i.e., they are quite small and they contain moieties such as Gal-GlcNAc, sialic acid, fucose and sulphate in addition to the mannose residue (see, for example, structure 4, Figure 17).

C. CYTOPLASMIC GLYCOSYLATION

Several years ago a novel form of protein glycosylation was revealed[30,31] when galactosyl transferase and radioactively labeled UDP-Gal were being used to probe for terminal GlcNAc residues in cell surface glycoproteins. The observation that the most heavily labeled proteins were located in the cytoplasm led to the discovery of *O*-linked GlcNAc in cytoplasmic proteins. *O*-GlcNAc is predominantly localised in the nuclear and cytoplasmic compartments of cells where it occurs on important nuclear pore proteins, chromatin proteins, transcription factors and cytoskeletal proteins. The functions of *O*-linked GlcNAc are not known but knowledge of the types of proteins to which it is attached has led to speculation[32] that it may be involved in regulation of gene transcription and nucleo-cytoplasmic transport, possibly via modulation of phosphorylation. Peptide sequencing studies have revealed common structural features in regions of *O*-GlcNAc attachment including the presence of proline and long stretches of Ser and/or Thr repeats. Structural studies to date have not revealed any functional groups attached to the *O*-linked GlcNAc, and elongation by glycosyl addition has not been observed except where exogenous enzymes and glycosyl donors have been employed.

D. GLYCOSYL-PHOSPHATIDYLINOSITOL PROTEIN ANCHORS

The glycosyl phosphatidylinositols (GPIs) of eukaryotic cells are functionally important components in the outer leaflet of the plasma membrane.[33] They act as membrane anchors for a diverse family of cell surface proteins (and also lipophosphoglycans in the case of single cell eukaryotes). The potential importance and unique structural features of these molecules warrant their inclusion in this chapter especially since FAB-MS is playing an increasingly important role in their characterization.

Although diverse in origin, all anchors have common structural motifs. At the heart of protein anchors whose structures have been defined is the sequence Manα1-2Manα1-6Manα1-4GlcNH$_2$-Inositol- which is attached via a phosphodiester linkage to a diacyl or an acyl/alkyl glycerol. The terminal mannose is connected at the 6-position to a phosphorylethanolamine moiety which is attached to the C terminal carboxyl group of the protein. The structure of the lipid anchor from the variant surface glycoprotein (VSG) of trypanosomes, the first anchor to be fully characterised,[34] is shown in Figure 18. Heterogeneity in the galactosyl arm which substitutes the mannosyl backbone is an important feature of the VSG anchor. Some anchors, including those from rat brain Thy-I[35] and human acetylcholinesterase,[36] have an additional phosphorylethanolamine moiety attached through a phosphodiester linkage to the glycan core. An additional novel feature of the latter anchor is the presence of an acyl group (palmitoyl) on one of the inositol hydroxyls.

Although the overall biosynthesis of GPI anchors has not been elucidated, several key steps have been defined.[33] Attachment of the GPI structure appears to occur cotranslationally and is concomitant with the removal of a C terminal hydrophobic peptide which may serve to retain the newly synthesised protein at the lumenal surface of the ER to facilitate linkage of the anchor. Synthesis of the anchor itself has been studied in a trypanosome cell free system and the first step has been shown[37] to be the transfer of GlcNAc to phosphatidylinositol and this is followed by rapid deacetylation to from GlcN-PI upon which the final structure is assembled.

E. PROTEOGLYCANS

Proteoglycans are glycoproteins that contain a protein core to which are covalently attached large numbers of glycosaminoglycans (GAGs) together with some *N*- and/or *O*-linked oligosaccharides.[38] GAGs are long linear polysaccharides constituted of tandem repeats of sulphated disaccharides one residue of which is an amino sugar. The second residue is either a uronic acid or galactose. GAGs are categorised according to the type of disaccharide repeat. The major classes are heparin, heparan sulphate, chondroitin sulphate, dermatan sulphate and keratan sulphate. With the exception of keratan sulphate, which is linked to Asn through the normal *N*-glycan core, the GAG chains of proteoglycans are linked to Ser or Thr through a xylose residue which is at the reducing end of a common core tetrasaccharide of sequence GlcAβ 1-3Galβ 1-3Galβ 1-4Xylβ-. 4-*O*-Sulphation of the galactose residue adjacent to the glucuronic acid has recently been described in some chondroitin sulphate linkage regions.[39] Superficially, GAGs may appear to be very uniform structures; there is no branching and only two types of residue are present in each GAG (except in the core). However, they are, in fact, very complex molecules due to heterogeneity in sulphation patterns and potent biological activities have been attributed to subdomains of proteoglycans such as heparin. Characterisation of these highly charged molecules poses a considerable challenge and FAB-MS is potentially a powerful means of providing important structural information.

III. FAST ATOM BOMBARDMENT MASS SPECTROMETRY

A. HISTORICAL PERSPECTIVE

Fast Atom Bombardment Mass Spectrometry (FAB-MS)[40,41] which was introduced at the beginning of the 1980s, is a mass spectrometric technique in which samples are ionised and desorbed from a liquid phase using a beam of accelerated particles. It is related to

Secondary Ion Mass Spectrometry (SIMS) which had been developed some years earlier to probe metal surfaces. In the SIMS experiment an accelerated beam of ions (the primary beam) was fired at a metal surface and metal ions which sputtered from the surface (the secondary beam) were recorded after mass separation in the analyser of the mass spectrometer. Prior to the introduction of FAB-MS, the SIMS technique had been applied to a limited number of organic samples which had been coated on the metal surfaces before primary ion bombardment. The data from these experiments was, however, of such poor quality that the method had not shown promise for biological applications. FAB-MS, although similar in many respects to SIMS, differed in two key ways: firstly an atom beam was used instead of an ion beam and, secondly, the sample was presented to the atom beam as a solution in a viscous liquid (the "matrix") rather than as a thin solid film on the metal surface. This combination, both components of which were believed to be essential, gave spectacular data from polar organic compounds including peptides, oligosaccharides and nucleotides. Subsequent research has revealed that an atom beam is not a necessary criterion for the successful analysis of organic substances and bombardment with ions gives analogous data. The key to success in the FAB experiment is, in fact, the matrix, which maintains the liquid nature of the sample in the high vacuum environment of the ion source and provides a surface which is constantly replenished with sample molecules for the duration of the experiment. Once it was firmly established that ion beams could be used in the FAB experiment, many researchers replaced their atom guns with ion guns (in particular the cesium ion gun[42]), which are more versatile, simpler to maintain, and importantly, can readily produce particle beams of much higher energy resulting in greater sensitivity. Consequently the acronym "FAB-MS" is somewhat of a misnomer for much of the current research in this area and the alternative designation "liquid SIMS" is used by some workers. However, we prefer the term FAB-MS, which is widely accepted in the biochemical literature, and in this chapter FAB-MS refers to all experiments involving the ionisation of samples from a liquid phase irrespective of the type of gun used or of the terminology chosen by the authors whose work is cited.

B. PRINCIPLES OF FAB-MS

In the FAB experiment, an accelerated beam of atoms (usually xenon) or ions (usually cesium) is fired from an atom or ion gun towards a small metal target which has previously been loaded with a viscous liquid such as glycerol or monothioglycerol in which is dissolved the sample to be analysed. When the atom or ion beam collides with the matrix, kinetic energy is transfered to the surface molecules, many of which are sputtered out of the liquid into the high vacuum of the ion source. A significant number of these molecules is ionised during the sputtering process. Thus gas-phase ions are generated without prior volatilization of the sample allowing the analysis of polar, involatile and thermally labile compounds. Both positive and negative ions are produced during the sputtering process, and either can be recorded by an appropriate choice of instrumental parameters. Molecules are ionised by the addition of a proton or cation such as sodium, potassium, ammonium etc. (positive ion formation) or by loss of a proton or addition of an anion such as chloride, bromide, thiocyanate etc. (negative ion formation). During ionization, some internal energy is imparted to the molecule resulting in fragmentation of labile bonds. The sputtered molecular and fragment ions are accelerated into the analyser region of the mass spectrometer where their mass/charge ratios are recorded to give the mass spectrum. Modern sector mass spectrometers are capable of focusing ions up to 15,000 Da at full sensitivity but the working mass range at the present time, for carbohydrate problems, is limited to about 6000 Da. Higher molecular weight samples are either too difficult to desorb from the matrix or do not produce a sufficient number of molecular ions to allow detection above background. In some cases, however, very large glycans or glycopeptides whose molecular ions cannot be detected, nevertheless produce abundant fragment ions which provide very useful structural information (see Section III.E.).

As a general rule glycans whose hydroxyl groups are protected by functional groups, such as methyl or acetyl, fragment more reliably than their native counterparts. Consequently derivatisation underpins most successful FAB-MS strategies in the glycoprotein area.[43] If there is insufficient fragmentation to allow sequence assignment, or if mixtures are being examined and fragment ions cannot readily be attributed to individual molecular ions, a mass spectrometric technique called Collisional Activation (CA) may afford useful data. Briefly CA involves the introduction of a collision gas into the flight path of the ion beam as it traverses the mass spectrometer. When the ions collide with the gas some of them will receive sufficient internal energy to fragment and the resulting so called daughter ions are then recorded at the detector. The daughter ions can be detected using a conventional double focusing two sector instrument (i.e., an instrument having one magnetic sector for focusing ions according to their masses and one electric sector for energy focusing) by introducing the collision gas immediately after the source and by performing a ''linked scan'' whereby the two sectors are scanned together in such a manner that their ratio remains constant. This is called a B/E scan.[44] Alternatively, two mass spectrometers can be joined in tandem to give a four sector instrument in which the collision gas is introduced after the first two sectors and the second mass spectrometer is used to focus and detect the daughter ions. This is often referred to as MS/MS. It is a much more expensive option than linked scanning on a single mass spectrometer. The main advantage of MS/MS over linked scanning is its greater parent ion resolution allowing molecular ions separated by only a few mass units to be separately examined. In the linked scanning experiment a window of ions is transmitted by the instrument and the daughters of all ions within the window will be observed.

C. FRAGMENTATION OF NATIVE AND DERIVATISED OLIGOSACCHARIDES AND GLYCOPEPTIDES[43]

The following general rules apply to the fragmentation behaviour of oligosaccharides and glycopeptides:

1. The most abundant fragment ions are formed by cleavage at the glycosidic linkages. Glycosidic cleavage is accompanied by a hydrogen transfer to the glycosidic oxygen (or sometimes a methyl or acetyl transfer in the case of derivatives) if the charge on the fragment ion is not specifically located at the point of cleavage.

2. Ring cleavage, when it occurs, is best rationalised as arising from the sequential movement of electron pairs around the ring resulting in the breakage of single bonds and the formation of double bonds (see Figure 19).

3. Fragment ions can sometimes be formed by two or more cleavage events occurring in different parts of the molecule. This phenomenon is more frequently observed in native samples than in derivatives. Unambiguous sequencing of native samples is often not possible if ''double cleavage'' ions are formed in abundance.

4. The fragment ions produced by glycopeptides are derived predominantly from cleavage of the glycosidic linkages. Fragmentation of the peptide moiety is usually minimal.

5. Permethyl and peracetyl derivatives fragment in a very predictable manner in contrast to native samples whose fragmentation behaviour is often unreliable.

6. If a permanent charge is present in the molecule (e.g., a sulphate moiety) then fragment ions produced by cleavage of labile bonds in the vicinity of the charge dominate the spectrum.

The major fragmentation pathways relevant to glycoprotein FAB-MS are summarised below (and illustrated in Figure 19):

Pathway A — Glycosidic cleavage yields an oxonium ion thereby locating the charge on the nonreducing fragment. The term A-type cleavage is normally applied to this type of fragmentation based on nomenclature used in electron impact mass spectrometry. A-type cleavage is the major mode of fragmentation of permethylated and peracetylated oligosaccharides. If HexNAc residues are present in the sequence, cleavage occurs predominantly (and sometimes exclusively) at the amino sugar residue. We will show later how this phenomenon is exploited in glycoprotein FAB-MS.

Pathway B — This involves cleavage of the glycosidic bond with a hydrogen transfer and charge retention on the reducing end moiety. Note that the charge is not necessarily located at the site of cleavage and that this fragmentation can occur in both positive and negative modes. It is a fragmentation pathway favoured by native oligosaccharides and glycopeptides and by oligosaccharides modified with a reducing end derivative where the charge can be localised. Unless samples are permethylated or peracetylated, fragment ions produced by pathway B have the same masses as molecular ions of molecules that are truncated at the nonreducing ends. The fragmentation occurring in pathway B is frequently referred to as beta cleavage.

Pathway C — This is analogous to pathway B except that cleavage occurs on the opposite side of the glycosidic linkage and charge is located on the nonreducing fragment. This is not a common pathway in the positive mode (where nonreducing end fragments are most frequently formed via pathway A) but is important in negative FAB-MS, especially when the negative charge is located towards the nonreducing end as it frequently is in sulphated *N*- or *O*-glycans.

Pathway D — This is a ring cleavage pathway which produces ions that are 28 U heavier than those obtained in Pathway B as a result of a formyl group, derived from the anomeric carbon and the ring oxygen, remaining attached to the glycosidic oxygen involved in the beta cleavage.

Pathway E — This is a ring cleavage pathway which can be rationalised using electron shifts analogous to those shown for Pathway D. In this case the charge is retained on the nonreducing end moiety and more than one fragment ion is possible depending on the linkage at the sugar residue undergoing ring cleavage. Figure 19 shows the two possible ring cleavage modes for a 4-linked residue. This mode of fragmentation is commonly observed in negative FAB spectra.

Some of the above cleavages may be accompanied by loss of water (native samples), methanol (permethylated derivatives) or acetic acid (peracetylated derivatives). Particularly informative is the loss of the substituent from the 3-position of A-type oxonium ions from permethylated samples which allows differentiation between type 1 (Galβ1-3GlcNAc) and type 2 (Galβ1-4GlcNAc) backbones in *N*- and *O*-glycans (see Figure 20).

D. FAB-MS: ANALYSIS OF GLYCOPROTEINS

FAB-MS can solve a wide range of structural problems in the glycoprotein area provided its strengths are exploited by employing appropriate experimental strategies.[45-48] In the following sections we show how FAB data acquired during an integrated chemical/enzymatic/FAB-MS experimental protocol can yield information on:

1. The type of glycosylation, e.g., high mannose, hybrid, complex-type, etc.
2. The position(s) in the protein sequence of the glycosylated residue(s)
3. The composition in terms of hexoses, aminohexoses, deoxyhexoses, sialic acids etc. of glycans or of the glycan chains in glycopeptides
4. The presence of lactosaminoglycan chains in *N*- or *O*-linked glycans
5. Branching patterns of *N*- and *O*-linked glycans
6. Complete sequences of glycans up to about 15 sugar residues in size

7. Sequences of all nonreducing structures in complex-type *N*-glycans, irrespective of their overall size and complexity
8. The number of antennae attached to the trimannosyl core of complex glycans
9. The number and type of modifying groups such as acetyl, phosphate, sulphate etc. and the location of modified residues in the carbohydrate sequence
10. The complexity of mixtures of heterogeneous oligosaccharides
11. The fidelity of the carbohydrate structures of recombinant glycoproteins as compared with the naturally occurring substances
12. Abnormalities in the biosynthesis of *N*-glycans resulting from enzyme deficiencies
13. The composition of both the glycan and the lipid moieties in GPI anchors
14. The progress of enzymatic and chemical reactions on glycopeptides or glycoproteins and the identification of by-products.

It is not our intention, in the following sections, to provide a comprehensive list of publications reporting applications of FAB-MS in the glycoprotein area. Instead we review a limited number of studies which highlight the above points and illustrate the FAB strategies which are best suited to glycoprotein analysis.

1. Asn-Linked Oligosaccharides
a. Screening Methods

Rapid, sensitive FAB protocols have been developed which address the following questions: Is this protein glycosylated and if so which consensus sequences are occupied and what types of *N*-glycans are present? The simplest approach is to acetolyse the sample and analyse chloroform soluble products by FAB-MS.[49] The chloroform fraction will contain acetylated oligosaccharides which were originally present in the biopolymer and were partially hydrolysed by the acetolysis reagent. The FAB mass spectrum contains molecular and fragment ions for each acetolysis product and the masses of these ions define the type of carbohydrate, e.g., high mannose yields hexose oligomers, complex structures are revealed by small oligosaccharides containing sialic acid, hexose, and *N*-acetylhexosamine, while lactosaminoglycans are defined by the presence of *N*-acetyllactosamine repeats.

Alternatively the protein or a tryptic digest of the protein is treated with peptide *N*-glycosidase F (an enzyme of broad specificity which hydrolyses the GlcNAc-Asn linkage of *N*-glycans) and the product mixture, without purification, is peracetylated and the derivatised oligosaccharides are separated from protein and salt by partitioning between chloroform and water.[47] The peracetylated oligosaccharides can either be analysed directly or they can be converted to the permethyl derivative prior to analysis. The latter step usually gives improved sensitivity of detection of large glycans. Each high mannose component will exhibit a molecular ion and a major fragment ion, the latter resulting from A-type cleavage at the first GlcNAc of the core. Hybrid structures will give analogous ions and, additionally, fragment ions resulting from cleavage at the GlcNAc residues in the complex-type branch attached to the trimannosyl core. Complex structures will give characteristic HexNAc cleavage ions (see below) and, additionally, molecular ions provided there is sufficient high mass sensitivity.

The identification of glycosylated consensus sequences is achieved by a modification[50] of the FAB mapping strategy which was originally introduced for screening recombinant proteins for mutations, ragged ends, etc.[51] This method relies on the ability of FAB-MS to produce molecular ion data from a complex mixture of peptides thereby providing a FAB "map" of the digested protein which can be matched with the predicted molecular weights of tryptic peptides. Glycopeptides are not usually observed in the FAB map partly because of their size and partly because they compete poorly with peptides for ionisation and desorption. Glycosylation sites are determined by comparing the FAB maps generated prior to and following treatment with peptide *N*-glycosidase F. New signals appearing in the second map will be derived from glycopeptides whose glycan chains were removed by the enzyme. During the enzyme reaction the Asn residue linked to the carbohydrate is converted to Asp and the tryptic peptide is therefore observed at a position 1 mass unit higher than that calculated from the protein sequence.

b. FAB-MS of Permethylated N-Glycans

Although many *N*-glycans can be analysed by FAB-MS without prior derivatisation the data afforded by permethylated derivatives are so superior that most glycoprotein studies incorporate permethylation as a key step (see, for example, References 18, 52-58). FAB spectra of permethylated oligosaccharides contain characteristic fragment ions resulting from A-type cleavage at each amino sugar residue.[59,60] Thus high mannose structures cleave at the chitobiose core to yield fragment ions whose masses define the number of mannosyl residues present, whilst complex structures afford fragment ions from each of the branches, in addition to chitobiose cleavage ions. The branch derived fragment ions define sites of sialylation and fucosylation, presence or absence of bisecting GlcNAc, branching patterns, and presence and extent of lactosamine repeat regions. The chitobiose cleavage ions give information on the number of antennae attached to the trimannosyl core. In the high mass region of the spectrum, provided sufficient sensitivity is maintained, molecular ions for the intact oligosaccharides are observed. These define the extent of fucosylation of the terminal GlcNAc residue.

Figure 21 shows the fragment ions that will be produced by a variety of complex-type structures. Relatively small oligosaccharides, e.g., biantennary complex, will afford some fragment ions derived from cleavages other than A-type cleavage at HexNAc (see Figure 22). Larger molecules fragment almost exclusively at amino sugar residues and this phenomenom is exploited in the characterisation of lactosaminoglycans which, despite their size and complexity, give readily interpretable data (see Figure 23).[18,52-55] Indeed, FAB-MS is the only currently available technique that can detect the presence of *N*-acetyllactosamine repeats in the antennae of minor components of mixtures. For example tetra-antennary structures containing three *N*-acetyllactosamine repeats in one of the branches were shown[26,58] to be present in Erythropoietin, a glycoprotein that stimulates the production of red blood cells, despite being only about 1% of the total *N*-glycan content of this molecule.

The above strategy is applicable to phosphorylated and sulphated *N*-glycans. Phosphodiester groups are converted to their methyl esters during the permethylation and phosphorylated residues are easily recognised by their unique mass. Sulphate groups are stable to the reaction conditions and are not methylated.[61] Sulphated *N*-glycans are therefore negatively charged after permethylation and afford excellent negative spectra which contain intense signals derived from glycosidic cleavages in which charge is retained on the sulphated fragment.[62] In the positive mode, nonsulphated branches give the most abundant ions whilst the sulphated branch gives fragments typical of complex glycans except that the sulphate moiety is lost via a beta cleavage to leave an unsubstituted hydroxyl group on the previously sulphated residue.

The fragmentation pattern afforded by a permethylated sample is very reproducible. Hence, a comparison of data obtained from two related glycoproteins, e.g., natural and recombinant, or normal and abnormal in the case of a genetic deficiency such as that in congenital dyserythropoietic anaemia type II (HEMPAS), will reveal whether the nonreducing sequences are identical, or, if not, the exact nature of the differences. For example, FAB-MS of permethylated *N*-glycans from Band 3 glycoprotein of human erythrocytes, showed that the polylactosaminoglycans present in normal individuals are replaced by truncated hybrid structures in HEMPAS patients indicating a defect in GlcNAc transferase II.[56] Further, the fragmentation pattern is unaffected by variations in the structure at the reducing end of the molecule. Thus it is possible to rapidly screen glycoproteins, or, indeed, cell surfaces, for glycosylation abnormalities by FAB-MS of the unpurified high molecular weight fraction obtained after pronase digestion of glycoproteins or cell membranes. Studies of *N*-glycans present on the cell surfaces of normal and transformed granulocytes provide an excellent illustration of the power of FAB-MS for mapping nonreducing structures in cell surface glycoproteins.[55,63] The *N*-glycans were stripped from the cell surface by pronase digestion of cell residues after organic extraction, and carbohydrate containing fractions were permethylated and analysed by FAB-MS. Normal granulocytes yielded A-type fragment ions indicating the presence of four unique branches each containing three *N*-acetyllactosamine repeats. These branches are unsubstituted, monosialylated, trifucosylated and monosialylated/monofucosylated, respectively. In contrast, chronic myelogenous leukemia glycans are composed of shorter branches and contain a unique sialylated/difucosylated antenna which gave a characteristic fragment ion at m/z 1622.

FAB-MS is an ideal procedure for monitoring the glycosidic reactions which are an integral part of classical carbohydrate sequencing methods. Comparison of spectra acquired before and after digestion with exoglycosidases, endoglycosidases or mild acid will reveal the extent and sites of trimming, thereby providing important structural information.[52-55,64] Similarly FAB-MS can be applied to biochemical investigations into how carbohydrates are biosynthesised and degraded.[47,65]

c. Sequencing High Mannose Structures

The permethyl derivative gives valuable information relating to the size heterogeneity of high mannose structures but usually minimal sequence data because of the selective cleavage at the GlcNAc of the core. Sequencing of high mannose structures therefore requires alternative strategies. One approach is to employ the peracetyl derivative which, in contrast to the permethyl derivative, usually gives fragment ions from Hex cleavages[66] (they may be of low abundance, however, in comparison with the HexNAc cleavage ion). Branching is defined by noting which combinations of Hex_n^+ are present (see Figure 24). The fragmentation data from this type of sample can sometimes be improved by collisional activation (see Section III.A.). Acetolysis experiments monitored by FAB-MS will also yield sequence data.[67] During acetolysis the 6-linkages are hydrolysed preferentially. Thus structure I (Figure 24) will give $Man_4GlcNAc_2$ as a major product after acetolysis while structure II will form $Man_3GlcNAc_2$.

High mannose structures have also been sequenced from the negative ion fragmentation pattern of samples which have been modified at the reducing end by reductive amination.[68] These give fragment ions from the nonreducing end which are formed via pathway E (see Section III.C.) together with reducing-end beta-cleavage ions. Double cleavages are observed but the single cleavage ions are significantly more abundant and therefore unambiguous sequencing is possible.

2. Ser/Thr-Linked Oligosaccharides

a. Screening for O-Glycans

O-Linked chains are selectively released from the protein backbone by reductive elimination and this methodology is exploited in FAB-MS strategies for rapidly screening proteins for putative glycosylation.[47] Information relating to oligosaccharide composition and heterogeneity is provided in the same experiment. Two protocols may be used depending on the subsequent experiments that are planned. The simplest involves peracetylation of the reductive elimination product mixture, without prior purification, and FAB-MS analysis of the chloroform soluble fraction which contains the peracetylated oligosaccharides. This method is rapid and gives a rough indication of the oligosaccharide composition. However, some hydrolysis of sialic acids probably occurs[69] so the procedure should only be used as a fast way of screening for *O*-glycosylation and/or gross differences in oligosaccharide content, and not for searching for minor components or for quantitation. The second protocol includes clean-up steps followed by permethylation of the mixture of oligosaccharides. Sub-microgramme quantities of permethylated *O*-glycans give excellent FAB spectra and a good indication of the sample complexity is afforded by the molecular ions observed. Figure 25 illustrates typical data from a mucin-type glycoprotein.[70]

In an alternative screening approach, *O*-glycans are released from the protein and per-methylated in a single step by treatment with sodium hydroxide and methyl iodide.[71] This procedure is especially attractive because of its speed and simplicity but there may be selective removal of different glycan structures. For example, in glycophorin samples containing a single GalNAc residue (the Tn antigen) attached at some sites only the larger oligosaccharides are released.[69,71]

b. Sequencing of O-Glycans

The most appropriate strategy for analysing *O*-linked glycans is dependent upon the complexity of glycosylation (few or multiple attachment sites) and the extent of knowledge of the protein structure. If the protein sequence has been established and only a limited number of glycosylation sites are present, a suitable strategy involves the preparation of glycopeptides by digestion at sites conveniently close to the glycosylated residues using a specific protease such as trypsin. For glycopeptides of molecular masses less than about 3500 Da, FAB-MS affords abundant molecular ions which define the carbohydrate com-position (in terms of Hex, HexNAc, NeuAc, etc.) and give an indication of heterogeneity.[58,72] Fragment ions derived from beta cleavages at glycosidic linkages are also normally present. Because these are indistinguishable from molecular ions of glycopeptides containing trun-cated carbohydrate chains, precise information on heterogeneity requires the examination of derivatives, for example after reaction with NaCH and methyl iodide.[71] If more than one potential glycosylation site is present in an *O*-linked glycopeptide, the exact site(s) of attachment can be determined by sequential Edman degradation using FAB-MS to define the molecular mass of the truncated peptide or glycopeptide at each step of the degradation.[58] Sialic acids and fucose are lost during the Edman cleavage but other linkages are stable.

Unambiguous sequencing of *O*-glycans released from glycoproteins is best achieved using permethyl derivatives which afford abundant molecular ions at high sensitivity (100 pmol, see, for example, References 72 to 79). These give very useful information on composition which is equally valuable for both typical mammalian *O*-glycans and the novel structures present in, for example, bacterial glycoproteins.[81] Well-defined fragmentation patterns result from A-type cleavage at each amino sugar residue. Sequences, branching and sites of sialylation, fucosylation, sulphation, etc. are defined by these fragment ions in the same way as for *N*-glycans (see Section III.D.1.) and this information can be unambiguously derived from the spectra of simple mixtures as well as from single components. Because of the sensitivity of analysis of permethylated samples by FAB-MS, the method lends itself to the screening of HPLC eluents for minor components with retention times very similar to those of more abundant relatives.[74] Furthermore, the speed and simplicity of the FAB-MS analysis of permethylated *O*-glycans permits detailed comparisons of structures present on the surfaces of related cell lines. For example, FAB-MS played a key role in the charac-terisation of the oligosaccharide chains on leukosialin isolated from normal granulocytes and two cancer cell lines.[73]

In some glycans, especially highly branched ones, the favoured HexNAc cleavages may not provide sufficient data for complete sequencing. One solution to this problem is FAB monitoring of controlled methanolytic degradation of permethylated samples.[82] The masses of the molecular ions of the hydrolytic fragments define the number of free hydroxyl groups present in each fragment and this information, together with the compositional data for each fragment, is used to assign sequences and branching patterns.

3. Glycosaminoglycans

The polysulphated polysaccharides of glycosaminoglycans (GAGs) are difficult to rigorously characterise and no proteoglycan has been fully sequenced to date. Small GAG-derived oligosaccharides (up to about 10 sugar residues in length), are, however, much more tractable, and FAB-MS is the most powerful tool currently available for their analysis. FAB-MS of native samples in the negative ion mode has successfully defined the size (i.e., the number of repeat disaccharides and the maximum number of sulphates) of oligosaccharides from a variety of proteoglycans.[83-86] In these experiments the majority of the ion current was carried by an array of molecular ions (differing in the number of cations present) and by abundant fragment ions formed by consecutive eliminations of sulphite moieties. The latter are indistinguishable from the molecular ions of corresponding oligosaccharides carrying fewer sulphate moieties which is why the method only unambiguously defines the maximum number of sulphates present. Precise information on sulphate heterogeneity requires derivative preparation. A sensitive permethylation strategy has been described[61] in which permethylated samples (which still carry the sulphate groups) are analysed by negative FAB-MS and, following an acetylation step in which sulphates are replaced by acetyl groups, by positive FAB-MS. The combined data from this FAB strategy provides precise compositional data and also sequence information including the positions in the sequence of sulphated residues.

4. Glycosyl-Phosphatidylinositol (GPI) Anchors

GPI anchors are constituted of a glycan chain terminating with a glucosamine which is connected to phosphatidylinositol. Attached to the glycan chain are one or more phosphorylethanolamine moieties and additional fatty acids may substitute the inositol (see Section II.D.). Thus these molecules are structurally very complex and their complete characterisation requires the implementation of several structural techniques incuding NMR, GC-MS, and chemical and enzymatic degradation. FAB-MS is a powerful adjunct to these techniques and it has, for example, assisted in the characterisation of the anchors for the Trypanosome VSG[87] and human acetylcholine esterase[36] and for GPIs from Leishmania.[88,89] The strategies used are exactly those which are most successful in other areas of glycoprotein analysis, namely negative ion FAB-MS for initial molecular weight assignment (the negatively charged phosphates result in excellent negative ion sensitivity) and for establishing the masses of chemically or enzymatically produced fragments, and positive FAB-MS of permethyl derivatives. If permethylation is preceded by N-acetylation of the unique amino sugar, then the major fragment ion observed in the permethylated spectrum is derived, as expected, from A-type cleavage at HexNAc. Omission of the N-acetylation step results in higher molecular ion abundances because the primary amino group of the glucosamine offers a site for facile protonation of an N-dimethylated product or for generation of a fixed positive charge if trimethylation of the amino group occurs. In fact both the di- and tri-methyl derivatives are formed and they give a pair of molecular ions of comparable abundance. These constitute a distinctive fingerprint for GPI anchors and allow them to be detected in fairly crude membrane extracts after permethylation.[90] Acyl groups are lost during permethylation but fatty alkyl groups on the glycerol are stable. Hence the chain lengths of the acyl and alkyl groups in an acyl/alkyl glycerol can be separately assigned by comparing the data from the native and permethylated samples.[89]

IV. SUMMARY AND FUTURE PROSPECTS

In this chapter we have focused upon FAB-MS of the glycan constituents of glycoproteins but the technique is, of course, equally valuable for defining the protein structure, if necessary. Further, although much of our discussion has centered around the ''normal'' oligosaccharide chains which are most commonly present in glycoproteins, the characterisation of novel structures (for example: bisected high mannose structures[91] and methylphosphomannosyl[92] residues in glycoproteins from *Dictyostelium discoideum*; oligosaccharides and glycopeptides from bacterial glycoproteins[81,93]) is always enormously assisted by precise molecular weight and sequence data. A particular strength of FAB-MS is its applicability to a wide range of structural types which means that many glycoconjugates can be examined intact, thus minimising the possibility that labile substituents remain unrecognised. An additional advantage of FAB-MS derives from its capacity for providing meaningful data from mixtures, thereby making it a powerful tool for the analysis of the heterogeneous array of oligosaccharides that is normally attached at each glycosylation site in a glycoprotein. With respect to sensitivity, FAB-MS competes very successfully with other structural techniques. With the FAB-MS instrumentation currently being used to solve problems of the type reviewed in this chapter, sample loadings between 0.5 μg (for *O*-glycans and glycolipids) and 5 μg (for most *N*-glycans and GAG derived oligosaccharides) are all that is necessary for good quality data. We anticipate that these sensitivities will be improved by at least an order of magnitude when array detector mass spectrometers,[94] which are being designed and constructed at the time of this writing, begin to be applied to glycoprotein problems. These instruments record all ions within a given mass range which arrive at the detector in a given time. They are therefore capable of much greater sensitivity than the previous generation of instruments which have a point detector at which ions of different masses are brought into focus by scanning the magnetic sector, resulting in ions of each mass value being detected for only a fraction of the time of the experiment.

During the past ten years FAB-MS has proven to be a very powerful tool for the structural characterisation of glycoproteins. Instrumental developments, such as array detection, should make it even more powerful, and if these developments are fully exploited by carbohydrate chemists and biochemists, many important and exciting problems should be solved in the second decade of FAB-MS.

ACKNOWLEDGMENTS

Anne Dell and Roy McDowell are grateful for Programme Grant support from the MRC.

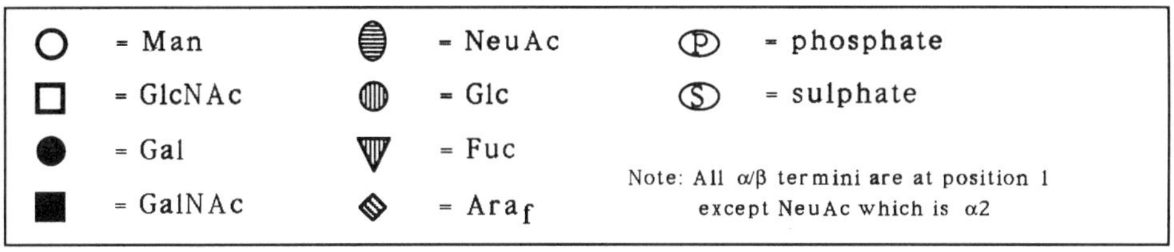

○	= Man	⊜	= NeuAc	Ⓟ	= phosphate
□	= GlcNAc	⊕	= Glc	Ⓢ	= sulphate
●	= Gal	▽	= Fuc		
■	= GalNAc	◈	= Ara$_f$		

Note: All α/β termini are at position 1 except NeuAc which is α2

Key to figures.

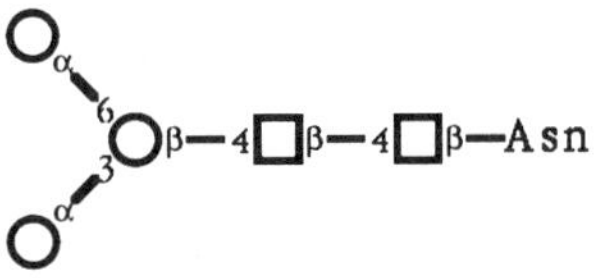

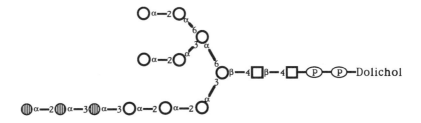

FIGURE 1. The pentasaccharide core that is common to all Asn-linked oligosaccharides; the GlcNAc-GlcNAc disaccharide is often called the chitobiose moiety.

FIGURE 2. The lipid-linked oligosaccharide which is the precursor of all Asn-linked oligosaccharides.

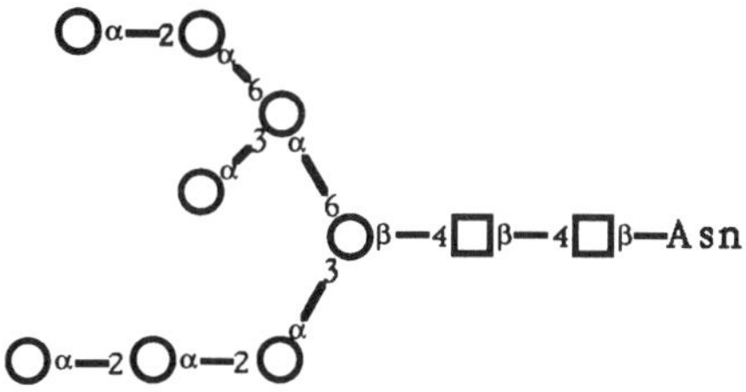

FIGURE 3. The glycan shown in Figure 2 is truncated by two glucosidases and a mannosidase in the ER to give this Man$_8$GlcNAc$_2$ glycan.

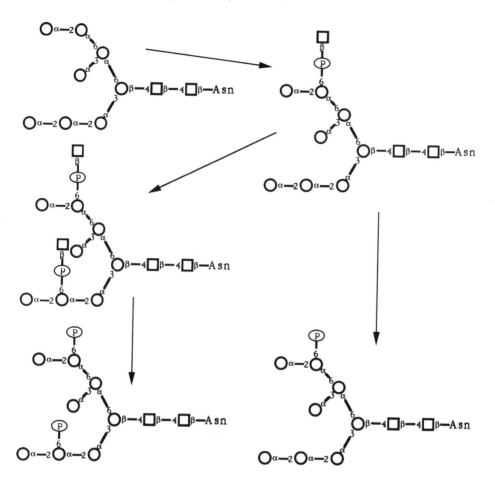

FIGURE 4.   Glycans containing mannose-6-phosphate are biosynthesised by the addition of GlcNAc-1-phosphate to the 6-positions of specific mannosyl residues in the $Man_8GlcNAc_2$ glycan shown in Figure 3. This is followed by hydrolysis of the phosphodiester linkages to yield mannose-6-phosphate residues. Major structures are shown in the figure. Trimming of the terminal mannose residues may occur before binding to the mannose 6-phosphate receptor.

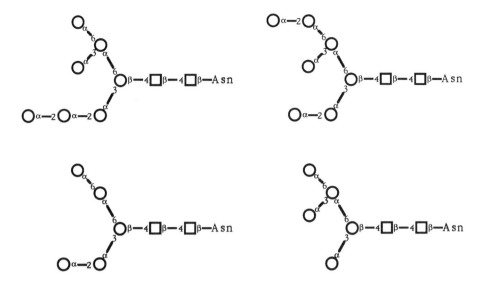

FIGURE 5.   Examples of high mannose structures.

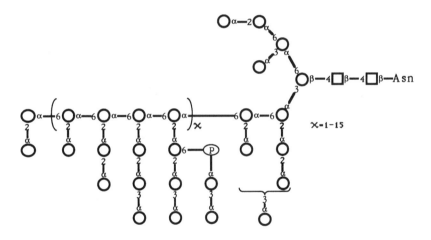

FIGURE 6.   Asn-linked glycans in yeast have structures of the type shown in this figure.

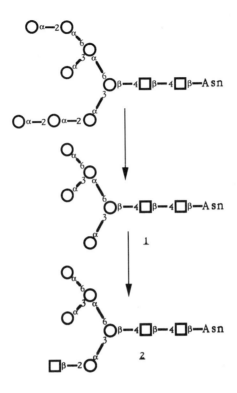

FIGURE 7.   Processing of the Man$_8$ glycan shown in Figure 3 yields structure $\underline{2}$ which is the precursor for hybrid and complex oligosaccharides.

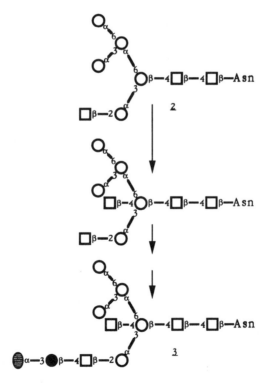

FIGURE 8.   The route to hybrid structures.

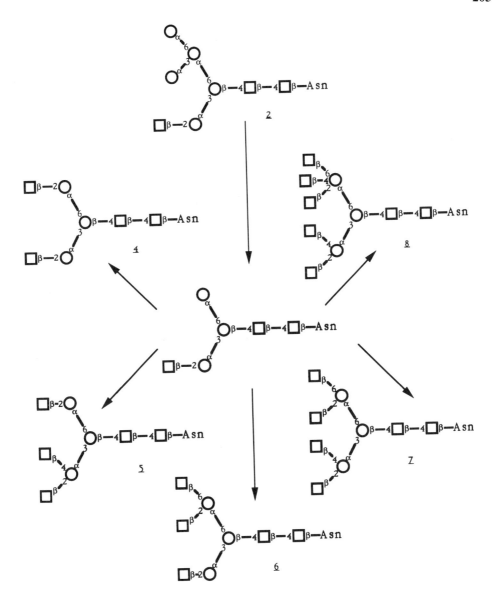

FIGURE 9. After alpha mannosidase trimming of structure 2 the product is acted upon by a series of GlcNAc transferases to give the core structures 4 to 8.

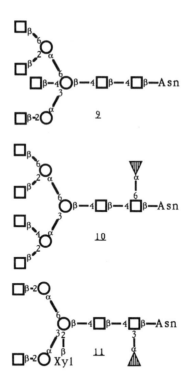

FIGURE 10.   In addition to the GlcNAc transferase reactions shown in Figure 9, the core can be modified by a bisecting GlcNAc transferase (e.g., structure 9) and by a fucosyl transferase (e.g., structure 10). Plant core structures often contain xylose as well as fucose (e.g., structure 11).

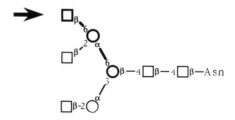

FIGURE 11.   The arrow shows the antenna which is preferentially acted upon by the elongating enzyme involved in polylactosaminoglycan production.

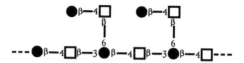

FIGURE 12.   Branching of polylactosaminoglycans can occur by GlcNAc addition to the 6-position of the in-chain Gal residues as shown here.

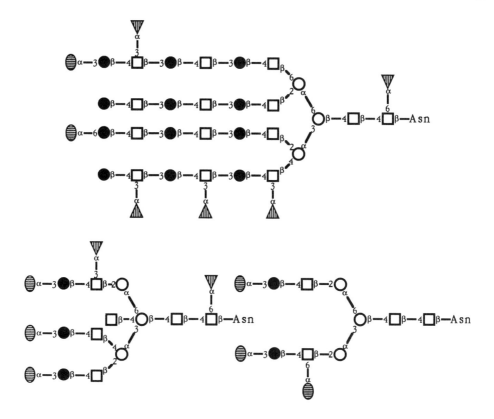

FIGURE 13.   Examples of complex glycans illustrating variations in branch number, branch length, and capping.

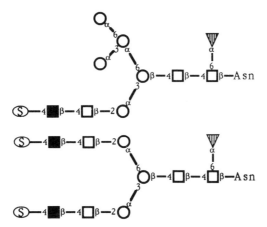

FIGURE 14.   Examples of glycans from pituitary hormones which contain unique sulphated GalNAc termini.

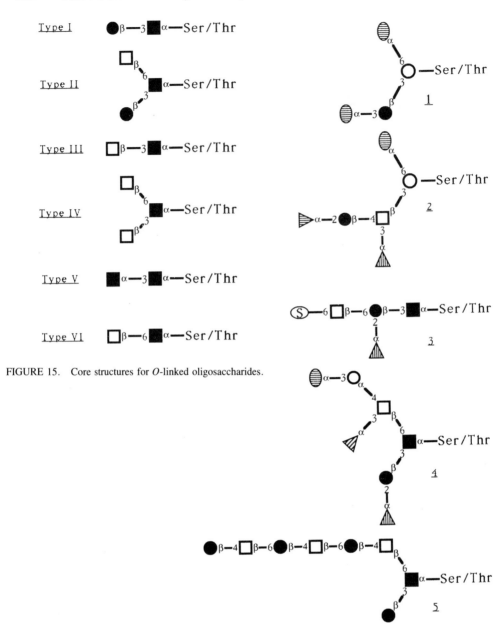

FIGURE 15.   Core structures for *O*-linked oligosaccharides.

FIGURE 16.   Examples of *O*-linked oligosaccharides illustrating variations in cores, branching and capping.

209

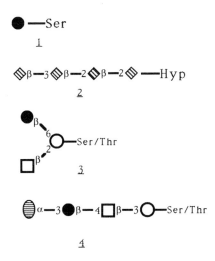

FIGURE 17. Examples of *O*-glycans not typical of mammalian systems; (1),(2) are common plant structures; (3),(4) are mannose-linked structures found in coral and brain, respectively.

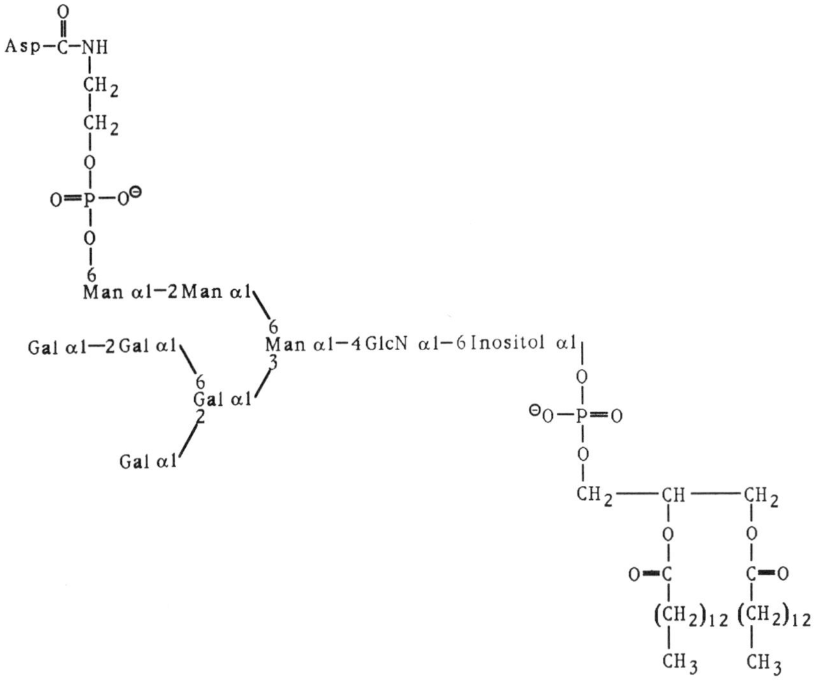

Asp—C—NH
║
O

CH₂
|
CH₂
|
O
|
O=P—O⁻
|
O
|
6
Man α1—2 Man α1

Gal α1—2 Gal α1

Man α1—4 GlcN α1—6 Inositol α1

Gal α1

Gal α1

O
|
⁻O—P=O
|
O
|
CH₂——CH——CH₂
| |
O   O
| |
O=C   C=O
| |
(CH₂)₁₂ (CH₂)₁₂
| |
CH₃   CH₃

FIGURE 18.   The GPI anchor of the VSG of Trypanosomes.

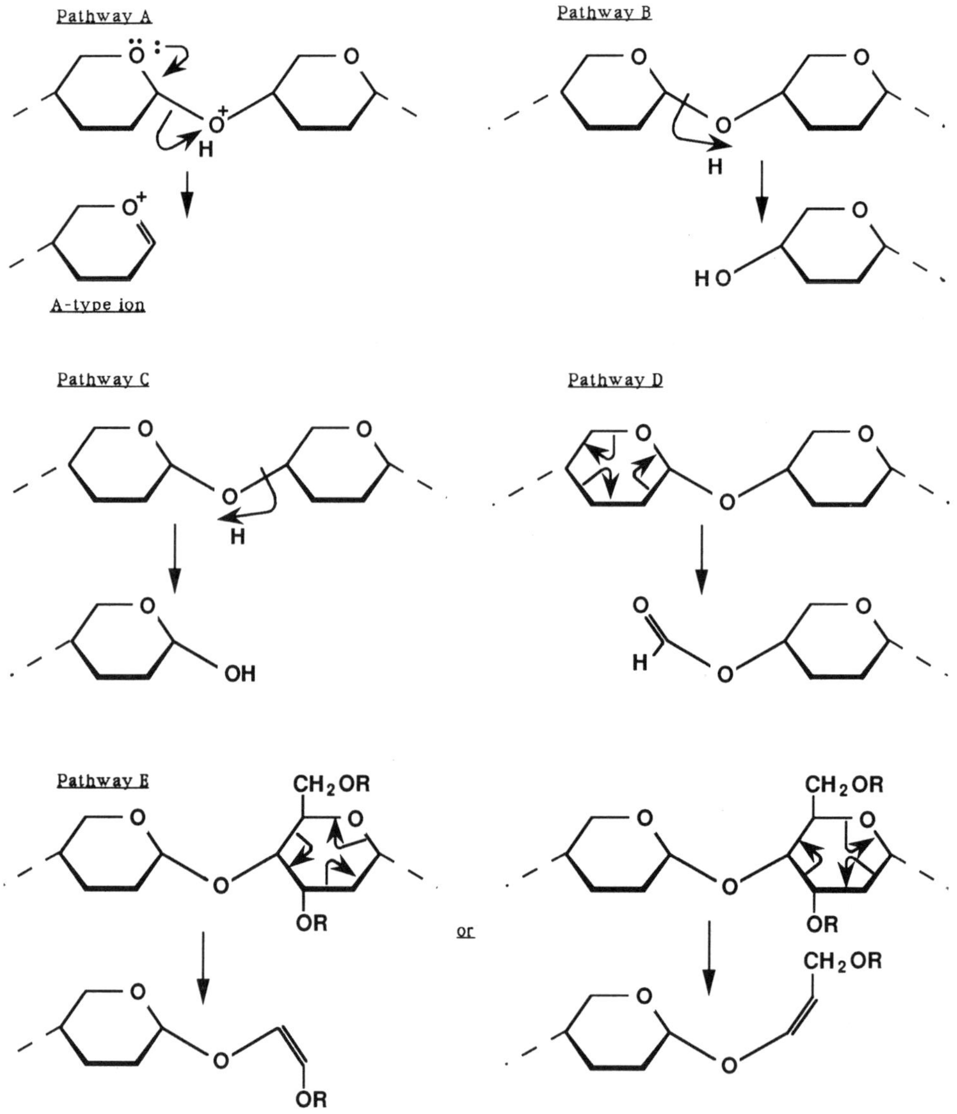

FIGURE 19. Major fragmentation pathways of oligosaccharides and glycoconjugates. Pathway A occurs in the positive ion mode only. Pathways B-E occur in both modes. In the negative ion mode ions derived from Pathways B and/or C are often accompanied by related ions resulting from loss of water (18 mass units lower) and oxidation of the hydroxyl group to a keto group (2 mass units lower).

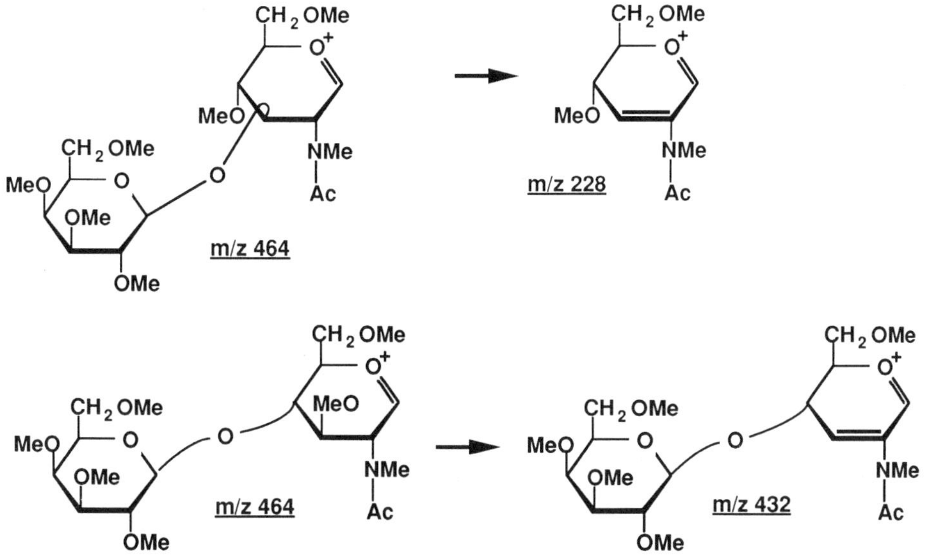

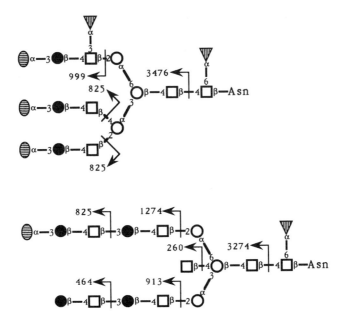

FIGURE 20. Elimination of the substituent at position 3 from the HexNAc A-type cleavage ion in permethylated samples allows assignment of Type I (upper scheme) or type II (lower scheme) backbones.

FIGURE 21. Specific A-type cleavage at each HexNAc gives characteristic ''maps'' of fragment ions which define all the nonreducing structures. This figure shows the major ions produced for two N-glycans.

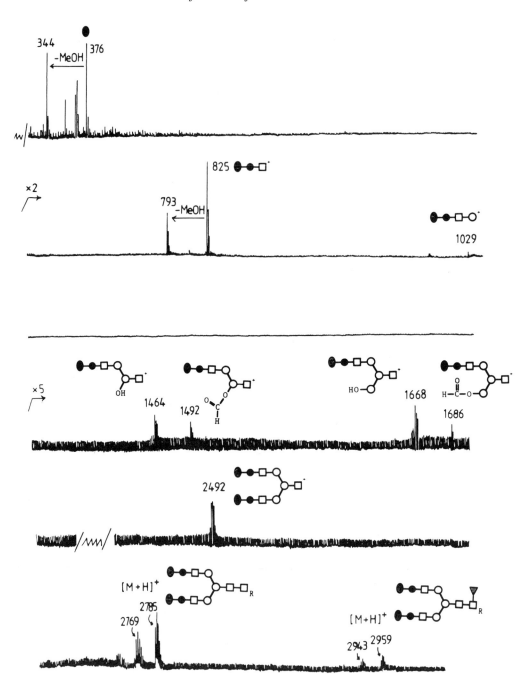

FIGURE 22.   FAB mass spectrum of reduced, permethylated disialylated biantennary oligosaccharide; fragment ions are assigned on the spectrum and arise via HexNAc A-type cleavage (major ions) or via beta-type cleavages (see Figure 19).

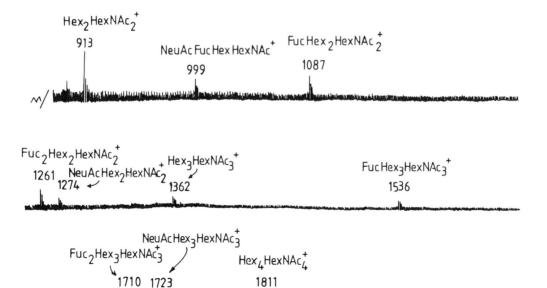

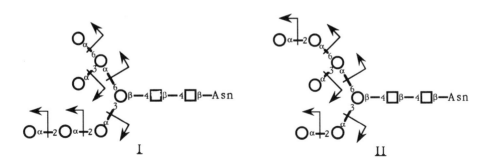

FIGURE 23. Part of the FAB mass spectrum of a permethylated sample of polylactosaminoglycan-containing oligosaccharides. All major fragment ions are formed by A-type cleavage at HexNAc and their compositions are shown on the figure.

FIGURE 24. Cleavage at each Man residue in peracetylated high mannose samples allows some isomers to be distinguished. For example, as shown here, structure I gives Hex_1, Hex_2, and Hex_3 but not Hex_4, while structure II gives Hex_1, Hex_2, and Hex_4, but not Hex_3. Fragment ions are often sodiated.

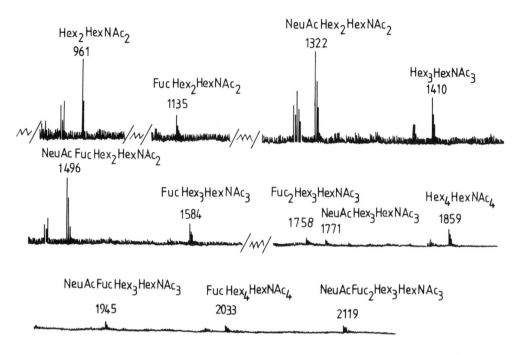

FIGURE 25.   FAB mass spectrum of the mixture of oligosaccharides released from 40 µg of a milk fat globule glycoprotein by reductive elimination. The sample was permethylated and the unseparated mixture analysed by FAB-MS. Molecular ion compositions are given on the figure.

REFERENCES

1. **Kornfeld, R. and Kornfeld, S.,** *Annu. Rev. Biochem.,* 54, 631, 1985.
2. **Sadler, J. E.,** in *Biology of Carbohydrates,* Vol. 2, Ginsburg, V. and Robbins, P. W., Eds., John Wiley and Sons, New York, 1984, 87.
3. **Kobata, A.,** *Biology of Carbohydrates,* Vol. 2, Ginsburg, V. and Robbins, P. W., Eds., John Wiley and Sons, New York, 1984, chap. 2.
4. **Schachter, H.,** *Biochem. Cell. Biol.,* 64, 163, 1986.
5. **Hirschberg, C. B. and Snider, M. D.,** *Annu. Rev. Biochem.,* 56, 63, 1987.
6. **Rademacher, T. W., Parekh, R. B., and Dwek, R. A.,** *Annu. Rev. Biochem.,* 57, 785, 1988.
7. **Kukurizinska, M. A., Bergh, M. L. E., and Jackson, B. J.,** *Annu. Rev. Biochem.,* 56, 915, 1987.
8. **Shaper, N. L., Shaper, J. H., Meuth, J. L., Fox, J. L., Chang, H., Kirsch, I. R., and Hollis, G. F.,** *Proc. Natl. Acad. Sci.,* 83, 1573, 1986.
9. **Tetteroo, P. A. T., de Heij, H. T., Van den Eijnden, D. H., Visser, F. J., Schoenmaker, E., and Geurts van Kessel, A. H. M.,** *J. Biol. Chem.,* 262, 15984, 1987.
10. **Nishikawa, Y., Pegg, W., Paulsen, H., and Schachter, H.,** *J. Biol. Chem.,* 263, 8270, 1988.
11. **Reichner, J., Whiteheart, S. W., and Hart, G. W.,** *J. Biol. Chem.,* 263, 16316, 1988.
12. **Paulson, J. C. and Colley, K. J.,** *J. Biol. Chem.,* 264, 17615, 1989.
13. **Pfeffer, S. R. and Rothman, J. E.,** *Annu. Rev. Biochem.,* 56, 829, 1987.
14. **Brockhausen, I., Hull, E., Hindsgaul, O., Schachter, H., Shah, R. N., Michnick, S. W., and Carver, J. P.,** *J. Biol. Chem.,* 264, 11211, 1989.
15. **Kaushal, G. P., Szumilo, T., and Elbein, A. D.,** in *The Biochemistry of Plants,* Vol. 142, Stumpf, P. K. and Conn, E. E., Eds., Academic Press, New York, 1988, chap. 2.
16. **Fukuda, M., Fukuda, M. N., Papayonnopoulou, T., and Hakomori, S.,** *Proc. Natl. Acad. Sci. U.S.A.,* 77, 3474, 1980.
17. **van den Eijnden, D., Koenderman, A. H. L., and Schiphorst, W. E.,** *J. Biol. Chem.,* 263, 12461, 1988.
18. **Fukuda, M., Delll, A., Oates, J. E., and Fukuda, M. N.,** *J. Biol. Chem.,* 259, 4792, 1984.

19. Spiro, R. G. and Bhoyroo, V. D., *J. Biol. Chem.*, 263, 14351, 1988.
20. Kamerling, J. P., Rijkse, I., Maas, A. A. M., van Kuik, J. A., and Vliegenthart, J. F. G., *FEBS Lett.*, 241, 256, 1988.
21. Freeze, H. H. and Wolgast, D., *J. Biol. Chem.*, 261, 127, 1986.
22. Baenziger, J. U. and Green, E. D., *Biochim. Biophys. Acta*, 947, 287, 1988.
23. Schachter, H., in *The Glycoconjugates*, Vol. 2, Pigman, W. and Horowitz, M., Eds., Academic Press, New York, 1978, 87.
24. Klein, A., Lamblin, G., Lhermitte, M., Roussel, P., Breg, J., Van Halbeek, H., and Vliegenthart, J. F. G., *Eur. J. Biochem.*, 171, 631, 1988.
25. Feeney, J., Frenkiel, A. T., and Hounsell, E. F., *Carbohydr. Res.*, 152, 63, 1986.
26. Sasaki, H., Bothner, B., Dell, A., and Fukuda, M., *J. Biol. Chem.*, 262, 12059, 1987.
27. Darvill, A., McNeil, M., Albersheim, P., and Delmer, D. P., in *The Plant Cell*, Tolbert, N. E., Ed., Academic Press, New York, 1980, 91.
28. Meikle, P., Richards, G. N., and Yellowlees, D., *J. Biol. Chem.*, 262, 16941, 1987.
29. Krusius, T., Reinhold, V. N., Margolis, R. K., and Margolis, R. U., *Biochem. J.*, 245, 229, 1987.
30. Torres, C. R. and Hart, G. W., *J. Biol. Chem.*, 259, 3308, 1984.
31. Hart, G. W., Haltiwanger, R. S., Holt, G. D., and Kelly, W. G., *Annu. Rev. Biochem.*, 58, 841, 1989.
32. Hart, G. W., Haltiwanger, R. S., Holt, G. D., and Kelly, W. G., in *Carbohydrate Recognition in Cellular Function*, 145th Ciba Found. Symp., Wiley, Chichester, 1989, 102.
33. Ferguson, M. A. J. and Williams, A. F., *Annu. Rev. Biochem.*, 57, 285, 1988.
34. Ferguson, M. A. J., Homans, S. W., Dwek, R. A., and Rademacher, T. W., *Science*, 239, 753, 1988.
35. Homans, S. W., Ferguson, M. A. J., Dwek, R. A., Rademacher, T. W., Anand, R., and Williams, A. F., *Nature*, 333, 269, 1988.
36. Roberts, W. L., Santikarn, S., Reinhold, V. N., and Rosenberry, T. L., *J. Biol. Chem.*, 263, 18776, 1988.
37. Doering, T. L., Masterson, W. J., Englund, P. T., and Hart, G. W., *J. Biol. Chem.*, 264, 11168, 1989.
38. Roden, L., in *The Biochemistry of Glycoproteins and Proteoglycans*, Lennarz, W. L., Ed., Plenum Press, New York, 1980, 267.
39. Sugahara, K., Yamashina, I., De Waard, P., Van Halbeek, H., and Vliegenthart, J. F. G., *J. Biol. Chem.*, 263, 10168, 1988.
40. Barber, M., Bordoli, R. S., and Sedgwick, R. D., in *Soft Ionisation Biological Mass Spectrometry*, Morris, H. R., Ed., Heyden, London, 1981, 137.
41. Barber, M., Bordoli, R. S., Sedgwick, R. D., and Tyler, A. N., *J. Chem. Soc. Chem. Commun.*, 325, 1981.
42. Burlingame, A. L., Maltby, D., Russell, D. H., and Holland, P. T., *Ann. Chem.*, 60, 294R, 1988.
43. Dell, A., *Adv. Carbohydr. Chem. Biochem.*, 45, 19, 1987.
44. Morris, H. R., Chatterjee, A., Panico, M., Green, B., Almeida da Silva, M. A. A., and Hartley, B. S., *Rapid Commun. Mass Spectrom.*, 3, 110, 1989.
45. Dell, A. and Thomas-Oates, J. E., in *Analysis of Carbohydrates by GLC and MS*, Biermann, C. J. and McGinnis, G. D., Eds., CRC Press, Boca Raton, FL, 1989, 217.
46. Egge, H., Dabrowski, J., and Hanfland, P., *Pure Appl. Chem.*, 56, 807, 1984.
47. Dell, A., Thomas-Oates, J. E., Rogers, M. E., and Tiller, P. R., 70, 1435, 1988.
48. Dell, A., in *Methods in Enzymology: Mass Spectrometry*, Academic Press, Orlando, FL, 193, 647, 1990.
49. Tsai, P., Dell, A., and Ballou, C. E., *Proc. Natl. Acad. Sci. U.S.A.*, 82, 4119, 1986.
50. Carr, S. A. and Roberts, G. D., *Ann. Biochem.*, 157, 396, 1986.
51. Morris, H. R., Panico, M., and Taylor, G. W., *Biochem. Biophys. Res. Commun.*, 117, 299, 1983.
52. Fukuda, M., Dell, A., and Fukuda, M. N., *J. Biol. Chem.*, 259, 4782, 1984.
53. Spooncer, E., Fukuda, M., Klock, J. C., Oates, J. E., and Dell, A., *J. Biol. Chem.*, 259, 4792, 1984.
54. Fukuda, M., Spooncer, E., Oates, J. E., Dell, A., and Klock, J. C., *J. Biol. Chem.*, 259, 10925, 1984.
55. Oates, J. E., Dell, A., Fukuda, M., and Fukuda, M. N., *Carbohydr. Res.*, 141, 149, 1985.
56. Fukuda, M. N., Dell, A., and Scartezzini, P., *J. Biol. Chem.*, 162, 4580, 1987.
57. Fukuda, M. N., Masri, K. A., Dell, A., Thonar, E. J., Klier, G., and Lowenthal, R. M., *Blood*, 73, 1331, 1989.
58. Sasaki, H., Ochi, N., Dell, A., and Fukuda, M., *Biochemistry*, 27, 8618, 1988.
59. Dell, A., Morris, H. R., Egge, H., von Nicolai, H., and Strecker, G., *Carbohydr. Res.*, 115, 41, 1983.
60. Fukuda, M. N., Dell, A., Oates, J. E., Wu, P. W., Klock, J. C., and Fukuda, M., *J. Biol. Chem.*, 260, 4110, 1984.
61. Dell, A., Rogers, M. E., Thomas-Oates, J. E., Huckerby, T. N., Sanderson, P. N., and Nieduszynski, I. A., *Carbohydr. Res.*, 179, 7, 1988.
62. Dell, A., Morris, H. R., Greer, F., Redfern, J. M., Rogers, M. E., Weisshaar, G., Hiyama, J., and Renwick, A. G. G., *Carbohydr. Res.*, 209, 33, 1991.

63. Fukuda, M., Bothner, B., Ramasamooj, P., Dell, A., Tiller, P. R., Varki, A., and Klock, J. C., *J. Biol. Chem.*, 260, 12957, 1985.
64. Dell, A., *Biochem. Soc. Trans.*, 17, 17, 1989.
65. Shao, M.-C. and Wold, F., *J. Biol. Chem.*, 264, 6245, 1989.
66. Lindberg, B., Leontein, K., Lindquist, U., Svenson, S. B., Wrangsell, G., Dell, A., and Rogers, M. E., *Carbohydr. Res.*, 174, 313, 1988.
67. York, W. S., Oates, J. E., Van Halbeek, H., Darvill, A. G., Albersheim, P., Tiller, P. R., and Dell, A., *Carbohydr. Res.*, 173, 113, 1988.
68. Hernandez, L. M., Ballou, L., Alvarado, E., Gillece-Castro, B. L., Burlingame, A. L., and Ballou, C. E., *J. Biol. Chem.*, 264, 11849, 1989.
69. Dell, A. and Thomas-Oates, J. E., unpublished.
70. Dell, A., Thomas-Oates, J. E., and Mather, I., unpublished.
71. Thomas-Oates, J. E. and Dell, A., *Biochem. Soc. Trans.*, 17, 243, 1989.
72. Robb, R. J., Kutny, R. M., Panico, M., Morris, H. R., and Chowdry, V., *Proc. Natl. Acad. Sci. U.S.A.*, 81, 6486, 1984.
73. Fukuda, M. N., Carlsson, S. R., Klock, J. C., and Dell, A., *J. Biol. Chem.*, 261, 12796, 1986.
74. Fukuda, M., Lauffenberger, M., Sasaki, H., Rogers, M. E., and Dell, A., *J. Biol. Chem.*, 262, 11952, 1987.
75. Hanisch, F. G., Uhlenbruck, G., Peter-Katalinic, J., Egge, H., Dabrowski, J., and Dabrowski, U., *J. Biol. Chem.*, 264, 872, 1989.
76. Sabharwal, H., Nilsson, B., Chester, M. A., Lindh, F., Gronberg, G., Sjoblad, S., and Lundblad, A., *Carbohydr. Res.*, 178, 145, 1989.
77. Martensson, S., Due, C., Pahlsson, P., Nilsson, B., Eriksson, H., Zopf, D., Olsson, L., and Lundblad, A., *Cancer Res.*, 48, 2125, 1988.
78. Wieruszeski, J. M., Michalski, J. C., Montreuil, J., Strecker, G., Peter-Katalinic, J., Egge, H., Van Halbeek, H., Mutsaers, J. H., and Vliegenthart, J. F. G., *J. Biol. Chem.*, 262, 6650, 1987.
79. Hanisch, F. G., Egge, H., Peter-Katalinic, J., and Uhlenbruck, G., *Eur. J. Biochem.*, 155, 239, 1986.
80. Scudder, P., Lawson, A. M., Hounsell, E. F., Carruthers, R. A., Childs, R. A., and Feizi, T., *Eur. J. Biochem.*, 168, 585, 1987.
81. Gerwig, G. J., de Waard, P., Kamerling, J. P., Vliegenthart, J. F. G., Morgenstern, E., Lamed, R., and Bayer, E. A., *J. Biol. Chem.*, 264, 1027, 1989.
82. Dell, A., Thomas-Oates, J. E., Rogers, M. E., and Tiller, P. R., *Biomed. Environ. Mass Spectrom.*, 16, 19, 1988.
83. Carr, S. A. and Reinhold, V. N., *Carbohydr. Chem.*, 3, 381, 1984.
84. Reinhold, V. N., Carr, S. A., Green, B. N., Petitou, M., Choay, J., and Sinay, P., *Carbohydr. Res.*, 161, 305, 1987.
85. McNeal, C. J., Macfarlane, R. D., and Jardine, I., *Biochem. Biophys. Res. Commun.*, 139, 18, 1986.
86. Scudder, P., Tang, P. W., Hounsell, E. F., Lawson, A. M., Mehmet, H., and Feizi, T., *Eur. J. Biochem.*, 157, 365, 1986.
87. Schmitz, B., Klein, R. A., Duncan, I. A., Egge, H., Gunawan, J., Peter-Katalinic, J., Dabrowski, U., and Dabrowski, J., *Biochem. Biophys. Res. Commun.*, 146, 1055, 1987.
88. Rosen, G., Pahlsson, P., Londner, M. V., Westerman, M. E., and Nilsson, B., *J. Biol. Chem.*, 264, 10457, 1989.
89. McConville, M. J., Homans, S. W., Thomas-Oates, J. E., Dell, A., and Basic, T., *J. Biol. Chem.*, 25, 7385, 1990.
90. Dell, A. and Khoo, K. H., unpublished.
91. Couso, R., Van Halbeek, H., Reinhold, V., and Kornfeld, S., *J. Biol. Chem.*, 262, 4521, 1987.
92. Gabel, C. A., Costello, C. E., Reinhold, V. N., Kurtz, L., and Kornfeld, S., *J. Biol. Chem.*, 259, 13762, 1984.
93. Paul, G., Lottspeich, F., and Wieland, F., *J. Biol. Chem.*, 261, 1020, 1986.
94. Cottrell, J. S. and Evans, S., *Ann. Chem.*, 59, 1990, 1987.

Chapter 6

GLYCOPROTEINS OF MAMMALIAN AND AVIAN CELLS

F. W. Hemming

TABLE OF CONTENTS

I. INTRODUCTION

Glycosylation is a common example of the posttranslational modification of proteins in eukaryotic cells. Although many biochemical details of the process are now understood, important new observations continue to be made. Despite some clear examples of important biological consequences of the process, in many cases its significance remains an enigma. In this chapter the current understanding of the structural diversity of glycoprotein oligo-saccharides (glycans) and of the mechanism and control of their assembly will be reviewed. This will be followed by a discussion of several examples of the biological consequences of protein glycosylation and of the possible significance of the completed glycan moiety. Many investigations have been carried out on mammalian and avian species and most general points can be illustrated by reference to these.[1-13] A recent conference report[14] on a model soluble glycoprotein, α-1 acid glycoprotein, exemplifies several of these points. Details of further complications presented by higher plants, yeasts, fungi, bacteria and other animals have been well reviewed.[3,4,8,10,15-17]

The scientific literature contains very many excellent papers, reviews, and books covering this area. It follows that a relatively short chapter must be highly selective and lack detailed evidence. To avoid an inordinately long list of references the reader is referred primarily to reviews and books where further information may be found.

```
(i)                        glycan--4GlcNAcβ1-N-Asn


(ii)                       glycan--4GalNAcα1-O-Ser


(iii)  PI-glycan-Man-6-P-ethanolamineNH-CO-CH(R)-NH


(iv)                       glycan--2Galβ1-O-HLys
```

FIGURE 1. Examples of glycan-protein links found in mammalian and avian glycoproteins:
(i) glycanyl *N*-asparagine, (ii) glycanyl *O*-serine, (iii) PI-glycanyl phosphoethanolamine ester,
(iv) glycanyl *O*-hydroxylysine. In most proteoglycans β-xylose replaces the *N*-acetylgalactosa-
mine residue of structure (ii). Recently, variations of (ii) have been described[19,20] in which (1)
an *N*-acetylglucosaminidic bond links short glycans to a serine (threonine) residue of a small
group of glycoproteins, and (2) mannose is present as a peptide chain, carrying also GAGs found
in the cytoplasm of neurones.[21,22] Structure (iv) has been found in collagen and the CIq component
of complement. These links are discussed in section IIA. See Table 1 for a list of monosaccharide
abbreviations.

II. STRUCTURAL FEATURES

A. THE GLYCAN-PROTEIN LINK

Glycoproteins carry one or more glycan chains linked glycosidically to the side chain
N atom of an asparagine residue or to the side chain O atom of a serine or threonine (or
hydroxylysine in collagen) residue.[5,8,9] A small number of glycoproteins (and proteins) have
recently been shown also to be esterified at the C terminus to the amino group of ethanolamine
phosphomannoglycan which in turn is linked to phosphatidyl inositol forming a GPI anchor.[18]
These structures are summarised in Figure 1.

Some proteins carry just one glycan but several have more than one and the latter can
include both N- and O-linked (and phosphoethanolamine-linked) glycans on the same peptide
chain. N-linked glycans are found only on asparagine residues that form part of the tripeptide
(sequon): Asn-X-Thr (Ser) in which X is any amino acid other than proline. The tripeptide
is usually part of a β-loop in the protein tertiary structure. On average approximately one
third of asparagine residues that meet these criteria are actually glycosylated.

No firm generalizations can be made about the amino acid environment of hydroxyamino
acids carrying O-linked glycans. However, several appear to have a proline residue adjacent
or nearby. In the case of proteoglycans a consensus sequence of two or three acidic amino
acids followed by a Ser-Gly-X-Gly sequence (sometimes a Ser-Gly repeat) appears to be
sufficient to ensure that the serine residue is glycosylated.[23] In α-1(II) and α-2(I) chains of
collagen the triplet Gly-X-Lys has been reported to be essential for hydroxylation of the
lysine residue and hence is a feature common to galactosylation sites in these molecules.
However, in α-1(III) chains the sequence about these sites appears to be quite different. Not
all hydroxylysine residues are glycosylated: in extensively glycosylated forms of α-1(II)
chains it has been estimated that just over one half remain without a sugar attachment.[8,24]

As isolation and analytical technology of protein-bound glycans has improved,[9] the
classification of glycans and their associated proteins has had to be modified. Initially the
protein chains of mucins were found to carry many O-linked glycans employing an *N*-
acetylgalactosaminyl *O*-serine (threonine) link. Subsequently, mucin-like glycans were found
on many glycoproteins often accompanied by N-linked glycans. Similarly, glycosamino-

TABLE 1
Common Monosaccharides of Animal
Glycoproteins with Abbreviations Used in
Descriptions of Glycan Structures

Glucose	Glc	N-acetylglucosamine	GlcNAc
Galactose	Gal	N-acetylgalactosamine	GalNAc
Mannose	Man	glucuronic acid	GlcUA
Fucose[a]	Fuc	iduronic acid[a]	IdUA
Xylose	Xyl	N-acetylneuraminic acid	NeuAc

[a]   Present in the L-form.

glycans (GAGs), polymers of repeating pairs of *N*-acetamido- and acidic monosaccharides were first seen solely as components of proteoglycans. A proteoglycan was characterized by the presence of many long GAG chains linked by a tetrasaccharide bridge to multiple serine residues in a relatively short core protein. More recently it has become established that keratan sulphates are linked differently: either through an N-linked pentasaccharide core or through a "mucin" type O-link. The protein core of several proteoglycans is also known to carry classical N-linked glycans (Section II.A.2). In addition it is now clear that a number of glycoproteins of the plasma membrane of epithelial cells carry one or a small number of GAG chains.[25-29]

Several aspects of structure and assembly of the GAGs clearly overlap with those of the more classical glycoprotein glycans. For these reasons it seems increasingly appropriate to treat proteoglycans as one end of a broad spectrum of glycoproteins. This is not to deny that some aspects of structure, biosynthesis and function are obviously peculiar to polyglycosaminoglycanated peptides.[23,25-29]

1. Monosaccharides Involved

The monosaccharides most commonly found in the glycan moieties of animal glycoproteins are listed in Table 1. The family of sialic acids[30,31] is restricted to animal glycoproteins and it is usually at the nonreducing end of a glycan chain either as a single residue or as part of a short oligosialyl chain. It is almost always attached to a galactose or *N*-acetylgalactosamine residue. A pKa value of close to 2.6 results in sialyl residues decorating glycoproteins with a strong negative charge at neutral pH (Figure 2).

2. N-Linked Glycans

All N-linked glycans contain a pentasaccharide core composed of three mannose and two *N*-acetylglucosamine residues (Figure 3i). Further substitution gives rise to characteristic peripheral structures. The high mannose (also called oligomannosyl and simple) glycans (Figure 3ii) carry further mannose residues to bring the total to a maximum of nine. Occasionally a chain of glucose residues (up to a maximum of three) may also be present.

The lactosaminyl (strictly *N*-acetylactosaminyl), also called complex N-linked glycans are characterized by the presence of a galactosyl β1-4 *N*-acetylglucosaminyl (*N*-acetylactosaminyl) substitutent on the branched core mannose residues (Figure 3iii). At their simplest the galactose residue may carry a sialic acid residue or it may remain unsubstituted. There may normally be one, two, three or four of these tri(di) saccharide chains giving rise to mono-, di-, tri or tetraantennary structures. Hybrid structures (Figure 3ii) have been described in which the α1-3 linked branch core mannose residue carries a lactosaminyl chain and the other only a mannose chain.

In some glycoproteins a polylactosaminyl chain (Figure 3v) is present on a branch core mannose giving rise to long chains (sometimes branched) of the type 2 (i and I) seen on glycolipids and O-linked glycans (see section 3a). Sometimes the polylactosaminyl chain

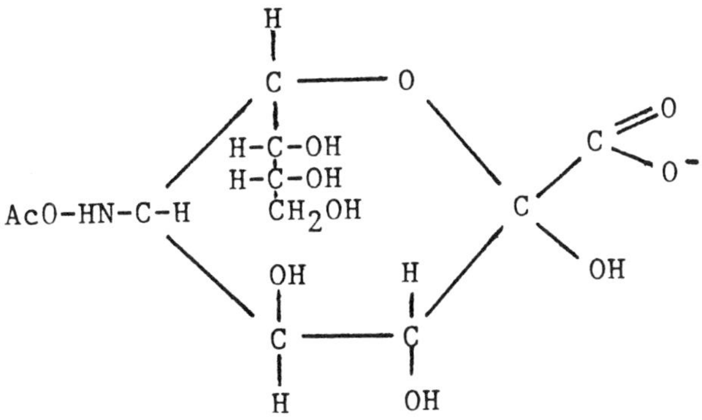

FIGURE 2. *N*-acetylneuraminate. The neuraminate skeleton is always *N*-acetylated or *N*-glycosylated and may be acetylated at one or more of the hydroxyl groups. This family of substituted neuraminic acids is called sialic acids.

bears a terminal sialic acid residue, like an extended version of a complex chain shown in Figure 3iii. Figure 3v shows the chain of keratan sulphate I (isolated from cornea) which is linked to the branch core mannose residues.

A picture emerges of a complex array of structures in which there is conservation, or at least constraint of diversity, at several levels. The structure of the core is invariant with respect to monosaccharide composition and arrangement and to the type of glycosidic bond between each of them. Considering the enormous potential for diversity the substitution pattern of this core appears to be limited to a relatively small range of structures. A feature of each structure on a particular glycoprotein from a particular cell type again is of a highly conserved monosaccharide composition and arrangement and detail of glycosidic linkages. In earlier work techniques did not allow elimination of the possibility that microheterogeneity, the detection of mixtures of "complete" and "incomplete" glycan chains on the same glycoprotein, was the result of random glycosylation or indeed due to artefacts of partial degradation. More recent technology provides evidence that strongly suggests a constancy of the range of structure of a glycan linked to a particular asparagine residue of a particular protein produced by a particular cell type under normal physiological conditions. Details of structure of any one glycan appear to be protein and cell-type specific. Clearly the range of glycan structures that can be elaborated on a protein will be limited by the glycosylation machinery of the cell producing that glycoprotein. To that extent there are common features in the glycans of several glycoproteins produced by one cell type. However, the extent of glycosylation and details vary from one glycoprotein to another sufficiently to make generalizations dangerous. Dwek's group[11] has coined the term "glycoforms" to describe the family of molecules formed by the presence on a glycoprotein of a range of N-linked glycans at any one glycosylation site. The term "glycotype" is used to describe the glycosylation pattern of a particular cell type.

3. O-Linked Glycans

In general, the type of *O*-glycosidic linkage, and in particular the nature of the linking monosaccharide is characteristic of the particular type of glycan involved. This gives rise to three types of glycan, (1) that linked to serine or threonine via *N*-acetylgalactosamine, (2) that linked to serine via xylose, and (3) that linked to hydroxylysine via galactose (see Figure 1). At present there is little evidence concerning the presence of sugar residues attached to the recently described[19,20] *N*-acetylglucosamine-serine (threonine) link (Figure 1).

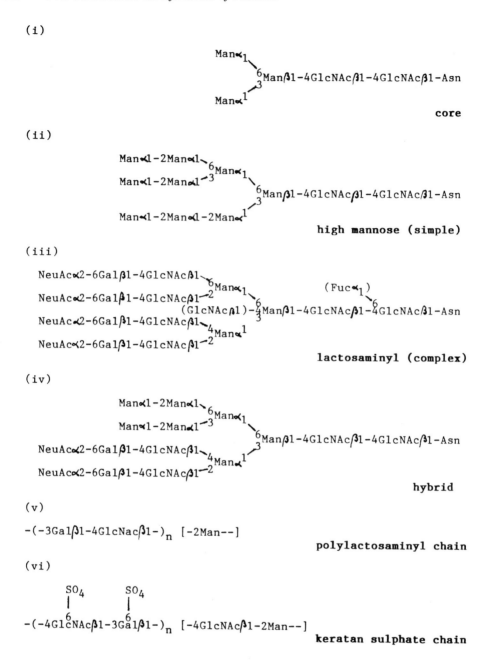

(i)

```
Man∝1
     6
      Manβ1-4GlcNAcβ1-4GlcNAcβ1-Asn
     3
Man∝1
```
core

(ii)

```
Man∝1-2Man∝1
            6
             Man∝1
Man∝1-2Man∝1-3      6
                     Manβ1-4GlcNAcβ1-4GlcNAcβ1-Asn
                    3
Man∝1-2Man∝1-2Man∝1
```
high mannose (simple)

(iii)

```
NeuAc∝2-6Galβ1-4GlcNAcβ1
                        6
                         Man∝1              (Fuc∝1)
NeuAc∝2-6Galβ1-4GlcNAcβ1-2    6                  6
                       (GlcNAcβ1)-4Manβ1-4GlcNAcβ1-4GlcNAcβ1-Asn
NeuAc∝2-6Galβ1-4GlcNAcβ1    3
                        4Man∝1
NeuAc∝2-6Galβ1-4GlcNAcβ1-2
```
lactosaminyl (complex)

(iv)

```
Man∝1-2Man∝1
            6
             Man∝1
Man∝1-2Man∝1-3      6
                     Manβ1-4GlcNAcβ1-4GlcNAcβ1-Asn
                    3
NeuAc∝2-6Galβ1-4GlcNAcβ1
                        4Man∝1
NeuAc∝2-6Galβ1-4GlcNAcβ1-2
```
hybrid

(v)

-(-3Galβ1-4GlcNacβ1-)$_n$ [-2Man--]

polylactosaminyl chain

(vi)

```
      SO4        SO4
       |          |
       6          6
-(-4GlcNAcβ1-3Galβ1-)n  [-4GlcNAcβ1-2Man--]
```
keratan sulphate chain

FIGURE 3.   Common types of N-linked glycans. All are derived by further glycosylation of the core pentasaccharide (i). The high mannose form (ii) may (rarely) carry 1-3 glucose residues as shown. The lactosaminyl form (iii) is shown as tetraantennary but they may involve fewer peripheral chains and they may not be complete (i.e., lacking the sialic acid or both sialic acid and galactose on one or more chains). The sialic acid may be linked α2-3 rather than α2-6. Some structures carry a fucose residue on the first *N*-acetylglucosamine of the core. Fucose may also replace sialic acid being linked α1-2 or 6 to galactose or it may be found linking α1,3 to the *N*-acetylglucosamine on a branch core mannose. Other structures have an *N*-acetylglucosaminyl residue (a bisecting GlcNAc) linked β1-4 to the β-mannose residue of the core. The hybrid type (iv) is a mixture of (ii) and (iii). The polylactosaminyl chain (v) is sometimes found instead of the single *N*-acetyllactosamine shown in (iii) and may be without a terminal sialic acid. The GAG of keratan sulphate I$_2$ is carried on a branched core mannose residue. The chain resembles a sulphated polylactosamine. Not all residues are sulphated. The structures in square brackets show the linkage to the core branch mannose.

$$NeuAc\alpha2-6GalNAc\alpha1-Ser(Thr)$$

$$
\begin{array}{c}
NeuAc\alpha2\diagdown_{6} \\
GalNAc\alpha1-Ser(Thr) \\
NeuAc\alpha2-3Gal\beta1\diagup^{3}
\end{array}
$$

FIGURE 4. Examples of simple O-linked glycans of the mucin type.

a. Glycans Linked to N-Acetylgalactosaminyl-O-Serine (Threonine)

A large number of structures have been described of this "mucin" type, ranging from simple di- and tetra-saccharides (Figure 4) to the very complex composed of up to 50 sugar residues. They usually contain galactose, N-acetylgalactosamine and sialic acid residues.

N-Acetylglucosamine may also be present often as part of a chain of repeating disaccharide units. Several of these chains are common to glycolipids. Type 1 is made up of repeating -[3Gal β1-3 GlcNAcβ1-] and type 2 consists of repeating -[3Galβ1-4 GlcNAcβ1-]. Type 2 chains give rise to the internal epitopes for the i and I antigens.[32,33] The former antigen consists of long unbranched repeats and the latter of long repeats occasionally branched through a GlcNAcβ1-6Gal link (Figure 5a). These also appear in glycolipid structures but whereas in man the glycolipids switch from i to I at about eleven years of age the I form predominates on glycoproteins even at birth.

The epitopes of several cell surface antigens are part of O-linked glycans of the mucin type. These epitopes include blood group determinants[34] (Figure 5b), stage-linked embryonic antigens[32] (Figure 5C) and several tumour antigens[35,36] (see Section V.B.7). The GAG of keratan sulphate II is also attached to proteoglycan core proteins of cartilage via this type of linkage (Figure 5d) (see, e.g., Hascall[25]).

b. Glycans Linked to Xylosyl-O-Serine

Only GAGs are attached to protein via this type of link. GAGs are made up of many repeating disaccharide units, each containing an acidic and an amino sugar which are often sulphated. The GAGs concerned are of chondroitin sulphate, heparan sulphate and dermatan sulphate. Summary structures are given in Figure 6. Further details are to be found elsewhere.[23,25-29]

c. Glycans Linked to Galactosyl-O-Hydrosylysine

This glycan has been found in collagen and the CIq component of complement. In most fibrous collagens it is simply the disaccharide unit Glcα1-2 Galβ- or even just the galactose residue. There is evidence for long chains of repeating disaccharide units on collagens (type iv) of basement membranes.

4. Phosphoethanolamine-Linked PI Glycans

An exciting development of recent years[18] has been the observation of a growing group of proteins anchored to the plasma membrane by phosphatidyl inositol (PI). Glycans bridge between a phosphoethanolamine on the protein and the PI to form a GPI link. The glycans share several of the features shown in Figure 7.

There are several unusual aspects of this structure. These include the ethanolamine phosphate linking the carboxyl end of the protein chain to the six position of a mannose residue, the presence of glucosamine with an unsubstituted amino group and the glycosidic bond between this sugar and a myoinositol residue to phosphatidyl inositol. (The presence of two myristate residues in the phosphatidyl moiety is also characteristic.)

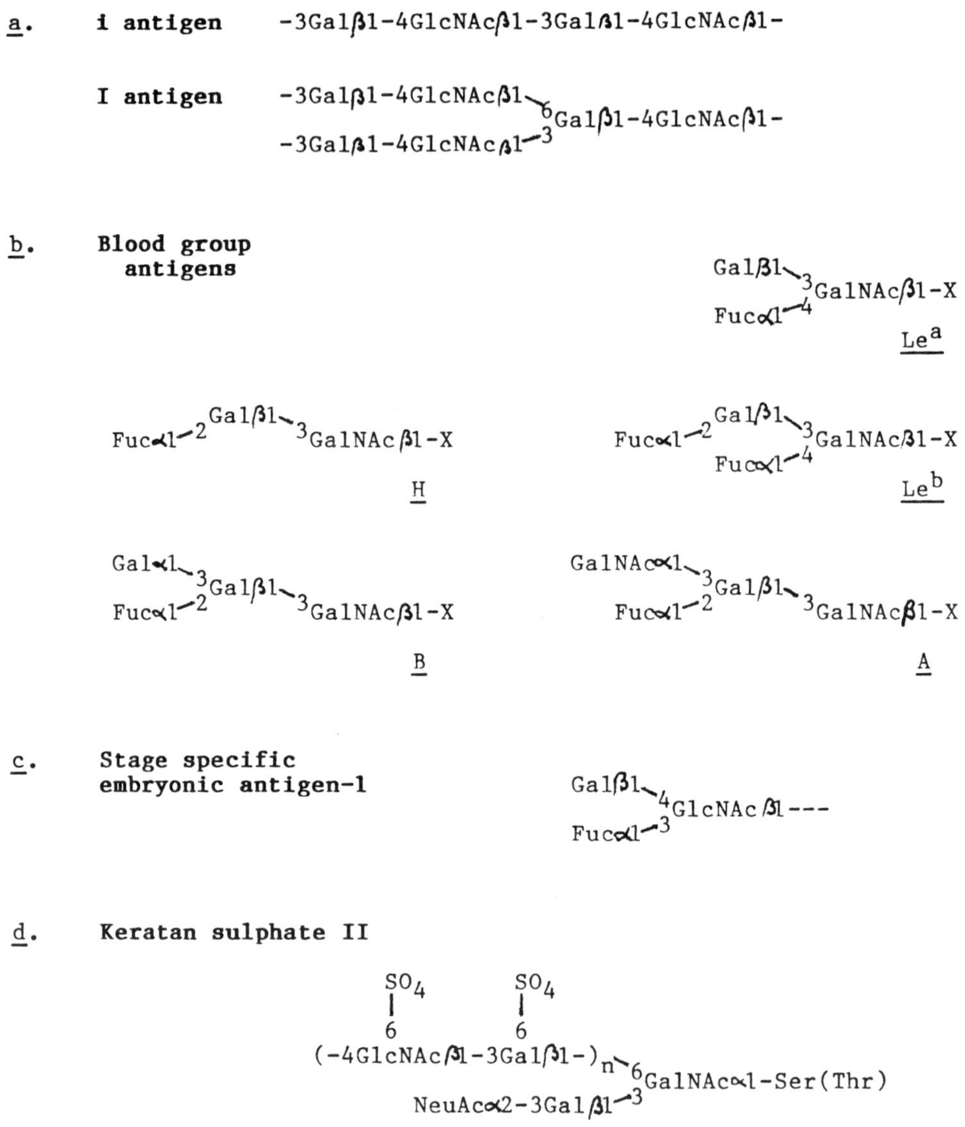

<u>a</u>.   **i antigen** $-3\text{Gal}\beta1-4\text{GlcNAc}\beta1-3\text{Gal}\beta1-4\text{GlcNAc}\beta1-$

I antigen

$$-3\text{Gal}\beta1-4\text{GlcNAc}\beta1\searrow_{6}$$
$$\text{Gal}\beta1-4\text{GlcNAc}\beta1-$$
$$-3\text{Gal}\beta1-4\text{GlcNAc}\beta1\nearrow^{3}$$

<u>b</u>.   **Blood group antigens**

$$\text{Gal}\beta1\searrow_{3}\text{GalNAc}\beta1-X$$
$$\text{Fuc}\alpha1\nearrow^{4}$$
$$\underline{\text{Le}}^{a}$$

$$\text{Fuc}\alpha1\nearrow^{2}\overset{\text{Gal}\beta1\searrow_{3}}{}\text{GalNAc}\beta1-X$$
$$\underline{\text{H}}$$

$$\text{Fuc}\alpha1\nearrow^{2}\overset{\text{Gal}\beta1\searrow_{3}}{\underset{\text{Fuc}\alpha1\nearrow^{4}}{}}\text{GalNAc}\beta1-X$$
$$\underline{\text{Le}}^{b}$$

$$\text{Gal}\alpha1\searrow_{3}$$
$$\text{Fuc}\alpha1\nearrow^{2}\overset{\text{Gal}\beta1\searrow_{3}}{}\text{GalNAc}\beta1-X$$
$$\underline{\text{B}}$$

$$\text{GalNAc}\alpha1\searrow_{3}$$
$$\text{Fuc}\alpha1\nearrow^{2}\overset{\text{Gal}\beta1\searrow_{3}}{}\text{GalNAc}\beta1-X$$
$$\underline{\text{A}}$$

<u>c</u>.   **Stage specific embryonic antigen-1**

$$\text{Gal}\beta1\searrow_{4}\text{GlcNAc}\beta1---$$
$$\text{Fuc}\alpha1\nearrow^{3}$$

<u>d</u>.   **Keratan sulphate II**

$$\overset{\displaystyle SO_4}{\underset{\displaystyle 6}{|}}\qquad\overset{\displaystyle SO_4}{\underset{\displaystyle 6}{|}}$$
$$(-4\text{GlcNAc}\beta1-3\text{Gal}\beta1-)_n\searrow_{6}\text{GalNAc}\alpha1-\text{Ser(Thr)}$$
$$\text{NeuAc}\alpha2-3\text{Gal}\beta1\nearrow^{3}$$

FIGURE 5.   Examples of complex peripheral glycans attached to the mucin type core O-linked to protein (see Figure 4).

Minor variations on the glycan structure concern the number of galactose residues (and their linkages) and the presence of up to two further phosphoethanolamine groups.

III. GLYCOSYLATION OF PROTEINS

A. GENERAL

Knowledge of the cell biology, biochemistry, and molecular genetics of the process of protein glycosylation is essential to an understanding of how the composition of the glycans concerned and their appearance on a protein is controlled. This leads to an appreciation of the consequences of aberration in the process. It also allows assessment of suitable targets for manipulation using inhibitors, mutants, and molecular biological techniques. This may be important in maintaining or changing the physicochemical and biological properties of

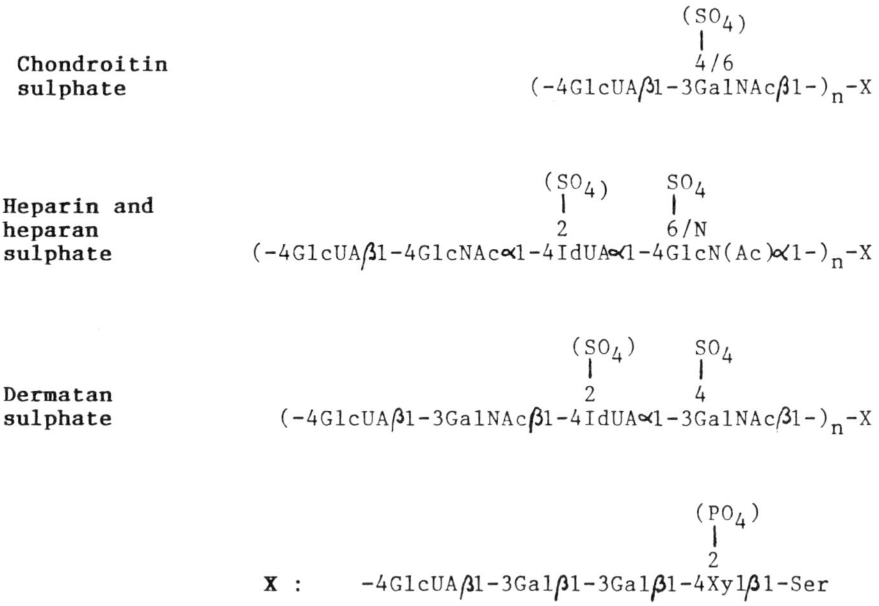

Chondroitin sulphate
(SO4)
4/6
(-4GlcUAβ1-3GalNAcβ1-)n-X
Heparin and heparan sulphate
(SO4)
SO4
2
6/N
(-4GlcUAβ1-4GlcNAcα1-4IdUAα1-4GlcN(Ac)α1-)n-X
Dermatan sulphate
(SO4)
SO4
2
4
(-4GlcUAβ1-3GalNAcβ1-4IdUAα1-3GalNAcβ1-)n-X
(PO4)
2
X :
-4GlcUAβ1-3Galβ1-3Galβ1-4Xylβ1-Ser

FIGURE 6. Summary structures of common glycosaminoglycans (GAGs) linked to protein via xylosyl-*O* serine. The repeating di (tetra) saccharide units are attached to xylose as in X forming a tetrasaccharide bridge to the serine. The presence of sulphate or phosphate in brackets indicates a substitution that is common but is not always found. The ratio of iduronic to glucuronic acid is distorted in these summary structures.

protein
ethanolamine
phosphate
6
Manα1-2 Manα1
(Galα1-2)Galα1
6Manα1-4GlcNH2α1-6myoinositol
3
6Galα1
2
(Galα1)
1
phosphate
6
dimyristoyl glycerol

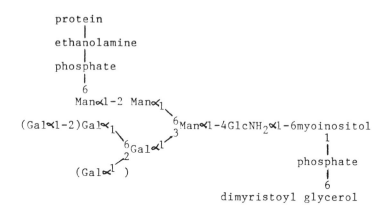

FIGURE 7. A glycanylphosphatidyl inositol (GPI) link. Structure proposed for the glycan bridge between the ethanolamine phosphate moiety of a variable surface glycoprotein (MIT at 1.4) and a PI anchor in the plasma membrane of *Trypanosomas brucei* (see also Figure 1, iii). Similar structures are found on mammalian cells.

key glycoproteins. Almost all mammalian and avian cell types seem to be capable of posttranslational glycosylation of proteins. In those cells actively secreting glycoproteins or subject to rapid membrane turnover or in need of lysosomal enzymes the process is particularly well developed. The obvious cell- and protein-related specializations of glycan assembly reflect differences in balance of features that are common to most cells. These features will be discussed in this section paying particular attention to *N*-glycosylation because knowledge is better developed in this area than in *O*-glycosylation or GPI assembly.

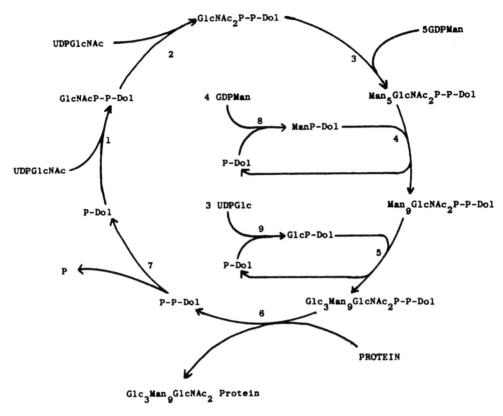

FIGURE 8. The phosphodolichol pathway of protein *N*-glycosylation. (From Hemming, F. W., *Biosci. Rep.*, 2, 206, 1982. With permission.)

B. GLYCOSYLATION IN THE ENDOPLASMIC RETICULUM

N- and *O*-glycosylation of proteins in animal cells differ in a number of ways that may have important biological consequences. In the first place *N*-glycosylation always precedes *O*-glycosylation: the former occurring in the rough endoplasmic reticulum, in many cases before the peptide chain is completely synthesized, whereas the latter is essentially a Golgi phenomenon. Both types of glycosylation are catalyzed by membrane-bound glycosyltransferases that show typical high specificity to monosaccharide, and its donor form, to acceptor and to the type of linkage formed (i.e., α- or β-, 1-2, 1-3, 1-4, etc.). However, in *N*-glycosylation the hydrophobic, membrane-bound phosphodolichol is used as an intermediary acceptor and donor of some monosaccharides and, as diphosphodolichol, for assembly and membrane anchor of a glycan (normally of fourteen monosaccharide residues) prior to its transfer *en bloc* to an asparagine residue in the peptide chain.[38]

As shown in Figure 8, three sugar donors formed in the cytoplasm are required for the assembly of this glycan: namely UDPN-acetylglucosamine, UDPglucose and GDPmannose. The first step is the reversible transfer of *N*-acetylglucosamine-1-phosphate from UDPN-acetylglucosamine to phosphodolichol. In effect one phosphate-phosphate bond is exchanged for another with little or no change in free energy. The next step is the first essentially irreversible and therefore committed step in the pathway. This second *N*-acetylglucosamine and the next five mannose residues are each donated directly from the appropriate nucleoside diphosphate sugar with the release of nucleoside diphosphate and the forging of a new glycosidic linkage. Bonds in both α- and β-configuration are formed. The next four mannose and three glucose residues are added indirectly from nucleoside diphosphate sugars via intermediary glycosyl phosphodolichol. There is an inversion of configuration at each step resulting in new glycosidic bonds in the α-configuration — the oligosaccharyl transferase

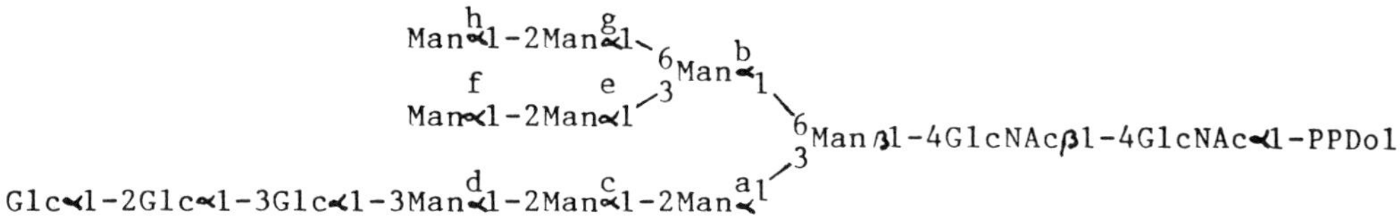

```
       h        g
    Manα1-2Manα1
                  \6   b
                    3Manα1
       f        e  /      \
    Manα1-2Manα1           \6
                             3Manβ1-4GlcNAcβ1-4GlcNAcα1-PPDol
                           /
                          /6
Glcα1-2Glcα1-3Glcα1-3Manβ1-2Manα1-2Manα1
                       d        c        a
```

FIGURE 9. The oligosaccharyl PPDol product of step 5 of the phosphodolichol pathway (Figure 8). Sugar residues transferred from phosphodolichol are underlined. The letters a-h indicate the most probable sequence of addition of mannose residues to the initial trisaccharide.

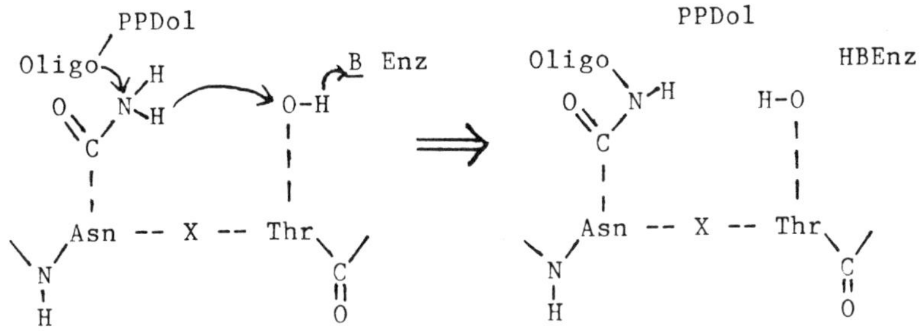

```
         PPDol                              PPDol
        /                                  /
 Oligo\   ,H           ,B  Enz      Oligo\   ,N-H      HBEnz
    O   N-H         O-H                O   N           H-O
     \\  \          |                   \\               |
       C            |                     C              |
       |            |          ⇒          |              |
       |            |                     |              |
    Asn -- X -- Thr\  /                 Asn -- X -- Thr\ /
     \  /           C                    \  /            C
      N             ||                     N             ||
      |             O                      |             O
      H                                    H
```

FIGURE 10. The mechanism of transfer of glycan to asparagine from oligosaccharyl diphosphodolichol. (After Bause, E. and Legler, G., *Biochem. J.*, 195, 639, 1981. With permission.)

catalyzing the *en bloc* transfer of the glycan to asparagine is most active with the three glucose residues (see Figure 9) present but it will transfer shorter glycans. The order of addition of monosaccharides, especially at steps 3, 4 and 5 is indicated in Figure 9. Further aspects of the biochemistry of these enzymes and the phosphodolichol pathway are discussed in subsequent sections.

The phosphodolichol pathway of *N*-glycosylation is thought to operate as in Figure 8 in all eukaryotic cells.[1-3,10,15] The necessity for asparagine to be part of the triplet -Asn-X-Thr(Ser)- for it to act as an acceptor in step 6 was mentioned in section II.A. Bause and Legler[39,40] have proposed the mechanism summarised in Figure 10 to explain this, on the basis of several *in vitro* studies on the oligosaccharyl transferase with short synthetic peptides as acceptors.

There are several consequences of the phosphodolichol pathway. One result is the presence of a pool of oligosaccharyl diphosphodolichol that allows rapid transfer of a bulky glycan to the new peptide chain rather than the more lengthy step by step assembly on the peptide. The advantage or necessity of this is not yet clear. It has also been suggested that the concentration of substrates and enzymes in a complex at the membrane surface contributes to the speed of glycosylation. Phosphodolichol has been implicated in the transfer of mannose and possibly glucose across the endoplasmic reticulum membrane (see section III.E.1). In some cells the very low concentration of phosphodolichol appears to be a rate-limiting factor in *N*-glycosylation (see section IV.E.3 and 5). Since *O*-glycosylation in animals does not involve phosphodolichol the pathway also allows *N*-glycosylation to be controlled independently of *O*-glycosylation.

Phosphodolichol is also involved in the *O*-mannosylation of yeast and fungal proteins. It will be of interest to learn if the incorporation of mannose into the PI linked glycan (see Figure 7 and section II 4) and also into the rare brain O-linked mannoproteins[21,22] of animal systems also involves phosphodolichol. Preliminary results from Hart's group[41-43] suggest that in trypanosomes glycan synthesis is initiated by transfer of *N*-acetylglucosamine from uridine diphosphate *N*-acetylglucosamine to common phosphatidyl inositol. Rapid deacetylation of this sugar then occurs followed by addition of mannose residues involving a lipid-soluble donor resembling mannosyl phosphodolichol. Ethanolamine phosphate is then added before the fatty acyl chains of the PI moiety are replaced by myristate. Since the formation of mannosyl phosphodolichol in animal cells appears to be carried out only in the endoplasmic reticulum it is likely that at least the early stages of synthesis of the GPI link also occur at this site.

In the case of most N-linked glycans of glycoproteins, processing commences while still in the endoplasmic reticulum (see Figure 11). This results from the action of two α-glucosidases to give the high mannose glycan (see Figure 3(ii)) which in some glycoproteins remains as an end product. Other glycans proceed along the processing pathway detailed in Figure 11 to produce the complex type of lactosaminyl chain shown in Figure 3(iii). The activity of most of these processing enzymes is expressed in the Golgi apparatus.

C. GLYCOSYLATION IN THE GOLGI APPARATUS

Glycosylation of lipids and proteins is a major activity of the Golgi apparatus. It is the site at which N-linked glycans are further processed (Figure 11) to produce the complex structures summarised in Figure 3. It is here that most *O*-glycosylation and glycosamino-glycan synthesis occurs. Reglycosylation of recycling, partially deglycosylated membrane glycoproteins is also probably a Golgi phenomenon.

Although small amounts of phosphodolichol can be isolated from Golgi membranes it appears that it does not function there as an intermediary carrier of either mono- or oligo-saccharide. In Golgi all glycosyltransferases catalyse the transfer of monosaccharide directly from the nucleotide donor to the final product so building up the new glycan one sugar residue at a time. In view of their high degree of specificity there are likely to be many different glycosyltransferases to be found.

Cytoplasmically-derived sugar donors include UDPgalactose, UDPglucuronic acid, UDPxylose, UDPN-acetylglucosamine, UDPN-acetylgalactosamine, GDPfucose and CMPsialic acids. All of these utilise carrier proteins to cross the Golgi membrane to reach the appropriate glycosyltransferase at the lumenal face of the membrane.[44] Glycosidases, phosphotransferases and sulphotransferases are also involved in this elaboration and processing of glycans. Understanding the arrangement and control of all of these proteins is a major challenge to modern molecular cell biology.

Progress on the distribution within the Golgi cisternae is an important step in this understanding. It is summarised in Figure 11. In the *cis*-cisternae the removal of α1-2 mannose residues begun in the endoplasmic reticulum, continues through the activity of Golgi α-mannosidase I. In the case of hydrolases destined for the lysosomes this does not occur but instead the glycan has added to the 6-position of some mannose residues *N*-acetylglucosamine-1-phosphate derived from UDPN-acetylglucosamine. This is catalysed by *N*-acetylglucosaminylphosphotransferase which is present only in Golgi *cis* cisternae. Further modification by *N*-acetylglucosamine-1-phosphodiester α-*N*-acetylglucosaminidase, also in the *cis* cisternae, leaves the glycan bearing just a mannose-6-phosphate group. The importance of this process will be discussed in section V.B.2.

Movement of glycoproteins, other than those destined for lysosomes, into the medial-cisternae exposes N-linked glycans to the action of *N*-acetylglucosaminyltransferase I. This catalyzes the addition of *N*-acetylglucosamine to the branch mannose of the pentasaccharide core linked α1-3. This is a prerequisite for the action of Golgi mannosidase II removing the final two "noncore" mannose residues. The medial-cisternae is also the site of action of *N*-acetylglucosaminyltransferase II which glycosylates the branch mannose linked α1-6 in the pentasaccharide core. Fucosyltransferase catalyzing the α1-6 fucosylation of the first *N*-acetylglucosamine residue in the core is found in both medial and trans cisternae.

The final elaboration of "antennae" of the complex N-linked glycans described in Figure 3 occurs in the trans cisternae where both galactosyltransferases and sialyltransferases are

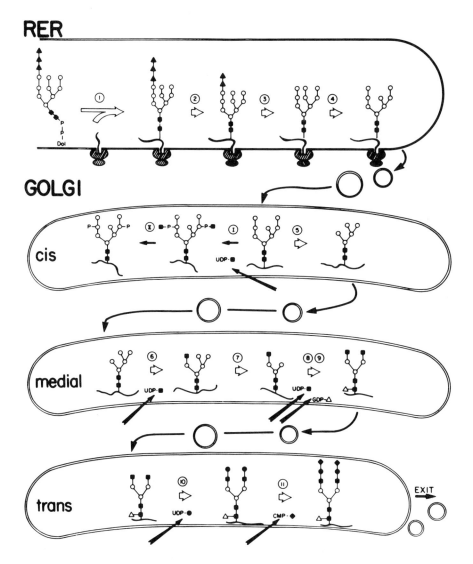

FIGURE 11.   The common pathway of processing of N-linked glycans. After transfer of the glycan to protein at step 1, the terminal glucose is removed by α-glucosidase-I (step 2). This is followed by α-glucosidase-II (step 3) which removes the remaining two glucose residues. Attack of the mannan commences in the rough endoplasmic reticulum through endoplasmic reticulum α1-2 mannosidase and continues in the *cis* cisternae of the Golgi through removal of three mannose by Golgi apparatus α-mannosidase-I (step 5). This then allows, in the medial cisternae of the Golgi, the addition of *N*-acetylglucosamine to the exposed core branch mannose (that linked α1-3 to β-mannose) through the activity of *N*-acetylglucosaminyltransferase-I (step 6). Golgi α-mannosidase-II removes the two remaining mannose residues on the α1-6 linked core mannose (step 7) to allow the addition of *N*-acetylglucosamine catalyzed by *N*-acetylglucosaminyl-transferase II (step 8). Step 9, the addition of fucose to the asparagine-linked *N*-acetylglucosamine is catalyzed by a fucosyltransferase before a galactosyltransferase in the *trans* cisternae of the Golgi catalyzes the addition of galactose (step 10) followed by the action of a sialyltransferase which catalyzes addition of the final sialic acid residues (step 11). I is the *N*-acetylglucosaminylphosphotransferase and II the *N*-acetylglucosamine 1-phosphodiester αN-acetylglucosaminidase involved in elaborating a mannose-6-phosphate signal on the glycan of lysosomal enzymes. (From Kornfeld, R. and Kornfeld, S., *Annu. Rev. Biochem.*, 54, 631, 1985. ©1985 by Annual Reviews, Inc. With permission.)

located. Whether or not, for example, tetraantennary (Figure 3) rather than diantennary (Figure 11) structures are formed depends on the multiplicity and specificity of *N*-acetyl-glucosaminyltransferases available at step 8 (Figure 11).

The synthesis of long repeating units such as those found in the polylactosaminyl regions of N- and O-linked glycans and in glycosaminoglycans raises particular questions of the relative dispositions and specificity of the glycosyltransferases concerned. In the case of GAGs these have been reviewed by Lindahl.[45,46] Reutter[47] has paid attention to the turnover of the glycan moieties of plasma membrane glycoproteins. The evidence suggests that the half life of the component monosaccharides varies, being shorter the closer they are to the nonreducing termini. This appears to be due to the action of exoglycosidases coupled with internalization of the glycoproteins through an endosome compartment to Golgi (possible primarily *trans* cisternae) where reglycosylation occurs before the glycoprotein is recycled to the plasma membrane.

D. GLYCOSYLATION IN OTHER REGIONS OF THE CELL
1. N-Linked Glycosylation
There has appeared over the last few years increasingly convincing evidence for protein glycosylation in nuclei, mitochondria, and cytoplasm.[19] There is little doubt that the major part of protein *N*-glycosylation commences in the rough endoplasmic reticulum and is completed in the Golgi apparatus. There is good evidence for the presence of phosphodolichol in other membranes but accurate, reliable assessment of this is clearly dependent upon the preparation of pure membranes uncontaminated with endoplasmic reticulum and Golgi membranes. This problem of membrane crosscontamination also complicates studies on the subcellular distribution of enzymes of protein *N*-glycosylation. However several reports[48,49] of very careful work attempting to eliminate these complications describe strong evidence for enzymes of the phosphodolichol pathway in mitrochondria and nuclei. The nature of the proteins so glycosylated and the importance of these observations remain to be assessed.

2. *N*-Acetylglucosaminylation of *O*-Serine Residues
Hart[41-43] and others have demonstrated quite recently the addition of single *N*-acetylglucosamine residues to serine residues of some (glyco) proteins. The process appears to occur on the cytoplasmic face of an unidentified membrane. The protein acceptors include components of nuclear pores, chromatin and the cytoskeleton.

E. MEMBRANE TOPOLOGY OF GLYCOSYLTRANSFERASES
1. Endoplasmic Reticulum
Detailed evidence for current views on the topological distribution of enzymes of the phosphodolichol pathway has been discussed by Lennarz[50] and by Hirschberg and Snider.[44] These are summarised in Figure 12 and explained briefly in the rest of this section.

Hirschberg and others have identified transporter (antiporter) systems for moving UDPN-acetylglucosamine and UDP glucose across the membrane from cytoplasm to lumen of rough endoplasmic reticulum. The distribution in the membrane of intermediates and enzymes of the phosphodolichol pathway has been probed with a glycosyltransferase, lectins or proteolytic enzymes.[50] A comparison of the effects of these treatments on whole microsomes with those on microsomes rendered leaky by dilute solutions of detergent has provided important results. Further evidence for the involvement of phosphodolichol and GDP mannose:phosphodolichol mannosyltransferase in the transport of mannose across the membrane comes from work in Tanner's laboratory[51] using reconstituted model membranes. The possibility that interpretation of some of the results of the experiments could be complicated by small areas of the rough endoplasmic reticulum being in a form more closely related to H^2 phase than to bilayer is discussed in Section III.F.2.

2. Golgi Apparatus
Membrane transporters for all of the nucleotide sugars needed for Golgi glycosylation have been reported to be present in Golgi membranes.[44] This is consistent with evidence

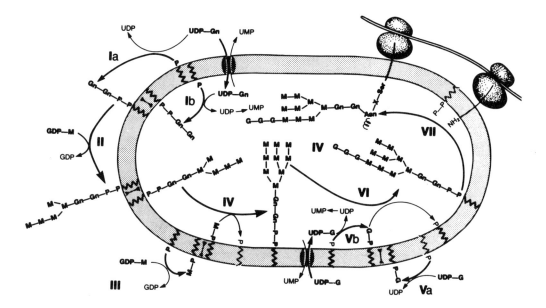

FIGURE 12. Membrane topology of enzymes of the phosphodolichol pathway. (From Hirschberg, C. B. and Snider, M. D., *Annu. Rev. Biochem.*, 56, 63, 1987. ©1987 by Annual Reviews, Inc. With permission.)

that all glycosyltransferases of this organelle are located at the cisternal (lumenal) surface. Sulphotransferase also function on this side of the membrane and an antiporter system has been proposed for PAPS.

F. GLYCOSYLTRANSFERASES

The large number of different glycosidic linkages catalyzed by glycosyltransferases and the wide variety of acceptors suggests the number of these enzymes to be very large. They are all membrane bound and because of the need for a concerted sequential activity it is likely that many of them exist as part of a multienzyme complex. In many cases it appears that the amount of enzyme present is very small. The hydrophobicity of regions of each enzyme causes them to bind strongly to other proteins and to aggregate if separated from phospholipids or solubilising detergent. All of these factors make it difficult not only to study in isolation the activity of any one enzyme using natural membranes but also to isolate pure, active glycosyltransferases. Some success has been achieved especially with those enzymes involved in adding peripheral sugar residues in the Golgi. In some cases this may have been aided by proteolytic release from the hydrophobic domain. However, in general in order to obtain sufficient quantities of enzymes in good condition for detailed study it will be necessary to take advantage of molecular biological techniques. Despite these problems some interesting aspects of these enzymes and of other sugar-binding proteins (see Section V.B.1) have emerged. Beeley[9] and Beyer et al.[52] have reviewed the properties of several Golgi glycosyltransferases extensively.

Not many of these enzymes have been isolated from the endoplasmic reticulum in a pure state. However two mannosyltransferases have been described by Schutzbach.[53] Both of these when pure are inactive but activity can be restored by addition of phospholipid. This reactivation is very specific to phospholipid structure showing a requirement for the H^2 phase-former phosphatidylethanolamine (especially with unsaturated fatty acids). Bilayer-forming phospholipids such as phosphatidyl choline will not reactivate these enzymes and indeed it is possible to titrate out their activity by adding increasing quantities of phosphatidyl choline to enzyme already activated by phosphatidyl ethanolamine. It is known that endoplasmic reticulum is enriched with phosphatidyl ethanolamine, compared with other natural membranes. It is also the main source of newly synthesised phosphodolichol and dolichol, both of which also tend to favour conversion of bilayer to H^2 phase. This raises the possibility that some enzymes of the phosphodolichol pathway function naturally in a membrane environment that may not always or entirely be in the form of a classical bilayer. A recent preliminary report on the enzyme catalyzing step 1 of the phosphodolichol pathway (Figure 8) suggests a preference for phosphatidylcholine.[54] It will be interesting to learn the phospholipid preferences of other enzymes of this pathway.

Progress has also been made on the isolation of oligosaccharyltransferase, primarily by Lennarz's group.[55]

IV. REGULATION OF PROTEIN GLYCOSYLATION

A. GENERAL

The previous section presents many possibilities of regulating protein glycosylation. The protein structure not only provides a site for glycosylation but carries other information that influences the enzymes modeling and processing the glycans to its final form. In addition control of glycosylation may involve regulation of the expression of the activity of glycosyltransferases, carrier proteins, glycosidases, or of enzymes influencing the availability of nucleotide sugars. In the case of N-glycosylation factors affecting the role of phosphodolichol may be crucial. Less directly, effectors of more general changes in the properties of membranes in which the enzymes function may well also affect enzyme activity. The importance of some of these factors is discussed in this section.

B. GENETIC CONTROL OF GLYCOSYLTRANSFERASES INVOLVED IN THE FORMATION OF BLOOD GROUP DETERMINANTS

Probably the best description of the genetic control of expression of individual glycosyltransferases is that concerning blood group determinants, summarised in Figure 13.[28]

C. MUTANTS
1. Human Mutants

No human genetic defects in O-glycosylation or in the phosphodolichol pathway of N-glycosylation have been described. Possibly mutations in the latter would be lethal. In fact the only mutation so far described leading to a clinical phentoype involves the enzyme N-acetylglucosaminyl-phospho-transferase — required for generation of the mannose-6-phosphate moiety on lysosomal enzymes (Figure 11, section IIIC).[56] There is a marked lack of this phosphotransferase activity in cells of patients with mucolipidosis II (I-cell disease) and a less marked deficiency in mucolipidosis III. Genetic heterogeneity is apparent and at least two complementation groups have been established. One variant of mucolipidosis III shows almost normal phosphotransferase activity with an artificial substrate but an almost complete lack of activity with a natural glycoprotein substrate. Presumably this mutant enzyme has a normal active site but has an aberrant further glycoprotein binding site. Such mutations may shed light on the mechanism of action of the natural enzyme.

2. Mutant Animal Cell Lines

Glycosylation mutations in animal cell lines have been fully reviewed by Stanley.[57] The resultant phenotypes show clonal inheritance and stability in absence of selection supporting a single mutation event. However there has been no molecular verification of a mutation either on the basis of DNA or protein analysis. All of those (just over 20) enzymes observed to be affected have been in the N-linked pathway. Approximately one third concern enzymes of the phosphodolichol pathway, approximately one half affect enzymes involved in elaboration of complex structures and a few are concerned with processing of the glycan. Mutants

Blood group antigen synthesis:

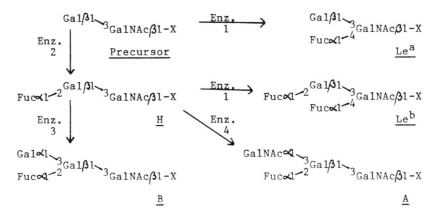

Enzyme 1: ∝1-4fucosyltransferase

Enzyme 2: ∝1-2fucosyltransferase

Enzyme 3: ∝1-3galactosyltransferase

Enzyme 4: ∝1-4N-acetylgalactosaminyltransferase

Blood group A requires expression of genes for enzymes 2 and 4

Blood group B requires expression of genes for enzymes 2 and 3

Blood group H requires expression of gene for enzyme 2

Blood group Le[a] requires expression of gene for enzyme 1

Blood group Le[b] requires expression of genes for enzymes 1 and 2

Appearance of blood groups on secreted glycoproteins requires expression of Se gene.

FIGURE 13.   Genes and glycosyltransferases controlling the elaboration of blood group determinants.

have been described that have an enzyme deficiency at steps 1, 4, 5, 6, and 9 in Figure 8 and steps 3, 5, 8, 9, 10, and 11 in Figure 11. Mutants with elevated activity at step 1 (Figure 8) and 9 and 11 (Figure 11) have also been reported.

3. Glycosylation Mutants in Yeast

The genetics of *S. cerevisiae* is in many respects better understood than for animal cell lines. It has been a useful organism for developing glycosylation mutants and because of the many common features of protein *N*-glycosylation across all eukaryotic organisms it is an excellent model for animal studies. Several mutants depleted in enzymes of the phosphodolichol pathway have been described[16,58] — the *alg* (*a*sparagine-*l*inked-*g*lycosylation) mutants. Relating to Figure 8 mutants are described for steps 1, 3 (1 each for 2 enzymes), 4, 5 (1 each for 2 enzymes) and 9.

Of the ALG genes, responsible for the enzymes deficient in the alg mutants, the ALG 7 gene codes for the *N*-acetylglucosamine-phosphate transferase of step 1. It is capable of rescuing cells from the lethality of prolonged treatment with tunicamycin, an inhibitor of step 1 (see section IV.E.3). The ALG 7 gene has been cloned, sequenced and its transcripts (two) have been studied. Two of the mutants, alg 1 and alg 2 (affecting mannosyltransferase I and II respectively at step 3, Figure 9) show temperature-sensitive conditional lethality.

The remaining mutants are not lethal but result in accumulation of incomplete oligosaccharyl diphosphodolichol. The alg 4 mutation resulted in 6 alleles, all defective in synthesis of oligo PPDol leading to accumulation of a range of products of intermediate size Man_5-Man_8 $GlcNAc_2$ PPDol. It is suggested that the ALG 4 gene product is a multifunctional protein of the endoplasmic reticulum and is not a transferase.

It appears from this that in yeast glycans of size Man_5 $GlcNAc_2$ and greater can be transferred from oligo PPDol to protein and that cell viability is retained. Smaller glycans such as $GlcNAc_2$ and Man_{1-2} $GlcNAc_2$ which accumulate in alg 1 and alg 2 mutants respectively, can also be transferred but lethality results. Also proteins other than transferases and which affect multiple steps are required for efficient transfer to protein. The transfer of Man_{1-2} $GlcNAc_2$ to protein is inconsistent with the topology model (Figure 12) suggesting, perhaps, that the distribution of intermediates across the membrane is not as clear cut as implied by the model, at least in yeast.

Several mutants (the mnn mutants) affect steps in the Golgi that extend N-linked oligosaccharides. However, these steps are not useful as a model for animal systems since they involve polymannosylation and some phosphorylation rather than elaboration of the "complex" chains of animal cells.

D. MOLECULAR BIOLOGY OF GYCOSYLTRANSFERASES

The enormous number and high specificity of glycosyltransferases has stimulated attempts to clone these enzymes. One objective is to obtain sufficient quantities of pure enzyme to allow detailed studies of structure-function relationships comparing, for example, carbohydrate recognition sites. Although progress in this area is still quite slow, several successful cloning experiments have now been reported. For example, cDNAs have been isolated and characterized for N-acetylglucosaminide β-1,4-galactosyltransferases[59-61] and for a β-galactoside α-2,6-sialyltransferase.[62] This allows transfection and overexpression of a glycosyltransferase in a cell to study the consequences of controlled endogenous enzymic manipulation of a glycan chain to be examined.[62]

E. INHIBITION OF PROTEIN GLYCOSYLATION
1. General

Inhibition of protein glycosylation has been an important approach by those attempting to unravel its biological importance. There is also good evidence for this being an important natural phenomenon. Endogenous protein inhibitors of a sialyltransferase,[63] of glucosyl- and galactosyltransferases[64] and of an N-acetylgalactosaminyltransferase[65] have been described. Some of these have been implicated in influencing both glycoprotein and glycolipid biosynthesis and their possible importance in the modulation of these activities during embryonic development has been discussed.[63] Some endogenous proteins activate enzymes. A long-known example of this is the influence of α-lactalbumin on a mammary gland galactosyltransferase.[66] In the absence of α-lactalbumin the transferase efficiently transfers galactose from UDP galactose to β-N-acetylglucosamine residues of glycoproteins but not to glucose. In the presence of α-lactalbumin the affinity of the enzyme for glucose is much increased such that lactose synthesis occurs. Other nonproteinaceous endogenous activators and inhibitors of glycosyltransferases ranging from lipids to metal ions have been described.

Various strategies have been used by researchers for blocking protein glycosylation. These are best exemplified by reference to N-protein glycosylation in which area most progress has been made. Several detailed reviews have appeared on inhibition of protein N-glycosylation,[66-69] and therefore this section will concentrate on general principles with examples, especially of those that have helped substantially our understanding of the regulation of protein N-glycosylation. It is convenient to consider these in terms of targeting different aspects of this biosynthesis process.

2. Targeting the Asn-X-Thr (Ser) Triplet

Using molecular biological techniques it is now possible to engineer unfavourable changes at all three sites in this triplet. It is unlikely that the triplet is part of the active site of biological function if it is also normally glysosylated (unless the glycan is involved directly in the function). Consequently substitution by an amino acid of similar type (e.g., glutamine for asparagine) should not be too disruptive of biological activity. Conversely, if greater N-glycosylation is required it should be possible to engineer an asparagine or threonine (serine) in a β-loop to create an extra available glycosylation site.

The technique of site-directed mutagenesis has been used[70] for example, to explore the need for the two N-glycans on the VSV G-protein for its proper export. Unlike the use of inhibitors this allowed investigation of each glycan separately. It became clear that either of these chains alone was sufficient for G-protein export. The same technique was used to introduce[71] new consensus triplets in the peptide chain which were duly glycosylated. This approach showed six of eight new sequences were actually glycosylated and that two of these was sufficient for function. It also showed that overglycosylation (i.e., at more than two sites) interfered with expression of wild-type protein.

An alternative approach using analogues of both asparagine and threonine has been used. For example, β-hydroxynorvaline instead of threonine was incorporated into the protein of cathepsin D by fibroblasts and found to block N-glycosylation such that aglycocathepsin D did not reach the lysosome and was rapidly degraded.[72] Threo-β-fluorasparagine also blocks N-glycosylation in some cells, probably by replacing asparagine in newly synthesized proteins.[73] Although the inhibitory effects of these analogues on N-glycosylation probably results from direct substitution in the Asn-X-Thr- sequon, their incorporation into numerous positions in nascent peptide chains may well result in other less specific effects.

3. Targeting the Phosphodolichol Pathway

Tunicamycin has been one of the most useful inhibitors in exploring the importance of N-glycosylation. The structure of this compound can be seen as a bisubstrate analogue of UDPN-acetylglucosamine:phosphodolichol N-acetylglucosaminyl phosphate transferase (see Figure 14), the enzyme catalyzing step 1 of the phosphodolichol pathway (Figure 8). It is a powerful, competitive and very specific inhibitor of this enzyme and of analogous enzymes in bacteria.[66,74] Tunicamycin is available commercially as a natural mixture by virtue of the presence of different fatty acyl chains, the composition of which appears to vary with strain and growth conditions of the *Streptomyces lysosuperificus* used as source. Some of these different fatty acyl derivatives are less specific inhibitors and influence protein synthesis. In general most tunicamycin mixtures if used with cell cultures or tissue slices at low concentrations (<1 μg/ml) for as short a period as possible are very specific to step 1. However, if being used with a new system (or if a new batch of tunicamycin) it is wise to test the inhibitor's effect on both protein synthesis and N-glycosylation. The use of tunicamycin to elucidate the role of N-linked oligosaccharides on different glycoproteins is summarised in Section V.C.1.

Several glycolipid analogues of tunicamycin have been described which are also strong, specific inhibitors of step 1 in Figure 8. One of these, corynetoxin, is of agricultural importance.[76,77] *Corynebacterium rathayi* is a common infection of nematodes which sometimes invade seedheads of ryegrass. The toxin, produced by the bacterium, is toxic to sheep that ingest the infected ryegrass by virtue of its inhibiting this particular form (N-glycosylation) of posttranslational modification.

Most inhibitors used to block other steps in the phosphodolichol pathway are less useful than tunicamycin because they can be used only in cell-free preparations, are required in higher concentrations, or are less specific. They are summarised in Table 2 and discussed in detail by Elbein[66] and by Schwarz and Datema.[67,68] Most of these inhibit the enzymes catalyzing the steps indicated. However, the sugar analogues 2-deoxyglucose (DG), 2-fluoro-

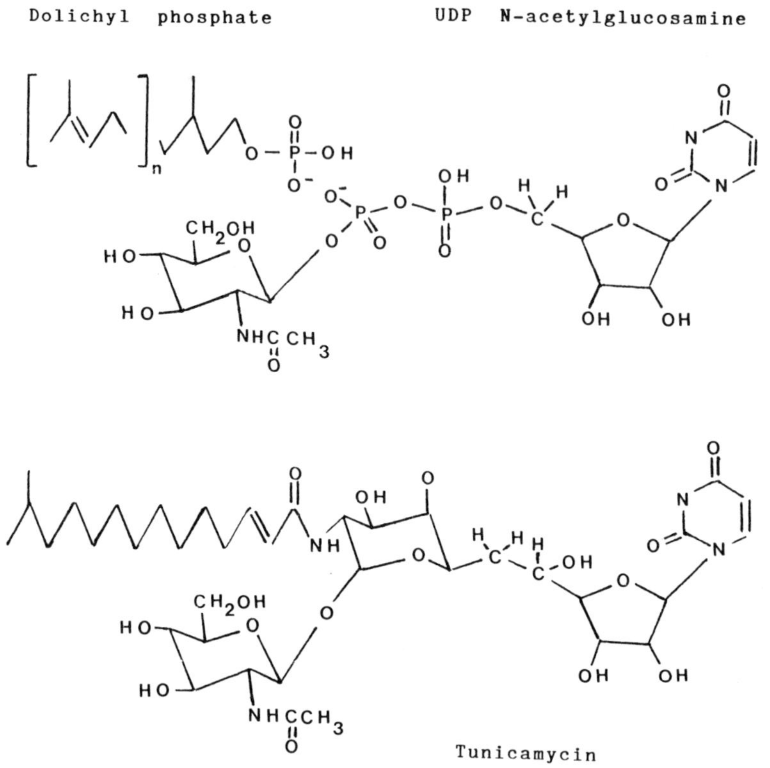

Dolichyl phosphate   UDP N-acetylglucosamine

FIGURE 14.   The structure of tunicamycin (v), phosphodolichol and UDPN-acetylglu-cosamine drawing attention to tunicamycin functioning as a bisubstrate analogue.

TABLE 2
Inhibitors of Steps in the Phosphodolichol Pathway (Figure 8)

Step 1   *tunicamycin*,[a,b,c,e] amphomycin tsushimycin),[a,d,f] diumycin[a,d,f]
Step 2   bis-(*p*-nitrophenyl)phosphate[a,d,g]
Step 3   2-deoxyglucose,[b,f] bis-(*p*-nitrophenyl)phosphate[a,d,g]
Step 4   ?glucosamine[b,f]
Step 7   *bacitracin*[a,d,f]
Step 8   *amphomycin*,[a,d,e] *diumycin*,[a,d,g] 2-deoxyglucose[b,f]
Step 9   *showdomycin*,[a,d,f] ?amphomycin,[a,d,e] bis-(*p*-nitrophenyl)phosphate[a,d,g]

Note: The italicizing of an inhibitor suggests that that step is its main point of action.

[a]   Active in cell-free preparations.
[b]   Active against whole cells.
[c]   Highly specific.
[d]   Moderately specific.
[e]   Active at 1 μM.
[f]   Active at 1 mM.
[g]   Active at >1 mM.

2-deoxyglucose, 2-fluoro-2-deoxymannose exert their inhibitor effects by metabolism and incorporation into intermediates of the phosphodolichol pathway. Thus DG is incorporated by cultured cells via GDPDG into DG-(GlcNAc)$_2$ PPDol and DGMan$_2$(GlcNAc)$_2$ PPDol, both of which cannot accept further sugar residues, and therefore accumulate, sequestering available phosphodolichol. It seems likely that DG will have several secondary effects through inhibition of glucose metabolism.

4. Targeting Protein Biosynthesis

Clearly blocking protein biosynthesis (e.g., with cycloheximide or actinomycin D) will deny the availability of nascent peptide for glycosylation and hence inhibit glycoprotein biosynthesis. In addition in whole cells, inhibition of protein synthesis has been reported to block the phosphodolichol pathway close to step 3 (Figure 8). It is suggested that this feedback regulation may be due to elevated levels of GTP.[66] This approach is not very selective towards glycosylation.

5. Targeting Phosphodolichol Biosynthesis

Phosphodolichol is formed from acetate by the pathway summarised in Figure 15.[38] In common with the biosynthesis of all compounds of isoprenoid origin this involves hydroxy-methylglutaryl coenzyme A (HMG CoA) being converted to mevalonate (MVA) through the activity of HMGCoA reductase. This step and the pathway as far as farneysl pyrophosphate (FPP) are shared with the biosynthetic routes to cholesterol and ubiquinone. In many cells the biosynthesis of cholesterol is responsible for 99% of the metabolic flux through FPP. The activity of HMGCoA reductase is rate limiting in the production of FPP and hence is a key enzyme in the control of the synthesis of all three products in Figure 16.

A cholesterol-rich diet leads to an influx of cholesterol into cells that initially causes a rapid and marked but incomplete loss of activity of HMGCoA reductase by, indirectly, turning down gene expression of this short lived enzyme. After two weeks on this diet inhibition of step B also occurs. In some instances this appears to cause a diversion of increased amounts of farnesyl pyrophosphate along route C resulting in an elevated concentration of phosphodolichol and increased protein N-glycosylation.[78]

Compactin and mevinolin are two powerful, highly specific inhibitors of HMGCoA reductase that can permeate cells and tissues at reasonable concentrations. Their presence can lead to a complete block of step A, uncomplicated by other effects. This, of course, inhibits bioysnthesis of all three produts in Figure 15. However, it can be made more specific to phosphodolichol and protein N-glycosylation by administering exogenous cholesterol (and ubiquinone) although the possibility of complications due to blocking isoprenylation of nucleotides and of ''MVA-derivatised'' substitution of proteins should not be ignored. An example of the use of compactin is given in section V.B.6.

The work of Lennarz[79] has shown that the sensitivity of N-glycosylation to the concentration of phosphodolichol is dependent upon the nature of the protein and cell type (Table 3). For example, exogenous phosphodolichol markedly stimulated pancreatic N-glycosylation of ribonuclease primarily because the extent of glycosylation of ribonuclease is normally less than 20% of maximum. On the other hand in oviduct slices, although the glycosylation of several glycoproteins does respond to exogenous phosphodolichol, the N-glycosylation of ovalbumin is not affected simply because it is normally fully N-glycosylated.

6. Targeting the Availability of Energy

The uncoupling of oxidative phosphorylation by 10 mM carbonyl-cyanide m-chlorophenylhydrazone shows the phosphodolichol pathway to be peculiarly sensitive to the energy charge of the cell concerned. In some cells,[80] the formation of ManPDol is blocked (step 8, Figure 8) but all other steps function resulting in the production of proteins carrying Glc_3 Man_5 $GlcNAc_2$ on asparagine residues (the ''alternative pathway''). The same treatment (also antimycin A or replacing air with nitrogen) of thyroid slices resulted in Man_8 $GlcNAc_2$ glycan.[81] Glucose starvation of cultured cells also brings about the alternative pathway and synthesis of N-linked Glc_3 Man_5 $GlcNAc_2$ glycan.[82,83]

It has been suggested that some of these effects may be due to a depletion of GTP and hence of GDPmannose. It has also been proposed that a lack of energy charge disrupts recycling to the plasma membrane of the glucose receptor — thus explaining the common result of glucose starvation and uncoupling of oxidative phosphorylation.

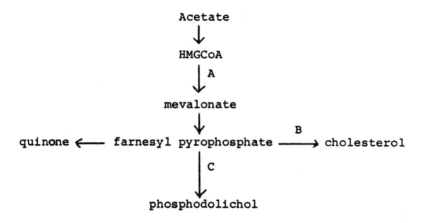

Acetate
↓
HMGCoA
↓ A
mevalonate
↓
quinone ⟵ farnesyl pyrophosphate ⟶ᴮ cholesterol
↓ C
phosphodolichol

FIGURE 15.   The biosynthetic pathway of phosphodolichol showing its relationship to cholesterol and ubiquinone synthesis. (From Hemming, F. W., *Biosci. Rep.*, 2, 215, 1982. With permission.)

TABLE 3
The Effect of Adding Phosphodolichol to Tissue Slices, an Abbreviated Summary Based on the Work of Lennarz's Group[70]

Bovine pancreas: [^{14}C-]amino acid into RNAse — no change
[^{3}H-]Man into RNAse — increase $4 \times$
(Normally bovine RNAse is only 10% glycosylated)

Oviduct: [^{35}S-]Met into general glycoproteins — no change
[^{3}H]Man into ovalbumin — no change
[^{3}H]Man into general glycoproteins — increase $2 \times$
(Normally ovalbumin is fully glycosylated)

7. Targeting Glycosidases of *N*-Glycan Processing

The inhibition discussed under sections IV.E.2 to 5 blocked production of the whole glycan: an all or nothing effect. Section 6 introduced the possibility of effecting a change in the structure of the glycan. This possibility has been taken much further by aiming inhibitors at steps in the processing of the glycan. The glycosidases concerned have been the prime target and following the discovery that two natural plant products are effective inhibitors of these, a number of other inhibitors have been synthesised chemically.

The first natural inhibitor identified was swainsonine, an alkaloid of *Swainsona* species in Australia, and of *Astralagus* species in the United States (reviewed by Elbein[66]). If these plants are eaten by cattle the swainsonine causes locoism characterized by unsteady gait and eventual death. Swainsonine is a powerful inhibitor of lysosomal α-mannosidase and of Golgi α-mannosidase II (step 7, Figure 11). It has no effect on other processing mannosidases. The N-linked glycan structure resulting from the use of swainsonine is usually of the hybrid type (Figure 16). The biological significance of this sort of change is discussed in section V.E.1.

It is suggested that swainsonine exists in a protonated form that structurally resembles the mannosyl cation formed during hydrolysis of mannosides and that this forms the basis of its inhibitory action. Chemically synthesized swainsonine is now available, as are several isomers. One of these, Glc-swainsonine, exists in a protonated form resembling the glucosyl cation. Interestingly, this compound did not affect lysosomal α-mannosidase or the processing glucosidases I or II of the endoplasmic reticulum but it did inhibit Golgi α-mannosidase II and fungal α-glucosidase.

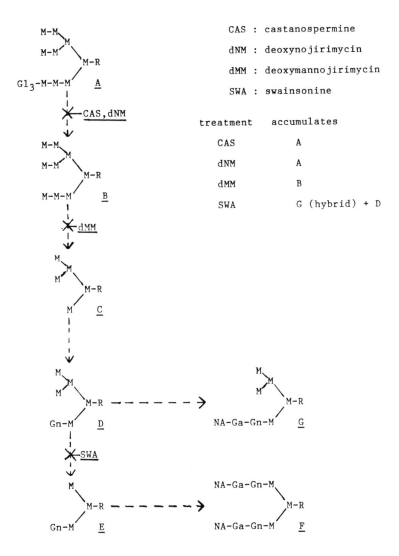

CAS : castanospermine

dNM : deoxynojirimycin

dMM : deoxymannojirimycin

SWA : swainsonine

treatment	accumulates
CAS	A
dNM	A
dMM	B
SWA	G (hybrid) + D

FIGURE 16.   N-linked glycan structures resulting from the use of processing inhibitors.

Another processing inhibitor of agricultural importance in Australia is the alkaloid castanospermine, isolated from the fruit of *Castanospermum australe*.[66] On ingestion of this material, animals suffer severe gastrointestinal symptoms and may die. Castanospermine is a powerful inhibitor of several α-glucosidases including intestinal sucrase, maltase and especially the processing α-glucosidases I and II of the endoplasmic reticulum (steps 2 and 3, Figure 11). This leads to a blocking of complex oligosaccharide formation from the N-linked glycan $Glc_3 Man_9 GlcNAc_2$ — although some processing to $Glc_3 Man_7 GlcNAc_2$ may occur (see Figure 16).

The glucose analogue containing a ring NH group instead of an O atom and called deoxynojirimycin is synthesised by *Bacillus* sp. and can also be made chemically. It is active against both processing α-glucosidases in cell free preparations and whole cells. It has a similar effect on glycan structure to that of castanospermine (reviewed by Elbein[66]). Deoxy-mannojirimycin, the mannose analogue of deoxynojirimycin, is an effective inhibitor of Golgi α-mannosidase I (step 5, Figure 11) giving rise to $Man_{8-9} GlcNAc_2$ glycans. More detailed descriptions of these and other inhibitors are provided by Elbein.[66]

8. Targeting Golgi Trafficking

Monensin and other ionophors are known to disrupt the movement of vesicles from *cis* to *trans* Golgi cisternae possibly through depletion of Ca^{++} levels required for fusion of vesicles with cisternal membranes. This causes glycoproteins to avoid the terminal glycosylation steps leading to complex glycan formation.[66]

9. Targeting *O*-Glycosylation

In contrast to the *N*-glycosylation and processing pathway, very few useful inhibitors of *O*-glycosylation have emerged that can be used with whole cells. Several potential exogenous acceptors will act as competitive inhibitors in cell free preparations but most of these fail to permeate the plasma membrane. One exception is the group of β-xylosides especially *p*-nitrophenyl-β-xyloside and 4-methylumbelliferyl-β-xyloside which at 1 m*M* effectively inhibit xylose-linked proteoglycan synthesis although the elaboration of GAG chains on the xyloside is enhanced.[84] Presumably the hydrophobic moiety of these compounds assists their permeation of the plasma and Golgi membranes. At much higher concentrations (40 m*M*) xylose is also inhibitory.

V. STRUCTURE-FUNCTION RELATIONSHIPS OF GLYCANS

A. GENERAL

Glycan chains of glycoproteins are generally bulky, hydrophilic structures. The range of possible glycosidic bonds between adjacent monosaccharides in a glycan is quite large giving rise to an enormous potential diversity of glycan structures. This has encouraged the concept of these structures carrying information extra to that in the peptide chain. In fact perhaps fewer than 10% of the potential structures are formed but the diversity and information content of these is still high. This may be of biological significance if suitable systems exist for interpreting the information.

Because of the hydrophilicity of the glycan in aqueous medium it will always be on the outside either of the folded protein chain of a soluble glycoprotein or on the outside of a cell membrane in the case of a plasma membrane glycoprotein. In either case any information in the glycan is clearly well exposed to be read. There have emerged a few but important examples of this sort of role of the glycan moiety.

It is also suggested that especially in the case of multiglycosylated proteins, the hydrophilic glycan groups may well influence the folding of the peptide chain and, in the case of plasma membrane glycoproteins, the orientation in the lipid bilayer. However, only in rare cases has this been shown to influence the activity of the active site of the protein.

The bulkiness of the glycan chains especially in multiglycosylated proteins has also been seen as protective against proteolytic enzymes. There are a number of observations of this role.

Finally, the negative charge on the sialo- and sulpho-glycans appears to play a role in maximizing the area of influence of the glycoprotein by causing it to occupy the maximum volume. These charges also order the structure of nearby water molecules causing gel formation and they sequester important cations such as Ca^{++} and Mg^{++}. Evidence has been described for all of these roles.

B. BIOMOLECULAR RECOGNITION OF GLYCANS
1. Lectins and Related Carbohydrate-Binding Proteins

Lectins are formally defined as proteins of nonimmune origin that precipitate glycoconjugates and agglutinate cells, these activities being inhibited by carbohydrate.[85-87] Since these activities depend on cross-linking it follows that lectins are multivalent having at least two binding sites. It is also a feature of lectins that these binding sites are specific to particular monosaccharide moieties of glycan structures. Some of the lectins with particular specificity

TABLE 4
The Sugar Specificity[a] of Several Plant Lectins

Source of Lectin		
Latin Name	**Common Name**	**Specificity**
Agaricus bisporus	Mushroom	Gal (of Gal 1-3GalNAc)
Arachis hypogaea	Peanut	Gal (of Gal 1-3GalNAc)
Bauhinia purpurea	Camel's foot tree	Gal (of Gal 1-3GalNAc)
Concanavalia	Jack bean meal	Man > Glc
Datura stramonium	Thorn apple	$GlcNAc_2$
Iberis amara	Candytuft	Gal
Lens culinaris	Lentil	Man > Glc
Phaseolus vulgaris	Red kidney bean	GalNAc
Phytolacca americana	Pokeweed	$GlcNAc_3$
Pisum sativum	Pea	Man
Ricinus communis	Castor bean	Gal, GalNAc
Sophora japonica	Japanese western pagoda tree	Gal
Triticum vulgaris	Wheat germ	$GlcNAc_2$, NeuAc
Vicia faba	Fava bean	Man > Glc
Vicia villosa	Hairy vetch	GalNAc
Wistaria floribunda	Japanese wisteria	GalNAc

[a] As determined by competition by simple sugars for binding with oligosaccharides.

to components of glycoprotein glycans are listed in Table 4. Although most known lectins are of plant origin several have been detected in animals.

Barondes[88] makes a case for broadening the definition to include any carbohydrate binding protein other than an enzyme or an antibody. He draws attention to the fact that several lectins are multivalent because they are multimeric proteins and that several closely related monomeric, and hence monovalent, proteins may be as biosynthetically and functionally important as the divalent forms. He also points out the increasing number of carbohydrate-binding proteins that appear to be multifunctional due to possessing a second site with a quite different biological activity.

The recognition of specific features of glycan structure by carbohydrate binding proteins is of importance in the biological phenomena discussed in the following sections (2 to 5).

Sequence work on some of these binding proteins reveals carbohydrate recognition domains which appear to be very similar.[89] The identification of homologous sequences in proteins hitherto not recognized as carbohydrate binding proteins adds a further dimension to the possible biological role(s) of these carbohydrate recognition domains.

The specific and reversible binding of lectins to glycoconjugates has formed the basis of a wide laboratory technology. This includes fractionation of lymphocytes in the treatment of leukemia, purification of glycoproteins by chromatography on columns of lectin on an inert support, blocking of the biological activity of glycoproteins (e.g., of receptors) and numerous applications of lectin complexes.[9,85,86,90,91] These include the use of ^{125}I-lectin, peroxidase-lectin, fluorescein-lectin to identify and label glycans on cell surfaces or on isolated glycoproteins following electrophoresis and drug-lectin to target drugs to cells. In addition the mitogenic properties of some lectins and the cytotoxic effects of others have proved very useful in biological and medical research.

2. A Lysosomal Address

A peripheral mannose-6-phosphate group on a high mannose N-linked glycan of a lysosomal enzyme is a clear signal in fibroblasts and other cells for its transfer to the lysosomes.[92-95] This address is initiated in the *cis*-Golgi membranes by a transferase (step I, Figure 11, Section III.E.2) that clearly must recognise a code in the peptide chain specific

TABLE 5

Removal of Ceruloplasmin from the Circulation by Rabbit Hepatocytes, an Abbreviated Summary Based on the Work of Ashwell's Group.[96,97]

	Half-life in the circulation
Ceruloplasmin	55 h
Asialoceruloplasmin	5 min
Asialo-agalactoceruloplasmin	55 h

to lysosomal enzymes as well as recognising the acceptor glycan and UDPN-acetylglucosamine. The address is completed by the action of a specific *N*-acetylglucosaminidase (step II, Figure 11). There is good evidence for a mannose-6-phosphate receptor in Golgi membranes that binds to the lysosomal enzymes and diverts them to vesicles that develop into the lysosomes. The development of a low pH causes the enzyme to dissociate from its mannose-6-phosphate receptor, allowing the latter to recycle back to the Golgi to repeat the process. Some mannose-6-phosphate receptor is present on the plasma membrane of cells possibly having escaped from the Golgi-lysosome shuttle system. This is able to bind extracellular lysosomal enzymes and transport them in to the lysosomes. Probably in normal cells there is some leakage of lysosomal enzymes from the Golgi into the secretion pathway despite the presence of a mannose-6-phosphate moiety. In fibroblasts from I-cell patients, which lack step I in Figure 11, considerably more of the enzyme is secreted. The lack of enzymes in the lysosome results in the accumulation of polymers that would normally be degraded, precipitating the symptoms of the diseased condition.

3. Removal of Glycoproteins and Cells from the Blood

The original observations establishing the influence of the glycan of serum glycoproteins in their removal from the blood stream were made by Ashwell and Morell[96,97] (Table 5). This work demonstrated the effect of exposed terminal galactose residues. Subsequent investigation showed that a specific galactoside-binding protein in the hepatocyte plasma membrane was responsible for removing asialoglycoproteins from the blood stream. After internalization of the asialoglycoprotein lysosomal degradation occurred in the hepatocyte.

This is now known to be a general phenomenon and binding proteins in several cell types for different terminal monosaccharides have been identified (see, e.g., Gross et al.,[98] Burgi[99]). A similar phenomenon has been proposed to be a factor in the elimination of aging red blood cells. It has been suggested that glycoproteins and cells in blood are exposed to exoglycosidases which result in the sequential exposure of new terminal monosaccharides. This is particularly apparent in aging glycoproteins and cells. The physiological significance of the phenomenon has sometimes been challenged. However, consistent with this being important at least in some medical conditions is the report that in crises of severe, viral or bacterial infection (e.g., in gangrene) the high level of viral or bacterial sialidase released into the serum is sufficient to stimulate elimination of serum proteins. This probably makes an important contribution to the massive drop in serum glycoproteins associated with these infections.

4. Binding of Bacteria to Cell Surface Glycoproteins

An important factor in several extracellular infections is the binding of the bacteria concerned to host plasma membranes. No internalization follows but the binding helps to stabilize the bacterial colony. An example of this is to be found in bladder infections by strains of *E. coli*.[86] The bacterial fimbriae contain mannose binding proteins which allow several anchors to be put down by a bacterium to the surface of bladder smooth muscle cells. The infection can be alleviated by administration of methylmannoside which competes for the binding protein, displacing the mannoglycoproteins of the plasma membrane. This releases the bacteria to be moved on by the flow of fluid past the cell surface.

5. Binding of Viruses to Cell Surface Glycoproteins

Glycans on cell surfaces appear to be important binding sites both for phage entering bacteria and for viruses entering host animal cells. In the case of some strains of *E. coli* it is a particular glycan moiety of the lipopolysaccharide that is concerned. In the case of animal cells it appears to be a link with a sialo-glycan, usually of a glycoprotein, that is involved. The virus tends to be very specific to the structure of the glycan it first binds. For example, the influenza C virus recognizes only 9-*O*-acetyl-*N*-acetylneuraminic acid. There is also a remarkable correlation of specificity with species of origin of the virus. Thus human isolates of Hong Kong influenza virus serotypes bind preferentially to red blood cells carrying α2,6 linked sialic acid, whereas avian and equine isolates prefer α2,3 linked sialic acid.[100-102] It is possible to isolate variants from human isolates that show avian specificity and vice versa. The selectivity of the viruses for specific glycan structures resides in the haemaglutinin glycoprotein. It has been shown that a single amino acid substitution at residue 226 occurs in these variants. In over 30 influenza viruses isolated the α2,3-preferring variant has been shown to have glutamine and the α2,6-preferring variant has leucine in this position.

Not all of the viral-binding glycans are protein bound; some are on glycolipids. Also of those on glycoproteins not all are part of N-linked oligosaccharides. Nevertheless there has been a lot of interest in the possibility of inhibitors of *N*-glycan processing such as castanospermine, swainsonine (section IV.E.7), turning off the generation of complex, sialic acid-containing, N-linked glycans, so reducing the infectivity of retroviruses such as, for example, the human immunodeficiency virus. The antiretroviral activity of castanospermine in mice infected with Friend leukaemia virus may also be in part due to reversing the virally induced damage to the numbers and function of CD4 and CD8 lymphocytes, without the immunosuppressive side effects of other drugs.[103] It may also potentiate alloantigen induced cytolytic activity of T-lymphocytes in the same mice.[104] To what extent this is due to inhibition of glycan processing is not yet clear.

6. Embryonic Development

The glycans on cell surface glycoproteins have been implicated directly, or indirectly, in recognition and adhesion phenomena at several stages of embryonic development.[32,105,106]

Antibodies have been used to distinguish progressive early stages of embryogenesis. Antigenic determinants in the glycan of surface glycoproteins and glycolipids of mouse embryos change in a characteristic manner. For example, embryos at the 1 to 4 cell stage display different glycan antigens from those at the 8-cell stage and at compaction etc. Several of these antigens appear to be derived from polylactosaminyl chains of N-linked glycans. These changes are also being explored using lectins. In general both approaches point to a progressive increase in complexity of glycan structure during the early developmental stages. However, changes in exposure of glycans at the cell surface due to other factors probably also influence their antigenicity.

The biological function of these glycan changes in the process of development is not clear. However, work with compactin and tunicamycin demonstrates the importance of *N*-glycosylation to gastrulation. In the case of sea urchin embryos both tunicamycin and compactin block gastrulation.[107] The effect of compactin (section IV.E.5) was countered by the presence of exogenous phosphodolichol but not by cholesterol or ubiquinone, consistent with the tunicamycin result which highlighted the importance of *N*-glycosylation. Similar results have been reported for the effect of compactin on *N*-glycosylation and compaction of the mouse blastomere at the 32 cell stage. The effect was reversed by mevalonate.[108]

Cell adhesion molecules (CAMs) have been shown to be important in embryological development.[109,110] Studies on the neural cell adhesion molecule (NCAM) have progressed

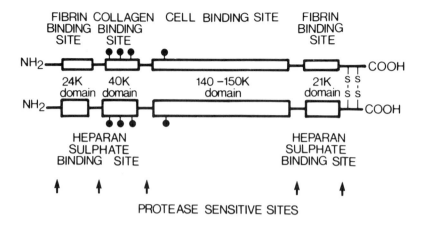

FIGURE 17.   The structure of fibronectin showing the various binding domains and sites of glycosylation. (From Hughes, R. C., *Glycoproteins,* Chapman and Hall, 1983. With permission.)

the furthest. NCAM is a cell surface glycoprotein that is involved in homophilic binding of neural cells (i.e., NCAM of one cell binds to NCAM of another). It exists in different forms in embryonic and adult tissues (E and A forms). The E form contains an unusually large amount (more than 100 moles/mole protein) of sialic acid linked 2-8 as a polysialo-chain whereas the A-form contains much less. The effect of tunicamycin suggests that the sialo-glycan is N-linked. Comparison of the binding properties of the E- and A-forms indicates that although the carbohydrate is not directly involved in the binding site the polysialo-chain reduces the strength of binding. This is probably a result of repulsion effects of the high concentration of negative changes plus steric effects of the bulky chain and possibly induced conformational changes in the protein chain. It appears that the change from E to A form is an important aspect of neural morphogenesis. The rate of change varies from one region of the brain to another. In the mouse cerebellar mutant, the staggerer, the change is delayed. This is consistent with a key role for NCAM in the development of a properly functioning brain.

Cell surface:extracellular matrix interactions are clearly also an important feature of embryonic morphogenesis. Fibronectin is thought to be involved in this, utilising its binding sites for cell surface receptors (integrins), collagen and heparan sulphate[112,113] (Figures 17 and 18). There is evidence indicating that the degree of glycosylation of fibronectin can influence its binding properties, presumably by inducing conformational changes. The possibility of this being of significance in cancer[35] and tuberous sclerosis[114] has been proposed. It has also been suggested that heparan sulphate moieties of cell surface glycoproteins may stabilize the binding of fibronectin to integrins. This may be of particular significance in epithelial cells where plasma membrane heparan sulphate is restricted to the basolateral cell surface which borders the extracellular matrix of the basement membrane. Since plasma membrane heparan sulphate is reported to have a short half-life (approximately 4 h) modulation of its biosynthesis may have a rapid effect on the interaction of cell with extracellular matrix.

7. Changes in Protein Glycosylation Associated with Tumors

Aberrant protein glycosylation is one of the most common correlates of the malignant phenotype.[35] Enhanced *N*-acetylglucosaminylation of a branched core mannose residue (step 8, Figure 11) is frequently observed leading to an increase in the number of antennae, many carrying terminal sialic acid residues. These changes seem to be much less antigenic than changes in the peripheral sections of both N- and O-linked chains.

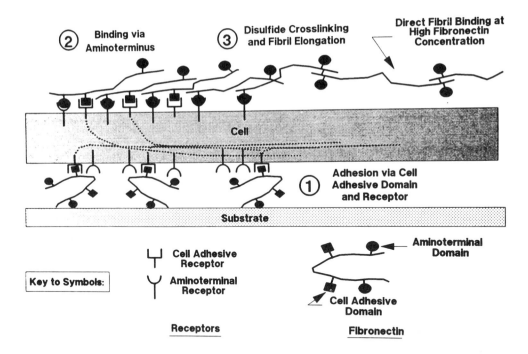

FIGURE 18.   The role of fibronectin in cellular:extracellular matrix interactions. (From McDonald, J. A., *Annu. Rev. Cell Biol.*, 4, 183, 1988. ©1988 by Annual Reviews, Inc. With permission.)

At the periphery changes frequently involve increased fucosylation and sialylation of O-linked chains. For example, type I chains (Section II.B.3a) are found with sialic acid linked α2-3 to the terminal galactose and linked α2-6 to the penultimate residue (*N*-acetylglucosamine) which already carries fucose-linked α1-4. Often, enhanced fucosylation (α1-4) is also found to give high levels of Le$_a$ blood group determinant (see Figure 5). These changes are found also in premalignant cells. Several workers[115,116] have observed an increased metastatic potential associated with these sorts of modifications. Interestingly, the increased fucosylation of hepatic glycoproteins in patients with cancer of tissues other than liver has also been reported.[117]

The relevance of aberrant glycan structures to the etiology of malignancy remains uncertain. However there is little doubt that the production of monoclonal antibodies to those cancers that generate tumor-associated antigens is proving useful in cancer research. Some of these antibodies may well develop into valuable diagnostic and prognostic markers. They also present possibilities of therapy by targeting attached cytotoxic drugs to malignant cells.

C. OTHER FUNCTIONAL CONSEQUENCES OF PROTEIN GLYCOSYLATION
1. Functions Responding to Inhibition of Glycosylation

Some of the effects of tunicamycin on biological phenomena or on the production or activity of glycoproteins are summarised in Table 6 (see, e.g., Elbein,[66] Hemming[69]). It is clear from this that there is a great variety in response and that it is not possible to generalize other than to conclude that the expression of cell surface glycoproteins is more likely to be blocked by tunicamycin than is the secretion or function of a secretory glycoprotein. The structure of the peptide chain as well as the cell type producing it are both important factors pertinent to the particular response to tunicamycin. Some of the disputed responses (labeled a in Table 6) might well be the result of using different cell types or of different sources of tunicamycin (see section IV.E.3). The expression of cell surface receptors is particularly sensitive to glycosylation of the receptor protein chain. Several examples of this sensitivity

TABLE 6
Summary of Observed Consequences of Inhibiting Protein *N*-Glycosylation by Tunicamycin

Marked Interference in Cell Biological Phenomena

Amino acid entry (some)
Cell cycle
Cell division in Tetrahymena
Cell spreading, morphology
Differentiation:
  Adipocyte
  Kidney
  Myoblast fusion
  Odontoblast

Embryonic development:
  Mouse, sea urchin
Fungal morphology
Glucose entry
Histogenesis
Mating in Tetrahymena
Metastasis, lung
Nutrient entry
Virus infectivity

Marked Interference in Biochemical Phenomena
(Usually Reduction in Amount of or in Activity of that Listed)

ACTH-endorphin:
  Precursor, secretion
α_1-Acute phase glycoprotein
α_1-Chymotrypsin inhibitor
Collagen, ex procollagen
Complement, component 4
Factor VIII
Fibronectin[a]
Glycoprotein secretion in
  hepatocytes, plants
Glycoprotein of:
  Chick embryo plasma membrane
  Viral envelope
Glycosidase, viral
HCG α-unit
Haemopexin
Hexose transporter protein
Hydrolases of fungi, plants
IgA[a], IgE, IgG[a], IgM[a]
Keratan sulphate
Invertase
Laminin, kidney

Lipase, heart
Lipoprotein lipase, activity
Phosphatase of L-cells
α_1-Protease inhibitor
Proteoglycan
Receptors:
  Ac. choline, musc. & nic.
  β-Adrenergic
  EGF
  Gonadotropin
  Gonadotropin RH
  IgM
  Insulin
  LDL
  NGF
  Opiate
  Prostacyclin
  Transferrin
Thyroglobulin
T-lymphocyte protein CD4
tPA
von Willebrand factor

No Marked Effect on Concentration or Activity of

ACTH
Apolipoproteins B, E, C
Asialoglycoprotein receptor
Endorphin
Fibronectin[a]
α-Foetoprotein
Glycophorin
Human leukemia antigen
IgA[a], IgG[a], IgD, IgM[a]

Interferon
Lipoprotein lipase, quantity
Pro-opiomelanocortin
Prothrombin
Rhodopsin
TSH
Transferrin
Trypanosome surface glycoprotein

[a] Disputed.

are summarised in Table 6. A discussion of a few examples of these will make the range of the effects of aberrant glycosylation more clear.

Experiments with tunicamycin reveal the intracellular accumulation of the aglyco insulin receptor.[118] Isolated aglyco receptor was also found to be devoid of insulin binding activity.

TABLE 7
Some Effects of Inhibitors of Processing of N-Linked Glycans

Inhibition by Swainsonine

No effect on production of:	Increased secretion of:	Decrease in:
α_1-Antitrypsin[a]	α_1-Antitrypsin[a]	Glucocorticoid stim. of osteoclasts
H2-D antigens	Ceruloplasmin	Immunosuppression
Hepatoma secretion of glycopro-	α_2-Macroglobulin	Mannoprotein uptake
tein	Transferrin	Melanoma colonization
Oncogene, erb B prod.		Trypanosome: macrophage interac-
Receptors:		tion
Asialoglycoprotein		Turnover: endocytosed glycopro-
EGF		teins
Insulin		
Viral infectivity		
Viral proteins		
von Willebrand protein		

Inhibition by Deoxynojirimycin or Castanospermine

No effect on production of:	Decrease in:
Alloantigen 1aK	Aminopeptidase N
Carboxypeptidase	α_1-Antitrypsin secretion
Complement C3	Cathepsin D
Histocompatability	EGF receptor
Antigen, class 1	IgD
IgM	Lymphocyte H2K
Influenza haemagglutinin	Mouse hepatitis viral E protein
Invertase	α_1-Proteinase inhibitor
Oncogene erb B product	Sindbis virus production
Oncogene vfms product	TSH
RSV glycoprotein-35, -85	VSV G-protein[a]
VSV G-protein[a]	

[a] Disputed.

Mutant cells deficient in N-glycan processing enzymes have also been shown to produce insulin receptors with different binding properties presumably due to the different glycan structures on the receptors.[119] Similarly, the ligand binding activity of c and d opiate receptors appears to be dependent not only on the presence of N-linked glycans but also on their glycan structure.[120]

Inhibitors of glycan processing (Table 7) also have a wide range of biological effects, even in the same cells. Again it is difficult to generalize about the effect of such an inhibitor on the biological properties of a glycoprotein.

That the glycan moiety in some cases protects against proteolysis is demonstrated in experiments using tunicamycin in the presence and absence of inhibitors of proteolysis. For example in the presence of tunicamycin alone the viral haemagglutinin of fowl plague virus could not be formed whereas if N-p-tosyl L-lysinchloromethyl-ketone, a protease inhibitor, was also present the formation of the viral aglyco-haemagglutinin could be detected quite readily.[121]

Similar results have been described for fibronectin, some immunoglobulins, TSH, histocompatability antigens HLA-A and HLA-B and for receptors for acetylcholine and asialoglycoprotein. It seems likely that the increased vulnerability to proteolysis of aglycopro- teins is due to changes in conformation. Indeed evidence for an effect on peptide conformation arose from studies on temperature sensitive mutants of VSV. At 30° one strain expressed aglyco G-protein at the cell surface when tunicamycin was present and replicated quite well. At 38° however the aglyco G-protein accumulated intracellularly. In addition the 30° product could be solubilized by Triton X-100 easily but the 38° product could not.[122]

2. Effects of Glycosylation of Enzymes

In most glycoprotein enzymes glycosylation appears to have little or no effect on the active site. Pancreatic ribonuclease is a prime example for it exists in several forms ranging from no glycosylation to full glycosylation. The K_m and V_{max} are essentially identical in each form. It appears that in the majority of cases any changes in conformation that accompany changed glycosylation affect areas of the protein away from the active site. Lipoprotein lipase[123] appears to be one of the few enzymes dependent upon *N*-glycosylation for expression of its catalytic activity.

Conformational changes may be more generally important in the routing of enzymes within the cell from the site of synthesis to the site of action and/or secretion. Changes in shape may also afford some protection from proteolysis.

A direct role of the glycan moiety is established for several other important aspects of some enzyme action. The targeting of some enzymes to their normal site of action, for example, lysosomal enzymes, has already been discussed (section V.B.2). Clearly glycan structure of glycoprotein enzymes in blood serum may also be critical to their elimination from the blood stream (section V.B.3). Bulky carbohydrate structures may protect against proteolytic enzymes.

The formation of heterologous mammalian proteins in bacteria has provided evidence supporting these general and other points. For example, the aglyco version, produced in bacteria, of tissue plasminogen activator expresses its full proteolytic activity in the absence of fibrin. Interestingly, the proteolytic activity of native tissue plasminogen activator, which differs from the bacterial product only in carrying an N-linked glycan, is dependent upon the presence of fibrin.[124]

3. Glycosylation of Peptide Hormones

Several pituitary peptide hormones are glycosylated.[125-127] Leucotropin (LH), follitropin (FSH) and thyrotropin (TSH) carry anionic N-linked glycans. In the case of LH and TSH the negative charge is due primarily to the presence of sulphate in the sequence SO_4-4GalNACβ1-4GlcNAC β1-2Man$_\alpha$. It is reported that the sulphotransferase is glycan specific but not protein specific. The selective addition of sulphate to LH and TSH seems to be due to the high specificity for protein structure (probably the α-unit) of the *N*-acetylgalactosamine transferase which competes with the galactosyltransferase that uses the same glycan acceptor but on other peptide chains.

The negative charge of FSH is due primarily to terminal sialic acid residues on di- and triantennary glycans. LH and FSH are synthesized in the same cell in the pituitary and it is suggested that the sulphated and sialylated glycans may be involved in directing them to different secretory granules.

Each hormone is composed of a common α-unit and a hormone-specific β-unit. Both carry N-linked glycans but those of the β-units are not involved in the function of the hormones. However the glycan on the α-unit has an important influence on function. On the one hand it tends to weaken the binding of hormone to its receptor. On the other hand, it is essential if the receptor-hormone complex is to stimulate production of cAMP. Deglycosylated hormone acts as antagonist to the normal hormone and there is some evidence that this competition may be of physiological significance.[128]

Human chorionic gonadotropin (hCG) carries two complex N-linked glycan chains on both α- and β-units, one having the rather rare monoantennary structure. As in FSH, the glycans of hCG are sialylated and show a similar relationship between glycosylation and function.

D. GLYCOSYLATION OF HETEROLOGOUS GLYCOPROTEINS

There is currently much interest in the possibility of using molecular biology to produce large quantities of pure animal glycoproteins, especially those of medical or veterinary/ agricultural importance, in cultured cells of bacteria, fungi, yeast and higher plants. The virtual absence of *N*- and *O*-glycosylation of proteins in bacteria, and the fact that non-animal cells do not produce the same complex glycan chains as animal cells (e.g., they lack sialic acid and rarely add *N*-acetylgalactosamine to serine residues) may present problems. Yeast systems have, for example, presented a number of interesting complications. One of these stems from the different form of serine-*O*-glycosylation found in these cells. This gives rise to short oligomannosyl chains instead of the classical "mucin-type" in animal cells (Section II.B.3.a). There is also evidence for mannosylation of serine residues in a protein by yeast systems that escape glycosylation completely in animal cells.[129] Hyperglycosylation of a different form may also occur, reflecting the tendency of yeast to add many mannose and some phosphate residues to the high mannose N-linked glycan.

A number of deleterious consequences of lack of or modification of glycosylation have been discussed in earlier sections. Several of these may be exaggerated in systems designed to over-express the products. Inappropriate secretion, changes in solubility, charge and stability to proteases could be of particular importance in a biotechnological production line. Once produced, differences in antigenicity, interactions with lectins and elimination from the blood stream may also be very relevant to the efficacy of the product if it is to be administered to animals. For these reasons the current interest in the control, manipulation and consequences of this particular form of posttranslational modification of proteins will continue to be of great commercial as well as academic and biomedical importance.

REFERENCES

1. **Lennarz, W. J.**, *Biochemistry of Glycoproteins and Proteoglycans*, Plenum Press, 1980.
2. **Hubbard, S. C. and Ivatt, R. J.**, Synthesis and processing of asparagine-linked oligosaccharides, *Annu. Rev. Biochem.*, 50, 555, 1981.
3. **Staneloni, R. J. and Leloir, L. F.**, The biosynthetic pathway of the asparagine-linked oligosaccharides of glycoproteins, *Crit. Rev. Biochem.*, 12, 289, 1982.
4. **Kobata, A.**, in *Biology of Carbohydrates*, Vol. 2, Ginsburg, V. and Robbins, P. W., Eds., Wiley-Interscience, 1987, 87.
5. **Hughes, R. C.**, *Glycoproteins*, Chapman and Hall, 1983.
6. **Ivatt, R. M.**, The Biology of Glycoproteins, Plenum Press, 1984.
7. **Sharon, N.**, Glycoproteins, *Trends Biochem. Sci.*, 9, 198, 1984.
8. **Stoddart, R. W.**, The Biosynthesis of Polysaccharides, Croom-Helm, 1984.
9. **Beeley, J. G.**, Glycoprotein and Proteoglycan Techniques, Elsevier, 1985.
10. **Kornfeld, R. and Kornfeld, S.**, Assembly of asparagine-linked oligosaccharides, *Annu. Rev. Biochem.*, 54, 631, 1985.
11. **Rademacher, T., Parekh, R. B., and Dwek, R.**, Glycobiology, *Annu. Rev. Biochem.*, 57, 785, 1988.
12. **Montreuil, J.**, Primary structure of glycoprotein glycans. Basis for the molecular biology of gycoproteins, *Adv. Carbohydr. Chem. Biochem.*, 37, 157, 1980.
13. **Montreuil, J.**, Glycoproteins, in *Protein Metabolism; Comprehensive Biochemistry*, Vol. 19BII, Neuberger, A. and van Deenen, L. L. M., Eds., Elsevier, 1982, 1.
14. **Baumann, P., Eap, C. B., Muller, W. E., Tillement, J. P., Eds.**, α1-acid glycoprotein: genetics, biochemistry, physiological functions and pharmacology, *Prog. Clin. Biol. Res.*, Liss, 300, 1989.
15. **Elbein, A. D.**, The role of lipid-linked saccharides in the biosynthesis of complex carbohydrates, *Annu. Rev. Plant Physiol.*, 30, 239, 1979.
16. **Kukurinska, M. A., Bergh, M. L. E., and Jackson, B. J.**, Protein Glycosylation in Yeast, *Annu. Rev. Biochem.*, 56, 915, 1987.
17. **Lechner, J. and Wieland, F.**, Structure and biosynthesis of prokaryotic glycoproteins, *Annu. Rev. Biochem.*, 58, 173, 1989.

18. **Ferguson, M. A. J. and Williams, A. F.,** Cell-surface anchoring of proteins via glycosylphosphatidyl-inositol structures, *Annu. Rev. Biochem.,* 57, 285, 1988.

19. **Hart, G. W., Holt, G. D., and Haltiwanger, R. S.,** Nuclear and cytoplasmic glycosylation, *Trends Biochem. Sci.,* 13, 380, 1988.

20. **Hart, G. W., Haltiwanger, R. S., Holt, G. D., and Kelly, W. G.,** Glycosylation in the nucleus and cytoplasm, *Annu. Rev. Biochem.,* 58, 84, 1989.

21. **Krusius, T., Reinhold, V. M., Margolis, R. K., and Margolis, R. U.,** Structural studies on sialylated and sulphated *O*-glycosidic mannose-linked oligosaccharides in the chondroitin sulphate proteoglycans of the brain, *Biochem. J.,* 245, 229, 1987.

22. **Scrisomsap, C., Richardson, K. L., Jay, J. C., and Marchase, R. B.,** Localization of the glucose phosphotransferase to a cytoplasmically accessible site on intracelluar membranes, *J. Biol. Chem.,* 263, 17792, 1988.

23. **Ruoslahti, E.,** Structure and biology of proteoglycans, *Annu. Rev. Cell Biol.,* 4, 229, 1988.

24. **Mayre, R. and Burgenon, R.,** Structure and Function of Collagen Types, Academic Press, 1987.

25. **Hascall, V.,** Proteoglycans: the chondroitin sulphate/keratan sulphate proteoglycan of cartilage, in *ISI Atlas of Science,* 1988.

26. **Evered, D. and Whelan, J.,** Function of Proteoglycans, *CIBA Foundation Symposium,* 124, Wiley, 1986.

27. **Wight, T. and Mecham, R.,** Biology of Proteoglycans, Academic Press, 1987.

28. **Poole, A. R.,** Proteoglycans in health and disease, *Biochem. J.,* 236, 1, 1986.

29. **Gallagher, J. T., Lyon, M., and Steward, M. P.,** Structure and function of heparan sulphate proteoglycans, *Biochem. J.,* 236, 313, 1986.

30. **Schauer, R.,** Sialic acids and their role as biological masks, *Trends Biochem. Sci.,* 10, 357, 1985.

31. **Schauer, R., Ed.,** *Sialic Acids. Chemistry, Metabolism and Function,* Springer-Verlag, 1982.

32. **Feizi, T. and Childs, R. A.,** Carbohydrate structures of glycoproteins and glycolipids as differentiation antigens, tumor-associated antigens and components of receptor systems, *Trends Biochem. Sci.,* 10, 24, 1987.

33. **Feizi, T. and Childs, R. A.,** Carbohydrates as antigenic determinants of glycoproteins, *Biochem. J.,* 245, 1, 1987.

34. **Watkins, W. M.,** Biochemistry and genetics of the ABO, Lewis and P blood group systems, *Adv. Hum. Genet.,* 10, 1, 1980.

35. **Hakomori, S.,** Aberrant glycosylation in tumors and tumor-associated carbohydrate advances, *Adv. Cancer Res.,* 52, 257, 1989.

36. **Yogeeswaran, G.,** Cell surface glycolipids and glycoproteins in malignant transformation, *Adv. Cancer Res.,* 38, 289, 1983.

37. **Hassell, J. R., Kimura, J. H., and Hascall, V. C.,** Proteoglycan core protein families, *Annu. Rev. Biochem.,* 55, 539, 1986.

38. **Hemming, F. W.,** Glycosyl phosphopolyprenols, in *Glycolipids,* Wiegandt, H., Ed., as cited in *New Comprehensive Biochem.,* Vol. 10, Elsevier, 1985, 4.

39. **Bause, E. and Legler, G.,** The role of the hydroxy amino acid in the triplet step during glycoprotein biosynthesis, *Biochem. J.,* 195, 639, 1981.

40. **Bause, E.,** Structural requirements of *N*-glycosylation of proteins, *Biochem. J.,* 209, 331, 1983.

41. **Doering, T. L., Masterson, W. J., Englund, P. T., and Hart, G. W.,** Biosynthesis of the glycosyl-phosphatidylinositol membrane anchor of the trypanosome variant surface glycoprotein, *J. Biol. Chem.,* 264, 11168, 1987.

42. **Masterson, W. J., Doering, T. L., Hart, G. W., and Englund, P. T.,** A novel pathway for glycan assembly: biosynthesis of the glycosyl-phosphatidylinositol anchor of the pathway for glycan trypanosome variant surface glycoprotein, *Cell,* 56, 793, 1987.

43. **Hart, G. W., Doering, T. L., Masterson, W. J., Roper, J., and Englund, P. T.,** Cell-free biosynthesis of the glycosyl-phosphatidylinositol membrane anchor of the trypanosome variant surface glycoprotein, *Glycoconj. J.,* 6, 374, 1989.

44. **Hirschberg, C. B. and Snider, M. D.,** Topography of glycosylation in the rough endoplasmic reticulum and Golgi apparatus, *Annu. Rev. Biochem.,* 56, 63, 1987.

45. **Lindahl, U. and Kjellen, L.,** Biosynthesis of heparin and heparan sulphate, in *Biology of Proteoglycans,* Wight, T. and Mecham, R., Eds., Academic Press, 1987, 59.

46. **Lindahl, U., Feingold, D. S., and Roden, L.,** Biosynthesis of heparin, *Trends Biochem. Sci.,* 11, 221, 1986.

47. **Tauber, R., Park, C.-S., and Reutter, W.,** Structural carbohydrates from liver and hepatoma, in *Structural Carbohydrates in the Liver,* Popper, H., Reutter, W., Gudat, F., and Kottigen, E., Eds., MTP Press, 1983, 333.

48. **Ardail, D., Gateau, O., Morelis, R., and Louisot, P.,** Glycosyltransferase activities in liver mitochondria. Phospholipid dependence of inner membrane mannosyltransferase, *Eur. J. Biochem.,* 149, 497, 1985.

49. **Fayet, Y., Galland, S., Degiuli, A., Got, R., and Frot-Coutaz, J.,** Glycoprotein mannosylation in rat liver nuclei, *Biochem. Int.,* 16, 429, 1988.

50. **Lennarz, W. J.,** Protein glycosylation in the endoplasmic reticulum: current topological issues, *Biochem.,* 26, 7205, 1987.

51. **Haselbeck, A. and Tanner, W.,** Dolichyl phosphate-mediated mannosyl transfer through liposome membranes, *Proc. Natl. Acad. Sci.,* 79, 1520, 1982.

52. **Beyer, T. A., Sadler, J. E., Rearick, J. I., Paulson, J. C., and Hill, R. J.,** Glycosyltransferases and their use in assessing oligosaccharide structure and structure-function relationships, *Adv. Enzymol.,* 52, 23, 1981.

53. **Schutzbach, J. S., Jensen, J. W., Lue, C.-S., and Monti, J.,** Membrane structure and mannosyl transferase activities: the effects of dolichols on membranes, *Chem. Scr.,* 27, 109, 1987.

54. **Ravi, K. and Elbein, A. D.,** Purification and properties of UDPGlcNAC:dolichyl-P GlcNAc-1-P transferase from pig aorta, *Glycoconjug. J.,* 6, 437, 1989.

55. **Kaplan, H. A., Naider, F., and Lennarz, W. J.,** Partial characterization and purification of the glycosylation site recognition component of oligosaccharyltransferase, *J. Biol. Chem.,* 263, 7814, 1988.

56. **Warner, T. G. and O'Brien, J. S.,** Genetic defects in glycoprotein metabolism, *Annu. Rev. Genet.,* 17, 395, 1983.

57. **Stanley, P.,** Glycosylation mutants of animal cells, *Annu. Rev. Genet.,* 18, 525, 1984.

58. **Snider, M. D., Huffaker, T. C., Conto, J. R., and Robbins, P. W.,** Genetic and biochemical studies of asparagine-linked oligosaccharide assembly, *Philos. Trans. R. Soc. London,* 300, 207, 1982.

59. **Shaper, N. L., Hollis, G. J., Douglas, J. G., Kirsch, I. R., and Shaper, J. H.,** Characterization of the full length cDNA for murine β1-4 galactosyltransferase, *J. Biol. Chem.,* 263, 10420, 1988.

60. **Nakazawa, K., Ando, T., Kimura, T., and Narimatsu, H.,** Cloning and sequencing of a full length cDNA of mouse *N*-acetylglucosamine (β1-4) galactosyltransferase, *J. Biochem.,* 104, 165, 1988.

61. **Masri, K. A., Appert, H. E., and Fukuda, M. N.,** Identification of the full length coding sequence for human galactosyltransferase (β-*N*-acetylglucosaminide, β-1,4-galactosyltransferase, *Biochem. Biophys. Res. Commun.,* 157, 657, 1988.

62. **Ujita, E., Roth, J., and Paulson, J. C.,** Alteration of terminal glycosylation sequences on N-linked oligosaccharides of Chinese hamster ovary cells by expression of β-galactoside α-2,6-sialyltransferase, *J. Biol. Chem.,* 264, 13848, 1989.

63. **Albarracin, T., Lassaga, F. E., and Caputto, R.,** Purification and characterization of an endogenous inhibitor of the sialyltransferase CMP-*N*-acetylneuraminate: lactosylceramide α2,6-*N*-acetylneuraminyltransferase (EC2.4.99.-), *Biochem. J.,* 254, 559, 1988.

64. **Constantino-Ceccarini, E. and Suzuki, K.,** Isolation and partial characterization of an endogenous inhibitor of ceramide glycosyltransferase from rat brain, *J. Biol. Chem.,* 253, 340, 1978.

65. **Quiroga, S., Caputto, B. L., and Caputto, R.,** Inhibition of the chicken retinal UDPGalNAc:GM$_3$, *N*-acetylgalactosaminyltransferase by blood serum and by pineal gland extracts, *J. Neurosci. Res.,* 12, 269, 1984.

66. **Elbein, A. D.,** Inhibitors of the biosynthesis and processing of N-linked oligosaccharide chains, *Annu. Rev. Biochem.,* 56, 497, 1987.

67. **Schwarz, R. T. and Datema, R.,** The lipid pathway of protein glycosylation and its inhibitors: the biological significance of protein-bound carbohydrates, *Adv. Carbohydr. Chem. Biochem.,* 40, 287, 1982.

68. **Schwarz, R. T. and Datema, R.,** Inhibitors of trimming: new tools in glycoprotein research, *Trends Biochem. Sci.,* 9, 32, 1984.

69. **Hemming, F. W.,** Control and manipulation of the phosphodolichol pathway of protein *N*-glycosylation, *Biosci. Rep.,* 2, 203, 1982.

70. **Machamer, C. E., Florkiewicz, R. Z., and Rose, J. K.,** A single N-linked oligosaccharide of either of the two normal sites is sufficient for transport of vesicular stomatitis virus G protein to the cell surface, *Mol. Cell Biol.,* 5, 3074, 1985.

71. **Machamer, C. E. and Rose, J. K.,** Influence of new glycosylation sites on expression of the vesicular stomatitis virus G protein at the plasma membrane, *J. Biol. Chem.,* 263, 5948, 1988.

72. **Hartin, G. and Boine, I.,** Inhibition of asparagine-linked glycosylation by incorporation of a threonine analogue into nascent peptide chains, *J. Biol. Chem.,* 255, 8007, 1980.

73. **Stern, A. M., Foxman, B. M., Tashjian, A. H., and Abeles, R. H.,** DL-threo-β-fluoroaspartate and DL-threo-β-fluoroasparagine: selective cytotoxic agents for mammalian cells in culture, *J. Med. Chem.,* 25, 244, 1982.

74. **Tamura, G.,** *Tunicamycins,* Tokyo, Japan Sci. Soc. Press, 1982.

75. **Tkacz, J. S.,** *Antibiotics, Modes and Mechanisms of Microbial Growth Inhibitors,* 6, 1, 1981.

76. **Vogel, P., Stynes, B. A., Coockley, W., Yeoh, G. T., and Patterson, D. S.,** Glycolipid toxins from parasitised annual ryegrass: a comparison with tunicamycin, *Biochem. Biophys. Res. Commun.,* 105, 835, 1982.

77. **Frahn, J. L., Edgar, J. A., Jones, A. J., Cockrum, P. A., Anderton, N., and Colvenor, C. C.,** Structure of the corynetoxins, metabolites of corynebacterium ruthayi, responsible for toxicity of annual ryegrass *(Lolium rigidum)* pastures, *Aust. J. Chem.,* 37, 165, 1984.

78. **Tavares, I. A., Coolbear, T., and Hemming, F. W.,** Increased hepatic dolichol and dolichyl phosphate-mediated glycosylation in rats fed cholesterol, *Arch. Biochem. Biophys.,* 207, 427, 1982.

79. **Carson, D. D., Earles, B. J., and Lennarz, W. J.,** Enhancement of protein glycosylation in tissue slices by dolichylphosphate, *J. Biol. Chem.,* 256, 11552, 1981.

80. **Datema, R. and Schwarz, R. T.,** Effect of energy depletion on the glycosylation of a viral glycoprotein, *J. Biol. Chem.,* 256, 11191, 1981.

81. **Spiro, R. G., Spiro, M. J., and Bhoyroo, V. D.,** Studies on the regulation of the biosynthesis of glucose-containing oligosaccharide lipids. Effect of energy deprivation, *J. Biol. Chem.,* 258, 9469, 1983.

82. **Rearick, J. J., Chapman, A., and Kornfeld, S.,** Glucose starvation alters lipid-linked oligosaccharide biosynthesis in Chinese hamster ovary cells, *J. Biol. Chem.,* 256, 6255, 1981.

83. **Baumann, H. and Jahreis, G. P.,** Glucose starvation leads in rat hepatoma cells to partially N-glycosylated glycoprotein including α1-acid glycoproteins, *J. Biol. Chem.,* 258, 3942, 1983.

84. **Schwartz, N. B.,** Regulation of chondroitin sulphate synthesis: effect of β-xylosides on synthesis of chondroitin sulfate proteoglycan, chondroitin sulfate chains and core protein, *J. Biol. Chem.,* 252, 6316, 1977.

85. **Lis, H. and Sharon, N.,** Lectins as molecules and as tools, *Annu. Rev. Biochem.,* 55, 35, 1986.

86. **Sharon, N. and Lis, H.,** *Lectins,* Chapman and Hall, 1989.

87. **Liener, I. E., Sharon, N., and Goldstein, I. J.,** *The Lectins: Properties, Functions and Applications in Biology and Medicine,* Academic Press, 1986.

88. **Barondes, S. H.,** Bifunctional properties of lectins: lectins redefined, *Trends Biochem. Sci.,* 13, 480, 1988.

89. **Drickamer, K.,** Demonstration of carbohydrate-recognition activity in diverse proteins which share a common primary structure motif, *Biochem. Soc. Trans.,* 17, 13, 1989.

90. **Green, E. D. and Baenziger, J. U.,** Characterization of oligosaccharides by lectin affinity high performance liquid chromatography, *Trends Biochem. Sci.,* 14, 168, 1989.

91. **Osawa, T. and Tsuji, T.,** Fractionation and structural assessment of oligosaccharides and glycopeptides by use of immobilized lectins, *Annu. Rev. Biochem.,* 56, 21, 1987.

92. **Goldberg, D., Gabel, C., and Kornfeld, S.,** in *Lysosomes in Biology and Pathology,* Dingle, J. T., Dean, R. T., and Sly, W. S., Eds., Elsevier, 1984, 45.

93. **von Figura, K. and Hasilik, A.,** Lysosomal enzymes and their receptors, *Annu. Rev. Biochem.,* 55, 167, 1986.

94. **Pohlmann, R., Nagel, G., Hille, A., Wendland, M., Waheed, A., Braulke, T., and von Figura, K.,** Mannose-6-phosphate specific receptors: structure and function, *Biochem. Soc. Trans.,* 17, 15, 1989.

95. **Creek, K. E. and Sly, W. S.,** in *Lysosomes in Biology and Pathology,* Dingle, J. T., Dean, R. T., and Sly, W. S., Eds., Elsevier, 1984, 63.

96. **Ashwell, G. and Morell, A. G.,** The role of surface carbohydrates in the hepatic recognition and transport of circulating glycoproteins, *Adv. Enzymol.,* 41, 99, 1974.

97. **Ashwell, G. and Morell, A. G.,** Membrane glycoproteins and recognition phenomena, *Trends Biochem. Sci.,* 2, 76, 1977.

98. **Gross, V., Steake, K., Tran-Thi, T.-A., Gerak, W., and Heinrich, P. C.,** Role of N-glycosylation for the plasma clearance of rat liver secretory glycoproteins, *Biochem. Soc. Trans.,* 17, 211, 1989.

99. **Burgi, W.,** The biological role of the carbohydrate moiety of human plasma glycoproteins in Alpha1-acid glycoprotein, *Progr. Clin. Biol. Res.,* Baumun, P., Eap, C. B., Muller, W. E., and Tillement, J.-P., Eds., *Progr. Clin. Biol. Res.,* 300, 1, 1989.

100. **Rogers, G. N., Herrler, G., Paulson, J. C., and Klenk, H. D.,** Influenza C virus uses 9-O-acetyl-N-acetylneuraminic acid as a high affinity receptor determinant for attachment of cells, *J. Biol. Chem.,* 261, 5947, 1986.

101. **Rogers, G. N. and Paulson, J. C.,** Receptor determinants of human and animal influenza isolates: differences in receptor specificity of the H3 haemagglutinin based on species of origin, *Virology,* 127, 361, 1983.

102. **Wiley, D. C. and Skehel, J. J.,** The structure and function of the haemagglutinin membrane glycoprotein of influenza virus, *Annu. Rev. Biochem.,* 56, 365, 1987.

103. **Bowlin, T. L. and Roberts, C. M.,** Castanospermine potentiates alloantigen induced cytolytic T lymphocyte generation from Friend leukemia virus infected mice *in vitro, Glycoconj. J.,* 6, 402, 1989.

104. **Bowlin, T. L., Sunkara, P. S., and Sjoerdsma, A.,** Castanospermine, an inhibitor of glucosidase-1, restores CD4 and CD8 positive lymphocyte frequency and function in Friend leukemia virus infected mice, *Glycoconj. J.,* 6, 401, 1989.

105. **Kimber, S. J.,** Changes in cell-surface glycoconjugates during embryonic development demonstrated using lectins and other probes, *Biochem. Soc. Trans.,* 17, 23, 1989.

106. **Aubery, M. and Codogno, P.,** Role of cell surface glycoproteins in embryo cell adhesion to extracellular matrix, *Biochem. Soc. Trans.,* 17, 27, 1989.

107. **Carson, D. P. and Lennarz, W. J.,** Inhibition of polysioprenoid and glycoprotein biosynthesis causes abnormal embryonic development, *Proc. Natl. Acad. Sci.,* 76, 5709, 1979.

108. **Surani, M. A. H., Kimber, S. J., and Osborn, J. C.,** Mevalonate reverses the developmental arrest of preimplantation mouse embryos by compactin, an inhibitor of HMGCoA reductase, *J. Embryol. Exp. Morphol.,* 75, 205, 1983.

109. **Cunningham, B. A.,** Cell adhesion molecules: a new perspective on molecular embryology, *Trends Biochem. Sci.,* 11, 4231, 1986.

110. **Edelman, G. M.,** Cell adhesion and the molecular processes of morphogenesis, *Annu. Rev. Biochem.,* 54, 135, 1985.

111. **Edelman, G. M.,** Cell adhesion molecules in the regulation of animal form and tissue pattern, *Annu. Rev. Cell Biol.,* 2, 81, 1986.

112. **Ruoslahti, E.,** Fibronectin and its Receptors, *Annu. Rev. Biochem.,* 57, 375, 1988.

113. **McDonald, J. A.,** Extracellular Matrix Assembly, *Annu. Rev. Cell Biol.,* 4, 183, 1988.

114. **Ellis, Z. K. and Hemming, F. W.,** Glycosylation of fibronectin and other proteins in tissues of patients with tuberous sclerosis, in *Neurocutaneous Syndrome,* Elsevier, 1990.

115. **Dennis, J. W., Laferte, S., and Vanderelst, I.,** Asparagine-linked oligosaccharides in malignant tumour growth, *Biochem. Soc. Trans.,* 17, 29, 1989.

116. **Passaniti, A. and Hart, G. W.,** Sialic acids and penultimate oligosaccharides on metastatic tumour cell surfaces, *Biochem. Soc. Trans.,* 17, 33, 1989.

117. **Thompson, S. and Turner, G. A.,** Elevated levels of abnormally-fucosylated haptoglobulins in cancer sera, *Br. J. Cancer,* 56, 605, 1987.

118. **Ronnett, G. V. and Lane, M. D.,** Post-translational glycosylation-induced activation of aglycoinsulin receptor accumulated during tunicamycin treatment, *J. Biol. Chem.,* 256, 4704, 1981.

119. **Rouiller, D. G., Sharon, N., McElduff, A., Podskalny, J. M., and Gorden, P.,** Lectins as probes of insulin receptor carbohydrates composition: studies in glycosylation mutants of Chinese hamster ovary cells with altered insulin binding, *Endocrinology,* 118, 1159, 1986.

120. **McLawhon, R. W., Berry-Kravis, E., and Dawson, G.,** Differential regulation of multiple neuroreceptors in a somatic cell hybrid by inhibitors of glycoprotein processing, *Biochem. Biophys. Res. Commun.,* 134, 1387, 1986.

121. **Schwarz, R. T., Rohrschneider, J. M., and Schmidt, M. F. G.,** Suppression of glycoprotein formation of Semliki Forest influenza and avian sarcoma virus by tunicamycin, *J. Virol.,* 19, 782, 1976.

122. **Gibson, R., Schlesinger, S., and Kornfeld, S.,** The nonglycosylated glycoprotein of vesicular stomatitis virus is temperature-sensitive and undergoes intracellular aggregation at elevated temperatures, *J. Biol. Chem.,* 254, 3600, 1979.

123. **Chajek-Shaul, T., Friedman, G., Knobler, H., Stein, O., Etienne, J., and Stein, Y.,** Importance of the different steps of glycosylation for the activity and secretion of lipoprotein lipase in rat preadipocytes studied with monensin and tunicamycin, *Biochem. Biophys. Acta,* 837, 271, 1985.

124. **Stump, D. C., Lijner, H. R., and Collen, D.,** Purification and characterization of single-chain urokinase-type plasminogen activator from human cell cultures, *J. Biol. Chem.,* 261, 1274, 1986.

125. **Sairam, M. R. and Bhargavi, G. N.,** A role for glycosylation of the α subunit in transduction of biological signal in glycoprotein hormones, *Science,* 229, 65, 1985.

126. **Calvo, F. O., Kentmann, H. T., Bergert, E. R., and Ryan, R. J.,** Deglycosylated human follitropin: characterization and effects on adenosinecyclic $3'5'$-phosphate production in porcine granulosa cells, *Biochemistry,* 25, 3938, 1986.

127. **Baenziger, J. U. and Green, E. D.,** Structure, synthesis and function of glycoprotein hormone oligosaccharides, in *Glycoconjugates,* Montreuil, Vergert, Spik, and Fournet, Eds., Lerouge Tourcoing, 1987, 12.

128. **Dahl, K. D., Bicsak, T. A., and Hsuch, A. J. W.,** Naturally occurring antihormones: secretion of FSH antagonists by women treated with GnRH analog, *Science,* 239, 72, 1988.

129. **Hard, K., Bitter, W., Kamerling, J. P., and Vliegenthart, F. G.,** *O*-mannosylation of recombinant human insulin-like growth factor 1 (IGF-1) produced in Saccharomyces cervisiae, *FEBS Lett.,* 248, 111, 1989.

INDEX